BIOMAT 2010

International Symposium on
Mathematical and Computational Biology

Rio de Janeiro, Brazil
24 – 29 July 2010

edited by

Rubem P Mondaini
Federal University of Rio de Janeiro, Brazil

World Scientific

NEW JERSEY • LONDON • SINGAPORE • BEIJING • SHANGHAI • HONG KONG • TAIPEI • CHENNAI

Published by

World Scientific Publishing Co. Pte. Ltd.
5 Toh Tuck Link, Singapore 596224
USA office: 27 Warren Street, Suite 401-402, Hackensack, NJ 07601
UK office: 57 Shelton Street, Covent Garden, London WC2H 9HE

British Library Cataloguing-in-Publication Data
A catalogue record for this book is available from the British Library.

BIOMAT 2010
International Symposium on Mathematical and Computational Biology

ISBN-13 978-981-4343-42-8
ISBN-10 981-4343-42-0

Printed in Singapore by World Scientific Printers.

Preface

The present book contains the papers selected for publication by the Referee Board of the BIOMAT Consortium (http://www.biomat.org). The papers have been selected among all submitted full papers of the BIOMAT 2010 International Symposium on Mathematical and Computational Biology. As a rule, the Editorial Board kept the level of acceptance on 20% of the submitted papers in order to preserve the rigour of analysis and critical evaluation of new scientific results and the soundness of reviews. The Editorial Board has also decided to proceed with a double-blind style for the work of the Referee Board on the stages of abstract and full paper. Due to the increasing number of submitted contributions every year, a third evaluation has been necessary for a final decision. This is done by invited Keynote Speakers during the presentation of the paper in the conference.

The BIOMAT Consortium has been working since 2001 on its fundamental mission of enhancing the number of practitioners on the Mathematical Modelling of Biosystems in Latin America and Developing Countries of five continents. The Consortium has been also working during these ten years of realization of international conferences for motivating the future scientific career of young graduate studies on the research areas of Mathematical and Computational Biology and Biological Physics.

The scientific seriousness of the Consortium and its leadership in the representation of Latin American countries in these research areas, was recognized by scientific societies and institutions worldwide. The BIOMAT Consortium has been invited to organize the 2010 Annual Meeting of the Society for Mathematical Biology (http://www.smb.org). This was the first meeting of the SMB Society in South America and it was held in Rio de Janeiro from 24th to 29th July 2010 in a joint conference with the BIOMAT 2010 International Symposium. Sixteen senior scientists have been invited by the BIOMAT Consortium, as Keynote and Plenary Speakers of the joint conference. Parallel sessions of the SMB 2010 Meeting have been mixed with sessions of contributed papers of the BIOMAT 2010 and SMB Mini-Symposium sessions. This joint conference has ended a cycle of organization of international conferences which was created on April 2001 in Rio de Janeiro, with the First BIOMAT Symposium. A new cycle is now starting in the year 2011, with the decision of organizing a number of BIOMAT symposia in other countries. Latin American countries are the candidates for the first symposia of this new cycle.

We acknowledge the Board of Trustees of three Brazilian sponsoring agencies - Coordination for the Improvement of Higher Education Personell - CAPES, The National Research Council for Scientific and Technological Development - CNPq, Foundation for Research Support of Rio de Janeiro State - FAPERJ. On behalf of the BIOMAT Consortium/ Institute for Advanced Studies of Biosystems, we thank their Directors and Authorized Representatives. Special thanks are due to Prof. José Oswaldo de Siqueira from CNPq for his expertise as a scientific administrator and his senior understanding of the scientific importance of the BIOMAT series of International Symposia.

Thanks are also due to the Commander-in-chief of the School of Naval War- Brazilian Navy, in Rio de Janeiro, to the Rector and the Dean of the Federal University of Rio de Janeiro State - UNIRIO, Prof. Malvina Tuttman and Prof. Luiz Amancio de Sousa Jr., respectively, and the Rector of the Federal University of Rio de Janeiro - UFRJ, Prof. Aloisio Teixeira. Thank you very much indeed for the available facilities on these institutions as well as for some financial help which was provided by UFRJ.

Last but not least, we thank our collaborators which have integrated the BIOMAT Consortium Administrative Staff during the Joint Conference SMB 2010 - BIOMAT 2010, Leonardo Mondaini, Felipe Mondaini and Sandro Pereira Vilela. The latter has also provided some help with the LaTeX version of the accepted papers. Other members of the Staff like Wanderson da Rocha, José Ricardo da Rocha, Janice Justo, Renata Figueiredo and Aline Coutinho should be also remembered for their good organizational skills during the conference.

Rubem P. Mondaini
President of the BIOMAT Consortium
Chairman of the SMB 2010 - BIOMAT 2010 Joint Conference

Rio de Janeiro, December 2010

Editorial Board of the BIOMAT Consortium

Contents

Population Dynamics

Population Biology

Theoretical Immunology

Computational Biology

Mathematical Aspects of Bioprocesses

Population Genetics

Systems Biology

USING DNA KNOTS TO ASSAY VIRAL GENOME PACKING

DE WITT SUMNERS

Department of Mathematics, Florida State University, Tallahassee, FL 32306 USA

E-mail: sumners@math.fsu.edu

Bacteriophages pack their double-stranded DNA genomes to near-crystalline density in viral capsids and achieve one of the highest levels of DNA condensation found in nature. Despite numerous studies some essential properties of the packaging geometry of the DNA inside the phage capsid are still unknown. Although viral DNA is linear double-stranded with sticky ends, the linear viral DNA quickly becomes cyclic when removed from the capsid, and for some viral DNA the observed knot probability is an astounding 95%. These observed DNA knots carry information about capsid packing geometry. This talk will discuss comparison of the observed viral knot spectrum with the simulated knot spectrum, concluding that the viral knot spectrum is non-random, writhe directed, and generated by local cholesteric interaction (juxtaposition at a small twist angle) between DNA strands

1. Introduction

DNA knots and links (catenanes) are of biological interest because they can detect and preserve topological information, notably information about the binding and mechanism of enzymes that act on DNA, and information about the geometry of viral genome packing in bacteriophage capsids. In the topological approach to enzymology, DNA substrate plasmids of known supercoiling density and knot/link type are incubated with purified enzyme *in vitro*, and changes from substrate to product in DNA topology/geometry are observed and quantified by gel electrophoresis and electron microscopy. Knotted and linked DNA molecules are very robust, and can survive experimental manipulation and visualization intact. Characterization of knotted products formed by random cyclization of linear molecules has been used to quantify important biochemical properties of DNA such as effective diameter in solution as a function of salt concentration [1,2]. DNA knots and catenanes obtained as products of site-specific recombination on negatively supercoiled substrate have been a key to understanding enzymatic binding and mechanism [3,4]. In all cases, the development of mathematical and computational tools has greatly enhanced analysis of the experimental results [5,6].

Significant numbers of endogenous DNA knots are found also in biological systems: in *Escherichia coli* cells harboring mutations in the GyrB or GyrA genes [7], bacteriophages P2 and P4 [8,9], and cauliflower mosaic viruses [10]. However, very little biological information about these systems has been inferred from the observed knots. In particular, interpretation of the experimental results for bacteriophages has been limited by the experimental difficulty in quantifying the complex spectrum of knotted products. These difficulties have paralleled those encountered in developing a theory for random knotting of ideal polymeric chains in cases where interactions with other macromolecules and/or confinement in small volumes have a significant function [11-16].

2. Bacteriophage Genome Packing

Bacteriophages are viruses that infect bacteria. They pack their double-stranded DNA genomes to near-crystalline density in viral capsids and achieve one of the highest levels of DNA condensation found in nature. When I was on sabbatical in Berkeley in 1989, Jim Wang described to me the problem of DNA packing in icosahedral viral capsids [8,9], and the high degree of knotting produced when the viral DNA is released from the capsids. Despite numerous studies some essential properties of the packaging geometry of the DNA inside the phage capsid are still unknown. Although viral DNA is linear double-stranded with sticky (cohesive) ends, the linear viral DNA quickly becomes cyclic when removed from the capsid, and for some viral DNA the observed knot probability is an astounding 95%. In the summer of 1998, my PhD students Javier Arsuaga and Mariel Vazquez spent 2 months in the laboratory of Joaquim Roca in Barcelona, supported by the Burroughs Wellcome Interfaces grant to the Program in Mathematics and Molecular Biology. In the Roca laboratory, they infected bacterial stock, harvested viral capsids and extracted and analyzed the viral DNA. They quantified the DNA knot spectrum produced in the experiment, and used Monte Carlo generation of knots in confined volumes to produce a random knot spectrum to be compared to the observed viral DNA knot spectrum. A subsequent experiment was performed in the Roca laboratory on a shorter viral genome [22]. A series of papers were produced as a result of these experiments [17-21, 23, 24], and I will describe some of the results from these papers, focusing on (and reproducing here) most of the discussion and analysis from [20], and a short discussion of the results from [24].

All icosahedral bacteriophages with double-stranded DNA genomes are believed to pack their chromosomes in a similar manner [25]. During phage morphogenesis, a procapsid is first assembled, and a linear DNA molecule is actively introduced inside it by the connector complex [26,27]. At the end of this process, the DNA and its associated water molecules fill the entire capsid volume, where DNA reaches concentrations of 800 mg/ml [28]. Some animal viruses [29] and lipo-DNA complexes used in gene therapy [30] are postulated to hold similar DNA arrangements as those found in bacteriophages. Although numerous studies have investigated the DNA packing geometry inside phage capsids, some of its properties remain unknown. Biochemical and structural analyses have revealed that DNA is kept in its B form [31-33] and that there are no specific DNA-protein interactions [34,35] or correlation between DNA sequences and their spatial location inside the capsid, with the exception of the cos ends in some viruses. Many studies have found that regions of the packed DNA form domains of parallel fibers, which in some cases have different orientations, suggesting a certain degree of randomness [32,33]. The above observations have led to the proposal of several long-range organization models for DNA inside phage capsids: the ball of string model [36], the coaxial spooling model [32,36,37], the spiral-fold model [38], and the folded toroidal model [39]. Liquid crystalline models, which take into account properties of DNA at high concentrations and imply less global organization, have also been proposed [33]. Cryo EM and spatial symmetry averaging has recently been used to investigate the surface layers of DNA packing [40]. In [20], the viral DNA knot spectrum was used to investigate the packing geometry of DNA inside phage capsids.

3. Bacteriophage P4

Bacteriophage P4 has a linear, double-stranded DNA genome that is 10-11.5 kb in length and flanked by 16-bp cohesive cos ends [41]. It has long been known that extraction of DNA from P4 phage heads results in a large proportion of highly knotted, nicked DNA circles [8,9]. DNA knotting probability is enhanced in P4 derivatives containing genome deletions [42] and in tailless mutants [43]. Most DNA molecules extracted from P4 phages are circles that result from the cohesive-end joining of the viral genome.

Previous studies have shown that such circles have a knotting probability of about 20% when DNA is extracted from mature P4 phages [8]. This high value is increased more than 4-fold when DNA is extracted from incomplete P4 phage particles (which we refer to as "capsids") or from non-infective P4 mutants that lack the phage tail (which we refer to as "tailless mutants" [8]). Knotting of DNA in P4 deletion mutants is even greater. The larger the P4 genome deletion, the higher the knotting probability [42]. For P4 vir1 del22, containing P4's largest known deletion (1.6 kb deleted [44]), knotting probability is more than 80% [43]. These values contrast with the knotting probability of 3% (all trefoil knots) observed when identical P4 DNA molecules undergo cyclization in dilute free solution [1,45]. These differences are still more striking when the variance in distribution of knot complexity is included. Although knots formed by random cyclization of 10-kb linear DNA in free solution have an average crossing number of three [1,45], knots from the tailless mutants have a knotting probability of 95% and appear to have very large crossing numbers, averaging about 26. [8,42,43].

The reasons for the high knotting probability and knot complexity of bacteriophage DNA have been investigated. Experimental measurements of the knotting probability and distribution of knotted molecules for P4 vir1 del22 mature phages, capsids, and tailless mutants was performed by 1- and 2- dimensional gel electrophoresis, followed by densitometer analysis. We will describe the Monte Carlo simulations to determine the effects that the confinement of DNA molecules inside small volumes have on knotting probability and complexity. We conclude from our results that for tailless mutants a significant amount of DNA knots must be formed before the disruption of the phage particle, with both increased knotting probability and knot complexity driven by confinement of the DNA inside the capsid.

In [20] it is shown that the DNA knots provide information about the global arrangement of the viral DNA inside the capsid. The distribution of the viral DNA knots is analyzed by high-resolution gel electrophoresis. Monte Carlo computer simulations of random knotting for freely jointed polygons confined to spherical volumes is performed. The knot distribution produced by simulation is compared to the observed experimental DNA knot spectrum. The simulations indicate that the experimentally observed scarcity of the achiral knot 4_1 and the predominance of the torus knot 5_1 over the twist knot 5_2 are not caused by confinement alone but must include writhe bias in the packing geometry. Our results indicate that the packaging geometry of the DNA inside the viral capsid is non-random and writhe-directed.

4. Knot Type Probabilities for Phage P4 DNA in Free Solution

The probability that a DNA knot K of n statistical lengths and diameter d is formed by random closure in free solution is given by $P_K(n,\ d) = P_K(n,\ 0)\, e^{-rd/n}$, where r depends on the knot type and equals 22 for the trefoil knot K and 31 for the figure 8 knot K^*[33]. The knotting probability of a 10-kb DNA molecule cyclized in free solution is 0.03 (25, 26), which implies an effective DNA diameter near 35Å. Because $P_K(34,\ 0) = 0.06$ and $P_{K^*}(34,\ 0) = 0.009$, then $P_K(34,\ 35) = 0.027(1/36$ times that of the unknot) and $P_{K^*}(34,\ 35) = 0.003(1/323$ times that of the unknot). These values were used to estimate the fractions of the knot K and the knot K^* generated for P4 DNA in free solution.

5. Monte Carlo Simulation

Knotting probabilities of equilateral polygons confined into spherical volumes were calculated by means of Markov-chain Monte Carlo simulations followed by rejection criteria. Freely jointed closed chains, composed of n equilateral segments, were confined inside spheres of fixed radius, r, and sampled: values of n ranged from 14 to 200 segments; r values, measured as multiples of the polygonal edge length, ranged from 2 to infinity. Excluded volume effects were not taken into account. Markov chains were generated by using the Metropolis algorithm [46]. The temperature, a computational parameter, was held at T = 300 K to improve the efficiency of the sampling algorithm. Other values of T produced similar results, thus indicating that the computation is robust with respect to this parameter. Chains contained inside the sphere were assigned zero energy. Chains lying partly or totally outside the confining sphere were assigned an energy given by the maximum of the distances of the vertices of the chain to the origin. Only chains with zero energy were sampled. A random ensemble of polygons was generated by the crankshaft algorithm as follows: (i) two vertices of the chain were selected at random, dividing the polygon into two subchains, and (ii) one of the two subchains was selected at random (with equal probabilities for each subchain), and the selected subchain rotated through a random angle around the axis connecting the two vertices. This algorithm is known to generate an ergodic Markov chain in the set of polygons of fixed length [47]. Correlation along the subchains was computed by using time-series analysis methods as described by Madras and Slade [48]. Identification of the knotted polygons was achieved by computing the Alexander polynomial $\Delta(t)$ [49-51] evaluated at t = -1. It is known that $\Delta(-1)$ does not identify all knotted chains; however, for polygonal chains not confined to a spherical volume, nontrivial knots with trivial $\Delta(-1)$ values rarely occur. This circumstance has been observed by using knot invariants, such as the HOMFLY polynomial, that distinguish between knotted and unknotted chains with higher accuracy than the Alexander polynomial. Computer simulations for small polygons ($<$ 55 segments) show that the knotting probabilities obtained by using $\Delta(-1)$ agree with those obtained by using the HOMFLY polynomial. Furthermore, Deguchi and Tsurusaki have reported that the value of $\Delta(-1)$ can almost always determine whether a given Gaussian polygon is unknotted for lengths ranging from 30 to 2,400 segments [51]. Each selected knotted polygon was further identified by evaluating its Alexander polynomial at t = -2 and t = -3. Although the Alexander polynomial is an excellent discriminator among knots of low crossing number and its computation is fast, it does not distinguish completely among some knotted chains [for example, composite knots $3_1\#3_1$ and $3_1\#4_1$ have polynomials identical to those of prime knots 8_{20} and 8_{21}, respectively [RW]. Evaluation of the polynomial at t = -2 and t = -3 is also ambiguous because the Alexander polynomial is defined up to units in Z[t^{-1}, t], and therefore the algorithm returns $(\pm n^{\pm m}\Delta(-n))$, where $n = 2$ or 3 and m is an integer. To deal with this uncertainty, we followed van Rensburg and Whittington [RW] and chose the largest exponent k such that the product $(\pm n^{\pm m}\Delta(-n))n^{\pm k}$ is an odd integer with $n = 2$ or 3. This value was taken as the knot invariant. To compute the writhe, we generated $>$ 300 regular projections and resulting knot diagrams for each selected polygon. To each of the projected crossings a sign was assigned by the standard oriented skew lines convention. The directional writhe for each diagram was computed by summing these values. The writhe was then determined by averaging the directional writhe over a large number of randomly chosen projections. To generate writhe-directed random distributions of polygons, we used a rejection method in which polygons whose writhe was below a positive value were not sampled.

6. Results and Discussion Knot Complexity of DNA Molecules Extracted from Phage P4

The 10-kb DNA from the tailless mutant of phage P4 vir1 del22 was extracted, producing 95% knotted molecules [18], and analyzed by a high-resolution two-dimensional gel electrophoresis [17]. This technique fractionates DNA knot populations according to their crossing number (i.e., the minimal number of crossings over all projections of a knot), as well as separating some knot populations of the same crossing number [52,53]. In the first dimension (at low voltage), individual gel bands corresponding to knot populations having crossing numbers between three and nine were discernible; knots with higher crossing numbers were embedded in a long tail. The second dimension (at high voltage) further resolved individual gel bands corresponding to knot populations with crossing numbers between six and nine. Although knot populations containing three, four, and five crossings migrated as single bands in a main arch of low gel velocity, knot populations containing six and more crossings split into two subpopulations, creating a second arch of greater gel velocity as in Fig. 1.

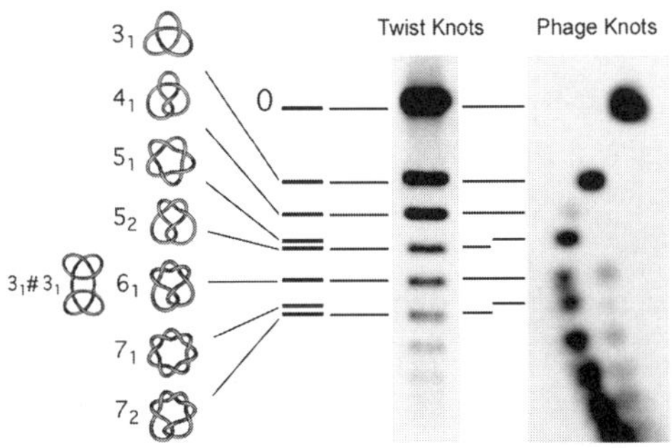

Figure 1. [20]. Fig. 1 shows the gel velocity at low voltage of individual knot populations resolved by two-dimensional electrophoresis (Right) is compared with the gel velocity at low voltage of the marker ladder of twist knots (3_1,4_1,5_2,6_1, and 7_2) of a 10-kb nicked plasmid (Center) and with known relative migration distances of some knot types. Geometrical representations of the prime knots 3_1,4_1,5_1,5_2,6_1,7_1 and 7_2 and of the composite knot $3_1\#3_1$ are shown. The unknotted DNA circle or trivial knot (0) is also indicated. Note that in the main arch of the two-dimensional gel and below the knots 3_1 and 4_1, the knot population of five crossings matches the migration of the torus knot 5_1, which migrates closer to the knot 4_1 than to the knots of six crossings. The other possible five-crossing knot, the twist knot 5_2, appears to be negligible or absent in the viral distribution. Note also that the knot population of seven crossings matches the migration of the torus knot 7_1 rather than the twist knot 7_2. In the secondary arch of the two-dimensional gel, the first knot population of six crossings has low-voltage migration similar to that of the composite knot $3_1\#3_1$.

We quantified the individual knot populations of three to nine crossings, which represented 2.2% of the total amount of knotted molecules. Densitometer readings confirmed the apparent scarcity of the knot 4_1 (the figure 8 knot) relative to the other knot populations in the main arch of the gel. It also made evident the shortage of the knot subpopulation of seven crossings in the second arch of the gel. The scarcity of the knot 4_1 relative to the knot 3_1 and to other knot populations is enhanced if we make the correction for DNA molecules plausibly knotted outside the viral capsid. Namely, if a fraction of the observed knots were formed by random cyclization of DNA outside the capsid, then, in the worst-case scenario, all observed unknotted molecules (no more than 5% of the total molecules extracted) would be formed in free solution. In such a case, one can predict that 38% of the total number of observed 3_1 knots and 75% of the observed 4_1 knots are formed by random knotting in free solution [1,2]. If all of the knots plausibly formed outside the capsid were removed from the observed knot distribution, the experimental values for knots 4_1 and 3_1 (1:18 ratio) would be corrected, resulting in a 1:44 ratio.

7. Identification of Specific Knot Types by Their Location on the Gel

Gel electrophoresis can distinguish some knot types with the same crossing number. For example, at low voltage, torus knots (such as 5_1 and 7_1) migrate slightly slower than their corresponding twist knots (5_2 and 7_2) (31, 32). We used this knowledge in conjunction with a marker ladder for twist knots (3_1, 4_1, 5_2, 6_1, and 7_2) to identify several gel bands of the phage DNA matching the migration of known knot types (Fig. 2). In the main arch of the gel, in addition to the unambiguous knots 3_1 and 3_1, the knot population of five crossings matched the migration of the torus knot 5_1. The other possible five-crossing knot, the twist knot 5_2 that migrates between and equidistant to the four-and six-crossing knot populations, appeared to be negligible or absent. The knot population of seven crossings matched the migration of the torus knot 7_1 rather than the twist knot 7_2, which has slightly higher gel velocity. Yet, we cannot identify this gel band as the knot 7_1, because other possible knot types of seven crossings cannot be excluded.

Several indicators led us to believe that the second arch of the gel consists of mainly composite knots. First, the arch starts at knot populations containing six crossings, and no composite knots of fewer than six crossings exist. Second, the population of six crossings matched the migration at low voltage of the granny knot $3_1\#3_1$. [54], although the square knot $3_1\#-3_1$. cannot be excluded. Third, consistent with the low amount of 4_1 knots, the size of the seven-crossing subpopulation is also reduced: any composite seven-crossing knot is either $3_1\#4_1$ or $-3_1\#4_1$. The increased gel velocity at high voltage (second gel dimension) of composite knots relative to prime knots of the same crossing number likely reflects distinct flexibility properties of the composites during electrophoresis [55].

8. Monte Carlo Simulations of Random Knot Distributions in Confined Volumes

Next, we asked whether the observed distribution of DNA knots could be compatible with a random embedding of the DNA inside the phage capsid. We used Monte Carlo simulations to model knotting of randomly embedded, freely jointed polygons confined to spherical volumes. Because the persistence length of the duplex DNA is not applicable in confined volumes (it is applicable in unbounded three-dimensional space), we considered freely jointed polygons as the zeroth approximation of the packed DNA molecule. Then, the flexibility of the chain is given by the ratio R/N, where N is the number of edges in the polygon and R is the sphere radius in edge-length units. When we computed random knot distributions for a range of chain lengths confined to spheres with a fixed radius, the probabilities of the knots 3_1,4_1 ,5_1, and 5_2 produced nonintersecting distributions, with simpler knots being more probable (Fig. 2A). That is, the knot 3_1 is more probable than the knot 4_1, and both are more probable than any five-crossing knot. In addition, the probability of the twist knot 5_2 is higher than that of the torus knot 5_1. Similar results had been observed for other random polymer models with/without volume exclusion and with/without confinement [52,13-16], indicating that this phenomenon is model-independent. All of the simulated distributions, showing the monotonically decreasing amounts of knotted products with increasing crossing number, highly contrasted with our experimental distribution, in which the probability of the knot 4_1 is markedly reduced and in which the knot probability of the knot 5_1 prevails over that of the knot 5_2 (Fig. 2B). These differences provide a compelling proof that the embedding of the DNA molecule inside the phage capsid is not random.

How can we explain the scarcity of 4_1 in the spectrum of viral knots? The knot 4_1 is achiral (equivalent to its mirror image). Random polygonal realizations of the 4_1 knot in free space and in confined volumes produce a family of polygons whose writhe distribution for any polygonal length is a Gaussian curve with zero mean (the writhe is a geometrical quantity measuring the signed spatial deviation from planarity of a closed curve) and whose variance grows as the square root of the length (Theorem 10). Therefore, we argue that the main reason for the scarcity of the knot 4_1 is a writhe bias imposed on the DNA inside the phage capsid. To test this hypothesis, we simulated polygons randomly embedded in spheres whose mean writhe value was gradually increased. To induce writhe in the sampling, we used a rejection method in which polygons of writhe below a cutoff value were not sampled. Then, we calculated the probabilities of the prime knots 4_1, 5_1, and 5_1 for each writhe-biased sampling. The results shown in Fig. 3 were computed with a freely jointed chain of 90 edges confined in a sphere of radius of 4 edge-length units. A drop of the probability of the knot 4_1, as well as an exponential increase of the probability of the torus knot 5_1 but not of the twist knot 5_2, readily emerged by increasing the writhe rejection value. The same results, but with knots of opposite sign, were obtained for knot distributions with the corresponding negative writhe values. These writhe-induced changes in the knot probability distribution are independent of the number of edges in the equilateral polygon and the sphere radius length. Accordingly, previous studies had shown that the mean writhe value of random conformations of a given knot does not depend on the length of the chain but only on the knot type and that these values are model-independent [56].

Because the writhe-directed simulated distributions approach the observed experimental spectrum of knots, we conclude that a high writhe of the DNA inside the phage is the most likely factor responsible for the observed experimental knot spectrum.

Consistent with the involvement of writhe in the DNA packing geometry, it is also the reduced amount of prime knots of six crossings visible in the main arch of the gels (Fig. 1). All prime knots of six crossings have a lower $(< Wr >)$ ($(< Wr >)$ of $6_1 = 1.23$, $(< Wr >)$ of $6_2 = 2.70$, and $(< Wr >)$ of $6_3 = 0.16$) than the torus knots of five and seven crossings ($(< Wr >)$ of $5_1 = 6.26$ and ($(< Wr >)$ of $7_1 = 9.15$). In contrast, the negligible amount of the twist knot of five crossings ($(< Wr >)$ of $5_2 = 4.54$) in the experimental distributions is striking. The apparent predominance of torus knots (5_1 and 7_2) over twist knots (5_2 and 7_2) in the experimental distribution suggests that writhe emerges from a toroidal or spool-like conformation of the packed DNA. Consistent with our findings, theoretical

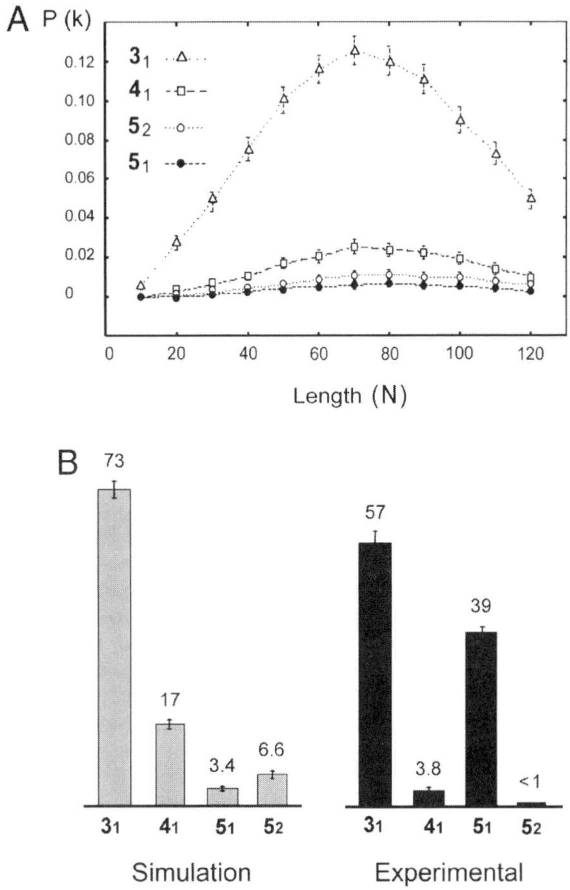

Figure 2. [20]. Fig. 2 shows the distribution probabilities $P(k)$ obtained by Monte Carlo simulations of the prime knots 3_1, 4_1, 5_1, and 5_2 for closed ideal polymers of variable chain lengths (n = number of edges) confined to a spherical volume of fixed radius ($R = 4$ edge lengths). Error bars represent standard deviations. (B) Comparison of the computed probabilities of the knots 3_1, 4_1, 5_1, and 5_2 (for polymers of length $n = 90$ randomly embedded into a sphere of radius $R = 4$) with the experimental distribution of knots. The relative amount of each knot type is plotted. Note that fractions of knots 3_1 and 4_1 plausibly formed in free solution are not subtracted from the experimental distribution. If these corrections are considered, the relative amount of knot 4_1 is further reduced.

calculations of long-range organization of DNA by Monte Carlo and molecular dynamics methods favor toroidal and spool-like arrangements for DNA packed inside the phage capsids. Calculations of optimal spool-like conformations of DNA in phage P4 already predicted a large nonzero writhe.

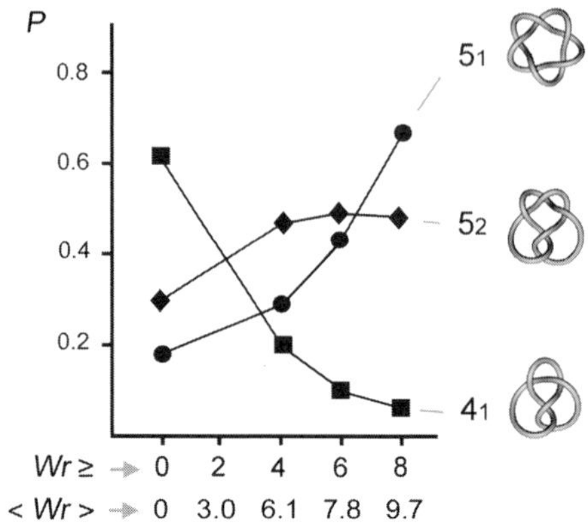

Figure 3. [20]. Fig. 3 shows the writhe of polygons of length n = 90 randomly embedded into a sphere of radius $R = 4$ were computed, and only conformations whose writhe values were higher than a fixed value (Wr = 4, 6, or 8) were sampled. The computed mean writhe value ($< Wr >$) of each sampled population is indicated. The ratios of the probabilities of the knots 4_1, 5_1, and 5_2 relative to that of the knot 3_1 for each writhe-biased sampling are plotted (P).

These studies gave an estimated writhe of 45 for the 10-kb DNA, which closely corresponds with the level of supercoiling density typically found in bacterial chromosomes [19].

The actual writhe value of the DNA packaged in the phage P4 capsid cannot be estimated in the present study. The phage P4 capsid has a diameter of 38 nm. If the parameters used to compute writhe-biased ensembles as in Fig. 3 ($n = 90$ and $R = 4$) were applied to a 10-kb DNA molecule, they would translate into 90 segments of 35 nm confined in a model capsid of radius 140 nm. Likewise, our study cannot argue for or against recent models that suggest that to minimize DNA bending energy, a spool conformation might be concentric rather than coaxial. Therefore, beyond the main conclusion of this work that the distribution of viral knots requires the mean writhe of the confined DNA be nonzero, the applicability of our simulations to other aspects of the DNA packaging in phage P4 is limited. We argue that further identification of the knotted DNA populations will provide more critical information for the packing geometry of DNA inside the phage.

Knots can be seen as discrete measuring units of the organizational complexity of filaments and fibers. Here, we show that knot distributions of DNA molecules can provide information on the long-range organization of DNA in a biological structure. We chose the problem of DNA packing in an icosahedral phage capsid and addressed the questions of randomness and chirality by comparing experimental knot distributions with simulated knot distributions. The scarcity of the achiral knot 4_1 and the predominance of the torus knot 5_1 in the experimental distribution highly contrasted with simulated distributions of random knots in confined volumes, in which the knot 4_1 is more probable than any five-crossing knot, and the knot 5_2 is more probable the knot 5_1. To our knowledge, these results produce the first topological proof of nonrandom packaging of DNA inside a phage capsid. Our simulations also show that a reduction of the knot 4_1 cannot be obtained by confinement alone but must include writhe bias in the conformation sampling. Moreover, in contrast to the knot 5_2, the probability of the torus knot 5_1 rapidly increases in a writhe-biased sampling. Given that there is no evidence for any other biological factor that could introduce all of the above deviations from randomness, we conclude that a high writhe of the DNA inside the phage capsid is responsible for the observed knot spectrum and that the cyclization reaction captures that information.

I will now briefly describe the results of more recent experiments [22] and analysis [24, 57, 58]. Recent experiments have shown that linear double-stranded DNA in bacteriophage capsids is both highly knotted with genome length half of previous studies [24], and with a highly structured surface structure [57,58]. In [24], evidence from stochastic simulation techniques suggests that a key element is the tendency of contacting DNA strands to interact at a small contact twist angle, as in cholesteric liquid crystals. This strand interaction promotes an approximately nematic (and apolar) local order in the DNA. The ordering effect dramatically impacts the geometry and topology of DNA inside phages. Accounting for this local potential allows the reproduction of the main experimental data on DNA organization inside phages, including the cryo-EM observations of surface ordering [57,58] and detailed features of the spectrum of DNA knots (Fig. 4) formed inside viral capsids. The DNA knots observed inside capsids were strongly delocalized, and were shown by simulation not to interfere with genome ejection out of the phage.

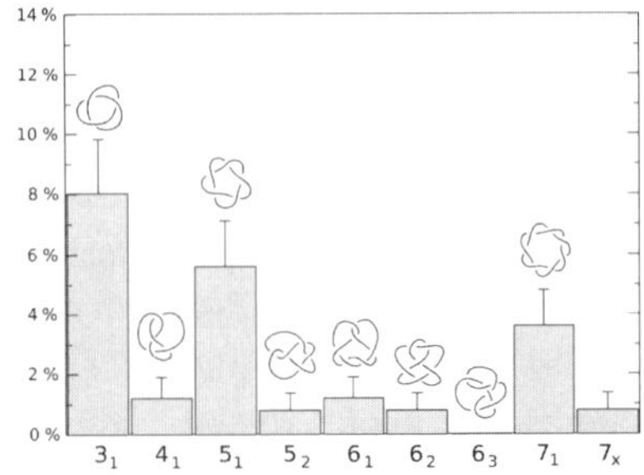

Figure 4. [24]. Knot spectrum (up to 7 crossings) produced by simulation of cholesteric interaction of a 4.7 *kb* genome inside the viral capsid. The error bar is calculated from Poissonian statistics. Chiral and torus knots prevail, as evidenced by the preponderance of 3_1, 4_1, 5_1, 6_1, and 7_1 knots. The index 7_x is used for the cumulative set of 7_2, 7_3, 7_4, and 7_5 knots.

References

1. V.V. Rybenkov, N. R. Cozzarelli, A.V. Vologodskii, *Proc. Natl. Acad. Sci. USA* **90**, 5307 (1993).
2. S.Y. Shaw, J.C. Wang, *Science* **260**, 533 (1993).
3. S.A. Wasserman, N.R. Cozzarelli, *Science* **232**, 951 (1986).
4. W.M. Stark, M.R. Boocock, D.J. Sherratt, *Trends Genet.* **5**, 304 (1989).
5. M.D. Frank-Kamentskii, A.V. Lukashin, V.V. Anshelevich, A.V. Vologodskii, *J. Biomol. Struct. Dyn.* **2**, 1005 (1985).
6. C. Ernst, D.W. Sumners, *Math. Proc. Camb. Phil. Soc.* **108**, 489 (1990).
7. K. Shishido, N. Komiyama, S. Ikawa, *J. Mol. Biol.* **195**, 215 (1987).
8. L.F. Liu, J.L. Davis, R. Calendar, *Nucleic Acids Res.* **9**, 3979 (1981).
9. L.F. Liu, L. Perkocha, R. Calendar, J.C. Wang, *Proc. Natl. Acad. Sci. USA* **78**, 5498 (1981).
10. J. Menissier, G. de Murcia, G. Lebeurier, L. Hirth, *EMBO J.* **2**, 1067 (1983).
11. M.L. Mansfield, *Macromolecules* **27**, 5924 (1994).
12. M.C. Tesi, J.J.E. van Resburg, E. Orlandini, S.G. Whittington, *Phys. A: Math. Gen* **27**, 347 (1994).
13. C. Micheletti, D. Marenduzzo, E. Orlandini, D.W. Sumners, *J. Chem. Phys.* **124**, 064903 (2006).
14. C. Micheletti, D. Marenduzzo, E. Orlandini, D.W. Sumners, *Biophys. J.* **95**, 3591 (2008).
15. D.W. Sumners, *in Lectures on Topological Fluid Mechanics, Springer-CIME Lecture Notes in Mathematics 1973, R. Ricca, ed.* **187** (2009).
16. D. Marenduzzo, C. Michelettti, E. Orlandini, *Physics Reports* (to appear 2010).
17. S. Trigueros, J. Arsuaga, M.E. Vazquez, D.W. Sumners, J. Roca, *Nucleic Acids Research* **29**, 67 (2001).
18. J. Arsuaga, M. Vazquez, S. Trigueros, D.W. Sumners and J. Roca, *Proc. National Academy of Sciences USA* **99**, 5373 (2002).

19. J. Arsuaga, K-Z Tan, M.E. Vazquez, D.W. Sumners, S.C. Harvey, *Biophysical Chemistry* **101-102**, 475 (2002).
20. J. Arsuaga, M.E. Vazquez, P. McGuirk, D.W. Sumners, J. Roca, *Proc. National Academy of Sciences USA* **102** (2005).
21. J. Arsuaga, J. Roca, D.W. Sumners, *in Emerging Topics in Physical Virology, P. Stockley and R, Trawick, eds, (World Scientific)* (to appear 2010).
22. S. Trigueros, J. Roca, *BMC Biotechnology* **7**, 94 (2007).
23. J. Arsuaga, Y. Diao, *J. Comp. Math. Meth. Med.* **9**, 303 (2008).
24. D. Marenduzzo, E. Orlandini, A. Stasiak, D.W. Sumners, L. Tubiana, C. Micheletti, *Proc. National Academy of Sciences USA* **106**, 22269 (2009).
25. W.C. Earnshaw, S.R. Casjens, *Cell* **21**, 319 (1980).
26. S. Rishov, A. Holzenburg, B.V. Johansen, B.H. Lindqvist, *Virology* **245**, 11 (1998).
27. C. Bustamante, *Nature* **413**, 748 (2001).
28. E. Kellenberger, E. Carlemalm, J. Sechaud, A. Ryter, G. Haller, *in Bacterial Chromatin, eds. Gualerzi, C. & Pon, C. L. (Springer, Berlin),* **11** (1986).
29. C. San Martin, R. Burnett, *Curr. Top. Microbiol. Immunol.* **272**, 57 (2003).
30. M. Schmutz, D. Durand, A. Debin, Y. Palvadeau, E.R. Eitienne, A.R. Thierry, *Proc. Natl. Acad. Sci. USA* **96**, 12293 (1999).
31. K. Aubrey, S. Casjens, G. Thomas, *Biochemistry* **31**, 11835 (1992).
32. W.C. Earnshaw, S. Harrison, *Nature* **268**, 598 (1977).
33. J. Lepault, J. Dubochet, W. Baschong, E. Kellenberger, *EMBO J.* **6**, 1507 (1987).
34. R. Hass, R.F. Murphy, C.R. Cantor, *J. Mol. Biol.* **159**, 71 (1982).
35. P. Serwer, *J. Mol. Biol.* **190**, 509 (1986).
36. K. Richards, R. Williams, R. Calendar, *J. Mol. Biol.* **78**, 255 (1973).
37. M. Cerritelli, N. Cheng, A. Rosenberg, C. McPherson, F. Booy, A. Steven, *Cell* **91**, 271 (1997).
38. L. Black, W. Newcomb, J. Boring, J. Brown, *Proc. Natl. Acad. Sci. USA* **82**, 7960 (1985).
39. N. Hud, *Biophys. J.* **69**, 1355 (1995).
40. W.Jiang, J. Chang, J. Jakana, P. Weigele, J. King, W Chiu, *Nature* **439**, (2006).
41. J.C. Wang, K.V. Martin, R. Calendar, *Biochemistry* **12**, 2119 (1973).
42. J.S. Wolfson, G.L. McHugh, D.C. Hooper, M.N. Swartz, *Nucleic Acids Res.* **13**, 6695 (1985).
43. M. Isaken, B. Julien, R. Calendar, B.H. Lindvist, *in DNA Topoisomerase Protocols, DNA Topology, and Enzymes, eds. M.A. Bjornsti, & N. Osheroff, (Humana, Totowa, NJ), Vol.* **94**, 69 (1999).
44. A. Raimondi, R. Donghi, A. Montaguti, A. Pessina, G. Deho, *J. Virol.* **54**, 233 (1985).
45. V.V. Rybenkov, C. Ullsperger, A.V. Vologodskii, N.R. Cozarelli, *Science* **277**, 690 (1997).
46. N. Metropolis, A.W. Rosenbluth, M.N. Rosenbluth, A.H. Teller, E. Teller, *J. Chem. Phys* **21**, 1087 (1953).
47. K. Millett, *in Random Knotting and Linking, Series of Knots and Everything,*

eds. D.W. Sumners & K.C. Millett, (World Scientific, Singapore), Vol. **7**, 31 (1994).
48. N. Madras, G. Slade, *The Self-Avoiding Walk, Birkhauser,* Boston (1993).
49. D. Rolfsen, *Knots and Link,* Publish or Perish (1976).
50. G. Burde, H. Zieschang, *Knots. De Gruyter, Berlin, New York* (1985).
51. C. Adams, *The Knot Book. W.H. Freeman and Co., New York* (1991).
52. T. Deguchi, K. Tsurusaki, *J. Knot Theor. Ramifications* **3**, 321 (1994).
53. A. Stasiak, V. Katrich, J. Bednar, D. Michoud, J. Dubochet, *Nature* **384**, 122 (1996).
54. A.V. Vologodskii, N.J. Crisona, B. Laurie, P. Pieranski, V. Katritch, J. Dubochet, A. Stasiak, *J. Mol. Biol.* **278**, 1 (1998).
55. R. Kanaar, A. Klippel, E. Shekhtman, J. Dungan, R. Kahmann, N. R. Cozzarelli, *Cell* **62**, 353 (1990).
56. C. Weber, A. Stasiak, P. De Los Rios, G. Dietler, *Biophys. J.* **90**, 3100 (2006).
57. E.J.J. van Rensburg, D.W. Sumners, S.G. Whittington, *in Ideal Knots, Series on Knots and Everything, eds. A. Stasiak, V. Katritch, L.H. Kauffman, (World Scientific, Singapore), Vol.* **19**, 70 (1999).
58. W. Jiang, J. Chang, J. Jakana, P. Weigele, J. King. W. Chiu, *Nature* **439**, 612 (2006).
59. L.R. Comolli, A.J. Spakowitz, C.E. Siegerist, P.J. Jardine, S. Grimes, D.L. Anderson, C. Bustamante, K.H. Downing, *Virology* **371**, 267 (2008).

CRUMPLED GLOBULE MODEL OF DNA PACKING IN CHROMOSOMES: FROM PREDICTIONS TO OPEN QUESTIONS

A.Y. GROSBERG

Department of Physics and Center for Soft Matter Research,
New York University,
4 Washington Place,
New York, NY 10003, USA
E-mail: ayg1@nyu.edu

Recent experiment[1] confirmed the theoretical prediction made long time ago [2,3] regarding fractal properties of DNA packed in the interphase chromosomes. This report summarizes the essence of both theoretical prediction and its experimental confirmation along with the explanation [4] given to the other observed properties of genome organization, such as chromosomal territories [5].

1. Introduction

A long time ago, theoretical prediction was formulated as to how DNA should be packed in the restricted volume of the cell nucleus [2,3]. Recently, this hypothesis received a strong impulse for revival from the experiment [1]. In this article, a brief review of the field is presented along with the formulation of some unresolved theoretical issues.

2. Melt of rings

Understanding the melt of unconcatenated and unknotted ring molecules is one of the surprisingly tough problems on the brink between polymer physics and topology. In purely mathematical form the problem can be formulated as follows. Consider a large segment of cubic lattice in 3D, with some $K \gg 1$ lattice sites inside. Self-avoiding random walks can be defined on such lattice in a standard way. Imagine now that there are some $M \gg 1$ distinct random walks, each of $N \gg 1$ steps, and each forming a loop, i.e., bond number N connects sites N and 1 in every loop. Furthermore, assume that every loop is a trivial knot (an unknot), and every pair of loops is a trivial link (an unlink). The "density" – fraction of occupied lattice sites, $\phi = N \times M/K$ is supposed to be of order unity. This simple purely mathematical description is believed to give a very reasonable model of a melt of unconcatenated ring polymers.

For simplicity the physics jargon will be used throughout this article, whereby every step of the random walk will be called "monomer" etc. It has to be emphasized however that a large fraction of the article treats really rather clean models in which mathematical formulation is straightforward.

The question is then as follows: There are obviously many ways to construct the above described system of closed unconcatenated random walks, and one wants to know some statistical properties. The typical properties in question might be:

- Consider a vector $\mathbf{R}_e$ connecting monomers 1 and $N/2$ on one of the loops; what is the mean squared average value of $\langle \mathbf{R}_e^2 \rangle$ and how does it depend on N?
- Consider vectors $\mathbf{r}(s)$ connecting monomers s' and $s' + s$; what is the mean squared value of $\langle \mathbf{r}^2(s) \rangle$ and how does it depend on s? The latter question is obviously related to the fractal dimension of the loops in the melt.
- Consider the so-called loop factor, which is the probability $P(s)$ that two monomers s' and $s' + s$ will be found next to each other in space; how does this probability depend on s?

Theoretical thinking about the melt of unconcatenated rings was dominated by the idea that the topological constraints imposed on a given ring by the surrounding ones can be treated using a sort of a "topological mean field", namely, by considering a single ring immersed in the lattice of immobile topological obstacles [6,7,8,9,10], with none of the obstacles piercing through the ring (such that the ring is unconcatenated with the whole lattice). The conformation of a ring in such lattice represents a lattice animal with every bond traversed by the polymer twice, in two opposite directions. We emphasize that the structure of branches in the lattice animal is completely annealed and subject to dynamics and equilibration. In the off-lattice case, the similar annealed randomly branched structure is expected, in which the chain follows itself twice along every branch only down to a certain scale d, similar to the distance between obstacles or the tube diameter in reptation theory. Such structure is reasonable as long as $b \ll d \ll R_g$, where b is the Kuhn length and R_g is the gyration radius of the loop.

3. Crumpled globule

Understanding the melt of unconcatenated and unknotted rings is also closely related to the problem of compact globular state of an unknotted loop. It was hypothesized a long time ago [2] that equilibrium of a compact unknotted loop is the state of the so-called crumpled globule, in which each subchain is condensed in itself, and, therefore, polymer backbone is organized as a fractal with dimension 3. The idea behind this prediction is that two crumples of the chain act pretty much like two unconcatenated rings, and the latter system obviously experiences at least some volume exclusion for topological reasons (on this subject see recent work [11] and references therein). A direct test of this prediction was attempted in the work[12] by modeling the closed Hamiltonian loops on the compact domains of the cubic lattice. Some evidence of segregation between crumples on the chain was observed, but overall the results were inconclusive because the simulated chains were too short. More detailed test was reported in the work[4].

Notice that the crumpled globule model[2] of the single collapsed unknotted ring is fully consistent with the lattice animal structure in the melt.

4. From territorial polymers to chromosome territories

Squeezing topologically constrained rings against each other in a melt, or squeezing one unknotted ring against itself in a restricted volume, is also a problem of great potential significance in application to the DNA organization in the cell nucleus. Chromatin fibers are packed *in vivo* at a rather high density, with volume fractions not dissimilar to those in a polymer melt. However, unlike what one would expect for the melt of linear chains, the different chromosomes in the nucleus (representing different chains) do not intermix, but stably occupy different distinct "territories": the image in which each chromosome is stained with a different color resembles a political map of some continent [5] (see also reviews [13,14]). This makes an obvious analogy with the melt of unconcatenated rings; indeed, if the rings in the simulated melt are shown in different colors, the image of a political map emerges [15]. The hypothesis becomes then appealing that chromosomes remain segregated for topological reasons, just like unconcatenated rings in the melt [4,16,17]. Why the topology of the chromatin fibers is restricted is a somewhat open question, but it might be that the DNA ends are attached to the nuclear envelope, or simply that the cell life time is not long enough for reptation to develop [18,19]. Territorial segregation of chromosomes appears a common feature for the cells of higher eukaryotes, including humans, but it is less well pronounced or absent altogether in lower eukaryotes such as yeast [14,20,21]. This observation is at least qualitatively consistent with the fact that topological segregation of unconcatenated loops is fully developed only when polymers are very long.

5. Chromatin loop factor

In recent years, more detailed information on chromatin fiber organization became available, due to the advent of novel experimental techniques, such as the ones nicknamed FISH[22] and HiC[23,24] [a], allowing to probe the large scale features of chromatin fiber fold inside a chromosome. In particular, the HiC study of human genome packing[1] revealed the signature of fractal folding pattern which was explicitly associated with the crumpled globule organization of DNA originally predicted on purely theoretical grounds[3]. Specifically, the measurement indicated that the loop factor $P(s)$ (the probability that two loci of genetic distance s base-pairs apart will be found spatially next to each other in chromosome) scales as $P(s) \sim s^{-\gamma}$ over the interval of s, roughly, 0.5 Mbp $\lesssim s \lesssim$ 7 Mbp, where the power γ is 1 or slightly higher. This scaling appears to be consistent with the crumpled globule model [25] (see below). There are also many more indications of self-similarity, or scale-invariance, in the chromatin folding (see recent review [26] and references therein). Current biological thinking about these issues also actively involves dynamics aspects, how this presumably crumpled self-similar structure can move to perform its functions. To this end, there are some observations regarding the structural difference between active part of chromatin fiber, currently transcribed, and the more densely packed, non-transcribing heterochromatin [1]. Not going into any further details, suffice it to say for the purposes of the present paper that chromosome studies necessitate the further and deeper understanding of the melt of unconcatenated rings, both in equilibrium and in dynamics, as a basic elementary model.

[a] FISH stands for Fluorescence In Situ Hybridization. HiC is a generalization of 3C, which means Capturing Chromosome Conformation[23], and 4C, which is 3C on a Chip [24].

6. Penetration of crumples

6.1. *An argument suggesting $1/s$ scaling of contacts between monomers*

Consider first the probability to meet for the two ends of homopolymer globule. The essence of globular state is that these ends are independent from each other, which means the chance for one of them to be inside the (small) volume v around the other is v/V, where $V \sim R^3$ is the total volume of the globule.

For equilibrium globule, this estimate works also for the two arbitrary monomers in the chain, provided however that the contour distance between them s exceeds $N^{2/3}$.

This estimate remains correct also in the case of a crumpled globule, for it is a globule. But for crumpled globule it allows also generalization for any pair of monomers, chemical distance s apart, provided s is larger than the minimal (microscopic) crumple scale. In this case $V = V(s) \sim R^3(s) \sim (s^{1/3})^3 \sim s$ which is the expected scaling.

In general, for any *fractal* conformation with $r(s) \sim r^{\nu}$, the probability for two monomers to meet should go as $v/r^3(s) \sim s^{-3\nu}$. This gives the well familiar result $s^{-3/2}$ for the Gaussian coil and for crumpled globule this yields $1/s$. This is what was found in [1] for chromosomes.

6.2. *Contacts between crumples*

Consider two crumples of the length g each, and suppose they are "close", i.e., the distance between their centers of mass is about their gyration radius or something like that. How many contacts are there between monomers of one crumple and monomers of the other? Let us say this number scales as g^{β}, and the goal is to find β.

Consider a polymer chain in terms of N/g blobs with g monomers each. Consider now pairs of blobs which are distance s monomers, or s/g blobs, apart along the chain. There are about N/g such pairs in the chain (assuming $s \ll N$).

Suppose now that among these pairs some fraction are in contact. This fraction is found to scale as a power of contour distance, i.e. it scales as $(s/g)^{-\gamma}$; according to the argument of section 6.1, $\gamma = 3\nu$, but let us continue the calculations for an arbitrary γ. Therefore, total number of blobs in contact is $(N/g)(s/g)^{-\gamma} = (N/s^{\gamma})g^{\gamma-1}$.

Here, we should digress to discuss what we mean by the "contact between blobs". Let us simply state that two blobs are contact if the spatial distance between them is smaller than $2r(g)$ (or $3r(g)$ or $1.5r(g)$ - the factor is not important), where $r(g)$ is the gyration radius of the blob of g monomers.

Now, let us consider monomer contacts instead of blob contacts. First, monomers do not contact if they belong to non-contacting blobs. Second, if two blobs are contact (in the above defined sense), then it is not obvious how many monomer contacts are there inside. Since we assumed the number of contacting monomer pairs inside two contacting g-blobs in a typical conformation to scale as g^{β}, knowing the number of contacts between blobs, we can find the number of contacts between monomers, it is $(N/s^{\gamma})g^{\gamma-1+\beta}$.

Finally, we have to realize that what we have counted is the number of contacting monomer pairs which are distance $s \pm g$ along the chain. The number of monomer contacts distance exactly s apart is g times smaller, i.e. $(N/s^{\gamma})g^{\gamma-2+\beta}$. This is the main result here.

Now, the number of monomer contacts cannot depend on how we counted them, that is, it cannot depend on g. Therefore, we arrive at

$$\gamma + \beta = 2 \ . \tag{1}$$

As an example, consider Gaussian polymer coil in 3D. In this case, $\gamma = 3/2$ and $\beta = 1/2$, so our formula (1) works. In general, for the Gaussian coil in d dimensions, we have $\beta = 2 - d/2$ and $\gamma = d/2$, so again it works.

For crumpled globule, as an independent argument (see above section 6.1) suggests, $\gamma = 1$, consistent with experiment [1]. Then formula (1) dictates that $\beta = 1$, which means crumples are in contact throughout their interior. This may be roughly consistent with the simulations [4], but it is in contradiction to [3] (where it was assumed that the number of contacts between crumples scales as $g^{2/3}$).

Note that the result (1) is quite independent of the nature of fractal globule. It seems a sort of general fractal statement. On the other hand, $\gamma = 1$ seems an inevitable consequence of the crumpled globule $r(g) \sim g^{1/3}$. If that is correct, then $\beta = 1$, i.e. "voluminous" overlap between crumples is also an automatic consequence of crumpled globule fractality – which seems to contradict the Peano curve example.

6.3. *Why $1/s$ scaling cannot be perfect*

The number of contacts for a given monomer is, on average,

$$\sum_s P(s) \ , \tag{2}$$

where $P(s)$ is the loop factor. If $P(s) \sim 1/s$, then this thing diverges and is determined by the upper limit, i.e., each monomer has about $\ln N$ contacts. This is impossible in the true limit $N \to \infty$.

In practice, this is an unimportant restriction, because "contact" in this topological game is on the level of the entanglement length N_e, therefore, the condition is $\ln N/N_e < N_e$ or $N < N_e^{N_e}$. This is a completely non-restrictive condition even when N_e is as small as 12 ($12^{12} \approx 10^{13}$), while usually N_e is closer to 50.

But as a matter of principle we have a problem here. The argument about $1/s$ must be flawed, there must be correlations between ends of s-piece.

6.4. *Speculation and conjecture*

Thus, the scaling argument suggests that neighboring crumples do not behave like solid impenetrable bodies. Instead, they penetrate each other quite strongly. This is consistent even with simple visual inspection of the image[15].

Peano curve can be easily constructed in 3D. Normal most trivial Peano curve is based on smooth volumed blocks, therefore, it has $\beta = 2/3$ and, then, $\gamma = 4/3$. The natural conjecture, which has to be proved or disproved, is that Peano curve can be constructed with any $\gamma > 1$ (not ≥ 1 but > 1).

7. Dynamics of crumples

An interesting question for the melt of rings is their dynamics. On the simplest level, one can address the dynamics of lattice animal type object in the topological mean field of fixed obstacles. The dynamics of such object is easy to visualize as an amoeba-like motion.

On a more realistic level, one aspect is overall motion of one ring among the others, it can be formulated in the following way.

Consider a single ring in the melt of other similar rings, and consider the position vector of its mass center as it changes in time $\mathbf{r}(t)$. In the very long time limit, there is no doubt that the behavior will be diffusive, $\langle \mathbf{r}^2(t) \rangle \simeq 6D_N t$, which opens two questions:

- What is the value of the diffusion coefficient of one ring, D_N, and how does it depend on the loop length N?
- What is time scale beyond which the diffusive behavior is established?

- What is the behavior on shorter time scales?

These questions were addressed in the recent simulation work [27], and we refer the interested reader to that work for many details. Here, we only announce a few relevant features.

- The dynamics of a single ring is subdiffusive for a significant period of time, before it crosses over to the diffusive regime.
- The cross over time is so long that the ring moves on average several times its own gyration radius before the correlations are forgotten and the diffusive asymptotics kicks in.

These observations are consistent with the idea that crumples penetrate significantly each other (with $\beta > 2/3$). It is conceivable that interpenetrating crumples take the form of invaginations which act as anchors slowing down the relative diffusion of crumples significantly beyond their apparent sizes measured by the gyration radii.

Another aspect of the dynamics in the melt of rings is the dynamics of a single ring, for instance, autocorrelation of its gyration tensor. These aspects were also addressed in the work [27].

In a more general theoretical prospective, the dynamics of rings in the melt represents a significant fundamental challenge, because the main workhorse in the dynamics of linear polymers – reptation theory – is definitely inapplicable directly and has to be modified in some unknown intricate way. Some estimates and suggestions on this subject have to do with the so-called primitive path analysis which is discussed for the rings systems in the works [28,27].

8. Sequence signatures

The attractiveness of crumpled globule hypothesis is largely due to the fact that it is based solely on very general robust polymer properties and does not employ anything specific. At the same time one can also speculate about the relation of the crumpled spatial structure and the DNA sequence, particularly the non-coding part, the introns. And this relation might be in either of the two directions:

- It is possible that the crumpled structure which exists for the general polymeric reasons may be made more stable by selecting proper sequences. Thus, in this scenario sequences are selected with the goal to make the more stable structure.

- It is possible that since sequences evolve while structure remains fractal, then sequences acquire some properties which are dictated by the structural framework.

Here we refer to the sequence-structure relationship of heteropolymers which was extensively studied in the protein context [29]. The use of these ideas in the DNA context is by no means obvious, but we think it should still be possible provided that one understands that heteropolymeric interactions in question are the effective renormalized interactions which are realized by the messenger function of the protein components, such as nucleosome positioning etc. In this sense, the situation is rather similar to that in the artificial heteropolymers whose sequence-structure relationship may be frustrated [30,31]. It is definitely premature to make any commitment in terms of the above mentioned causality (sequence because of structure versus sequence for the structure), but in any case it is important that certain fractal signatures are possible in the sequence depending on the type of structure [32]. It remains to be seen whether or not these correlations are related to the ones reported in the work [33] (see also further discussion and references in[26]).

9. Conclusion

In this article we presented some thoughts on the application of the crumpled globule model developed from purely theoretical perspective to the issue of chromatin folding. It seems safe to conclude that the investigation of this model at least opens up some theoretical and experimental questions which are definitely useful for the understanding of these important phenomena.

Acknowledgments

It is a great pleasure to acknowledge here useful discussions with R.Everaers, M.Imakaev, J.Halverson, K.Kremer, L.Mirny, J.-M.Victor, E.Yeramian, C.Zimmer and particularly with Y.Rabin.

Appendix A. Is there entropic segregation of polymers without topological constraints?

There are also attempts in the literature [37,38,39] to explain territorial segregation of chromosomes by the entropic factors related to polymeric nature of chromatin fiber but unrelated to its topology, that is, by considering the linear chains in a confined volume. Some of the authors [37,38] concentrate on the related problem of DNA packaging in bacterial (prokaryote) cells, others look at interphase chromosomes of eukaryotes[39]. In any case it is useful to discuss under which conditions such segregation unrelated to topology can take place.

Consider therefore a regular polymer with excluded volume confined in a closed volume in the shape of a cylinder of length H and diameter D. The question is this: is the polymer fills the volume randomly, or there is a correlation between coordinate s along the chemical (genetic) distance and the coordinate z along the cylinder axis?

To address this, we should look at the correlation distance ξ which determines the character of polymer conformation. The value of ξ can be established based on the general principles of polymer physics[34,35,36]. Indeed, the correlations are broken by the two mechanisms – by the walls of the confining volume and by the collisions between parts of the chain. These two factors are related to the confinement blobs and concentration blobs, respectively. Therefore, we can estimate the two correlation radii independently and then the real ξ will be the smallest of the two.

Confinement correlation length is nothing else but the tube diameter D. Corresponding confinement blob includes the number of monomer g_{conf} which is determined from $bg_{\text{conf}}^{\nu} \sim D$. It is useful to define the length H_N which the polymer would occupy along the tube if there were not longitudinal restrictions along the tube. This length is given by $H_N = (N/g_{\text{conf}})bg_{\text{conf}}^{\nu} \sim Nb\,(b/D)^{(1-\nu)/\nu}$.

Concentration blob is determined by the condition that monomer concentration inside the blob matches that averaged over the entire volume, which gives

$$\frac{g_{\text{conc}}}{\left(bg_{\text{conc}}\right)^3} \sim \frac{N}{HD^2} \ .$$

It is a matter of simple algebra to show now that

$$\frac{\xi_{\text{conc}}}{\xi_{\text{conf}}} \sim \left(\frac{g_{\text{conc}}}{g_{\text{conf}}}\right)^{\nu} \sim \left(\frac{H}{H_N}\right)^{\frac{\nu}{3\nu-1}} \ . \tag{A.1}$$

We can draw the conclusions now. The correlations imposed by the confinement dominate if $H \gg H_N$. This corresponds to a bacteria which is so very long that DNA occupies only a small fraction of its length. In this situation there is no doubt that indeed DNA in the bacteria is "segregated" in the sense that, for instance, the 1/3 of DNA contour next to one of the DNA ends will occupy about 1/3 of the occupied place, also close to the corresponding end. However, most of the length will remain free of DNA anyway. This is perhaps not very realistic and definitely not very interesting limit, for in this limit DNA hardly can be called truly "confined" in the sense used in biological literature.

In the opposite limit, when $H \ll H_N$, the DNA is truly strongly confined, compressed in all directions. In this case correlations imposed by the concentration completely dominate, and, therefore, the tube shape has marginal effect on the DNA placement. That means, the probability to find one end of the DNA in some place along the tube is hardly at all affected by the placement of the other end. Therefore, there is no territorial segregation of any kind.

There is, of course, also the cross-over regime, when H_N happens to be just the same as H. In this case, one can speak of a marginal confinement and there is at the same time marginal segregation.

References

1. E. Lieberman-Aiden, N.L. van Berkum, L. Williams, M. Imakaev, T. Ragoczy, A. Telling, I. Amit, B.R. Lajoie, P.J. Sabo, M.O. Dorschner, R. Sandstrom, B. Bernstein, M.A. Bender, M. Groudine, A. Gnirke, J. Stamatoyannopoulos, L.A. Mirny, E.S. Lander and J. Dekker, *Science* **326**, 289 (2009).
2. A.Y. Grosberg, S.K. Nechaev and E.I. Shakhnovich, *Journal de Physique (France)* **49**, 2095 (1988).
3. A.Y. Grosberg, Y. Rabin, S. Havlin and A. Neer, *Europhysics Letters* **23**, 373 (1993).
4. T. Vettorel, A.Y. Grosberg and K. Kremer, *Physical Biology,* **6**, 025013 (2009).
5. T. Cremer and C. Cremer, *Nature Reviews Genetics* **2**, 292 (2001).
6. M.E. Cates and J.M. Deutsch, *Journal de Physique (France)* **47**, 2121 (1986).
7. A.R. Khokhlov and S.K. Nechaev, *Physics Letters A* **112**, 156 (1985).
8. M.Rubinstein, *Phys. Rev. Lett.* **57**, 3023 (1986).
9. S.K. Nechaev, A.N. Semenov and M.K. Koleva *Physica A* **140**, 506 (1987).
10. S.P. Obukhov, M. Rubinstein and T. Duke, *Phys. Rev. Lett.* **73**, 1263 (1994).
11. M. Bohn, D. Heermann, O. Lourenço and C. Cordeiro, *Macromolecules* **43**, 2564 (2010).
12. R. Lua, A.L. Borovinskiy, and A.Y. Grosberg, *Polymer* **45**, 717 (2004).
13. T. Cremer, M. Cremer, S. Dietzel, S. Müller, I. Solovei and S. Fakan, *Current Opinion in Cell Biology* **18**, 307 (2006).
14. K.J. Meaburn and T. Misteli, *Nature* **445**, 379 (2007).
15. T. Vettorel, A.Y. Grosberg and K. Kremer, *Physics Today* **62**, 72 (2009).
16. T. Blackstone, R. Scharein, B.Borgo, R.Varela, Y.Diao and J.Arsuaga, *Journal of Mathematical Biology,* , (2010).
17. J. Dorier and A. Stasiak, *Nucleic Acids Research* **37**, 6316 (2009).

18. J.-L. Sikorav and G. Jannink, *Biophysical Journal* **66**, 827 (1994).
19. A. Rosa and R. Everaers, *PLoS Computational Biology* **4**, e1000153 (2008).
20. Z.Duan, M. Andronescu, K. Schutz, S. McIlwain, Y.J. Kim, C. Lee, J. Shendure, S. Fields, C.A. Blau and W.S. Noble, *Nature* **465**, 363 (2010).
21. P. Therizols, T. Duong, B. Dujon, C. Zimmer and E. Fabre, *Proc. Nat. Ac. Sci.* **107**, 2025 (2010).
22. I. Solovei, A. Cavallo, L. Schermelleh, F. Jaunin, C. Scasselati, D. Cmarko, C. Cremer, S. Fakan and T. Cremer, *Experimental Cell Research* **276**, 10 (2002).
23. J.Dekker, K.Rippe, M.Dekker, and N.Kleckner, *Science* **295**, 1306 (2003).
24. M. Simonis, P. Klous, E. Splinter, Y. Moshkin, R. Willemsen, E. de Wit, B. van Steensel and W. de Laat, *Nature Genetics* **38**, 1348 (2006).
25. A. Rosa, N.B. Becker and R. Everaers, *Biophysical Journal* **98**, 2410 (2010).
26. A. Bancaud, C. Lavelle, S. Huet and J. Ellenberg, in preparation (2010).
27. J.D. Halverson, Won Bo Lee, G.S. Grest, A.Y. Grosberg and K.Kremer, *Bull. Amer. Phys. Soc. APS* **55**, No. 2 (2010).
28. N. Uchida, G.S. Grest, and R. Everaers, *The Journal of Chemical Physics* **128**, 044902 (2008).
29. E.I. Shakhnovich, *Chemical Reviews* **106**, 1559 (2006).
30. Y. Takeoka, A.N. Berker, R. Du, T. Enoki, A.Y. Grosberg, M. Kardar, T. Oya, K. Tanaka, G. Wang, X. Yu and T. Tanaka, *Phys. Rev. Letters* **82**, 4863 (1999).
31. T. Enoki, T. Oya, K. Tanaka, T. Watanabe, T. Sakiyama, K. Ito, Y. Takeoka, G. Wang, M. Annaka, K. Hara, R. Du, J. Chuang, K. Wasserman, A.Y. Grosberg, S. Masamune and T. Tanaka, *Phys. Rev. Letters* **85**, 5000 (2000).
32. E.N. Govorun, V.A. Ivanov, A.R. Khokhlov, P.G. Khalatur, A.L. Borovinsky and A.Y. Grosberg, *Phys. Rev. E* **64**, 040903 (2001).
33. C.-K. Peng, S.V. Buldyrev, A.L. Goldberger, S. Havlin, F. Sciortino, M. Simons and H.E. Stanley, *Nature* **356**, 168 (1992).
34. P.-G. de Gennes, *Scaling Concepts in Polymer Physics*, Cornell Univ. Press (1979).
35. M. Rubinstein and R. Colby, *Polymer Physics*, Oxford University Press (2005).
36. A.Y. Grosberg and A.R. Khokhlov, *Statistical Physics of Macromolecules*, American Institute of Physics (1994).
37. S. Jun and B. Mulder, *Proc. Natl. Acad. Sci. USA* **103**, 12388 (2006).
38. S. Jun, *Quantitative Biology. arXiv.org, Cornell University Library. Available at: http://arxiv.org/abs/0808.2646* (2008).
39. P.R. Cook and D. Marenduzzo, *The Journal of Cell Biology* **186**, 825 (2009).

INTERNAL SYMMETRIES AND CLASSIFICATION OF TUBULAR VIRAL CAPSIDS

R. KERNER

Laboratoire de Physique Théorique de la Matière Condensée
Université Pierre et Marie Curie,
Tour 23, B.C. 121, 4 Place Jussieu, 75005 Paris, France
E-mail: richard.kerner@upmc.fr

We present a generalization of the analysis of viral capsid symmetries which was quite successfully applied to icosahedral capsids. It consists in the extension of the same method to prolate and tubular capsids formed with similar protein building blocks, namely the five-fold and six-fold capsomers displaying different internal symmetries than those forming strictly icosahedral structures. The dimensions and shapes of tubular capsids are analyzed and evolutionary relationships are suggested by structural affinities.

1. Introduction

Mathematicians often investigate algebraical or geometrical structures without special incitement or demands coming from other branches of human knowledge. Sometimes the structures are considered just for fun, or for their exceptional beauty and simplicity. This was probably the case when Coxeter found an infinite series of icosahedral shells constructed with twelve perfect pentagons and certain amount of perfect hexagons with the same side length ([1]). Each such icosahedron is characterized by two non-negative integers, p and q, and the *triangular number* $T = p^2 + pq + q^2$. The total number of hexagons is given by simple formula

$$N_6 = 10\,(T - 1)$$

The Euler theorem, obviously valid in this particular convex polyhedron case states that

$$N_F - N_E + N_V = 2,$$

where N_F is the number of faces, N_E the number of edges, and N_V is the number of vertices. Here we have

$$N_F = 10\,(T-1) + 12, \qquad N_E = \frac{1}{2}\,(12 \times 5 + 10(T-1) \times 6),$$

$$N_V = \frac{1}{3}\,(12 \times 5 + 10(T-1) \times 6), \tag{1}$$

because two adjacent polygons share their edges, and three adjacent polygons meet at each vertex. It is easy to see that the total number of hexagons is irrelevant, because one always has

$$N_6 - \frac{1}{2}(6 \times N_6) + \frac{1}{3}(6 \times N_6) = (1 - 3 + 2)N_6 = 0,$$

whereas the number of pentagons must be equal to 12 in order to satisfy Euler's relation: indeed from the same counting principle, if the number of pentagons is N_5, we must get

$$N_5 - \frac{1}{2}(5 \times N_5) + \frac{1}{3}(5 \times N_5) = 2,$$

which leads to $N_5 = 12$.

The first few Coxeter polyhedra with icosahedral symmetry correspond to triangular numbers $T = 3$, $T = 4$, $T = 7$, $T = 9$, and so forth. The smallest one is used as the most popular soccer ball model. It is amazing that a few decades later new molecular structures of pure carbon discovered by Smalley, Kroto and Kretschmar ([7]) turned out to be a realization of Coxeter's schemes. Closer investigations led to the discovery of prolate and tubilar structures, or the capped tubules, in which the two extremities were made of halves of the C_{60}, separated by belts made exclusively of hexagons. The corresponding formulas are C_{70}, C_{80}, etc.

Similar icosahedral structures were observed since a long time in many species of viruses. Their protecting shells, called capsids, display pentagonal or hexagonal symmetry, even when they are composed of smaller units. Each icosahedral capsid has exactly twelve pentamers and the number of hexamers respecting Euler's rule and falling into Coxeter's scheme, which was introduced for viral capsids by Caspar and Klug in 1962 ([2]). In ([14]) we gave a more detailed classification of icosahedral viral capsids based on their internal symmetries, of which there are four distinct classes.

The aim od this article is to show how the classification of icosahedral capsids can be generalized to the case of prolate and tubular viral capsids. By these we mean capsids which are constructed with the same coat proteins as the icosahedral ones. We shall not consider here the very long tubular capsids made of one type of protein, repeating the same structure, like what is observed in the *tobacco mosaic* virus.

From a purely geometrical point of view one can construct an infinite series of prolate and tubular capsids starting from an icosahedral capsid, whose two halves would become tubular capsids' caps. The cylindrical part in between is made exclusively by hexamers, whose number depends on the type of the caps and the length of the tube. Apparently, only a small fraction of possible structures of this type is realized in nature, which was already the case with icosahedral capsids. On the other hand, we cannot claim the ultimate knowledge of all existing viruses - new discoveries are yet to come, as well as spontaneous mutations which occur sometimes.

One can conjecture that the prolate and tubular structures produced with pentamers and hexamers use for their two caps the same scheme as the corresponding icosahedral viruses do. The main difference resides then in the extra hexamer types that alter the assembly rules leading to the prolate capsids. In what follows, we shall classify all possible prolate capsids constructed on the base of corresponding icosahedral capsids, at least for the first few triangular numbers.

Even if we knew for sure that certain prolate structures do not occur in nature, the question remains: why do they not appear ? The answer may be found by considering relative volume gain, rigidity and robustness of structure, and some other parameters that can influence survival capacities of a given virus species. Another interesting question concerns the evolutionary patterns that relate icosahedral and prolate and tubular capsid viruses: how many mutations, and of what kind were needed in order to transform one species into another. We believe that detailed knowledge of structural parameters characterizing all possible capsids may shed some light on these questions.

2. Capsomer types in icosahedral capsids

In order to understand internal symmetries and construction scheme of prolate capsids with icosahedral caps, we must first recall the classification of icosahedral capsids.

Since Caspar and Klug [2] introduced simple rules predicting a sequence of observed viral capsids, several models of growth dynamics of these structures have been proposed, e.g. A. Zlotnick's model [8] published in 1994.

In capsids, the building blocks made of *coat proteins* are called *monomers*, *dimers*, *trimers*, *pentamers* and *hexamers*, according to their shape, the bigger ones usually being assembled from smaller ones prior to further agglomeration into capsid shells [10] .

In some cases, the similarity with the fullerene structure is striking: for example, the TRSV capsid is composed of 60 copies of a single capsid protein (56 000 Da, 513 amino acid residues) [11], which can be put in a one-to-one correspondence with 60 carbon atoms forming a fullerene C_{60} molecule; the aforementioned *Cowpea* viruses provide another example of the same type (see Fig. 1).

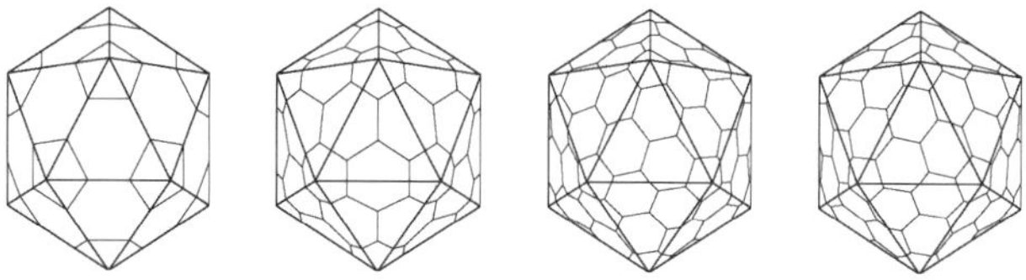

Figure 1. Schematic representation of capsids with T-numbers 3, 4 and 7, right- and left-handed (Courtesy of VIPERdb)

Capsids are built progressively from agglomerates of giant protein molecules displaying pentagonal or hexagonal symmetry, or directly from smaller units (*monomers* or *dimers*). It also seems that there is no such thing as universal assembly kinetics: the way the capsids are assembled differs from one virus to another.

Quite strikingly, viruses use almost 100% of pentamers and hexamers at their disposal to form perfect icosahedral capsid structures, into which their DNA genetic material is densely packed once the capsid is complete. This means that the initial nucleation ratio of pentamers versus hexamers is very close to its final value in capsids in order to minimize the waste. Secondly, the final size of the capsid must depend on particular assembly rules, which can be fairly well deduced from the unique final form of given capsid. These assembly rules can be summarized in a square table, or a matrix, describing the affinity between proteins forming pentagonal or hexagonal building blocks. Denoting different capsid proteins by letters a, b, c, etc., the square table will display 1 at the crossing of line and row corresponding to two proteins that are bound to stick together, and 0 otherwise. In order to make the agglomeration rules free of any ambiguity, each row and each column can display only one non-zero entry. The resulting assembly rules are like a user's description for a kit model, giving precise instructions as to which parts have to be put together, and by which sides.

Similar assembly rules are apparently valid for the prolate capsids as well; but to make our method clear we start with a short reminder of how the isocahedral capsids are assmebled.

From symmetry considerations (and confirmed by chemical analysis) it results that the pentamers are composed from five identical dimers, so that their five edges are perfectly equivalent (identical), because 5 is a prime number, and any division into parts will break the symmetry. Concerning the hexamers, as 6 is divisible by 2 and 3, one can have the following four situations:

- All 6 sides equivalent, $(aaaaaa)$
- Two types of sides, disposed as $(ababab)$
- Three types of sides, disposed as $(abcabc)$
- Six different sides, $(abcdef)$

The hexamers are also *oriented*, with one face becoming external, and the other one turned to the interior of the capsid. The three differentiated hexamers are represented in Figure 2 below.

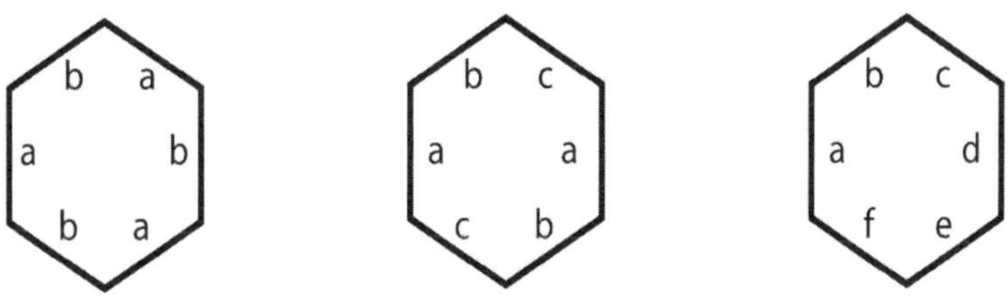

Figure 2. Three differentiations of hexamers

Let us denote pentamers' sides by symbol p, whereas two different kinds of sides on hexamers' edges will be called a and b (Fig. 2). Suppose that a hexamer can stick to a pentamer with only $(p + a)$-combination; then two hexamers must stick to each other only through a $(b + b)$ combination, with both $(p + b)$ and $(a + b)$ combinations being forbidden by chemical potential barrier.

Similarly, with a more differentiate hexamer scheme, $(abcabc)$, and with the assembling rules allowing only associations of $p + a$ and $b + c$, we get with a 100% probability the $T = 4$ capsid, as shown in Fig. 3.

Note that in both cases we show only one of the "basic triangles" forming the capsid, which is always made with 20 identical triangles sticking together to form a perfect icosahedral shape. One can also observe that in many cases the capsids are made of smaller building blocks, called *dimers* and *trimers*, containing only two or three proteins; however, the final symmetry is exactly the same as in the case of hexamers and pentamers, like the examples in Fig. 3 clearly show.

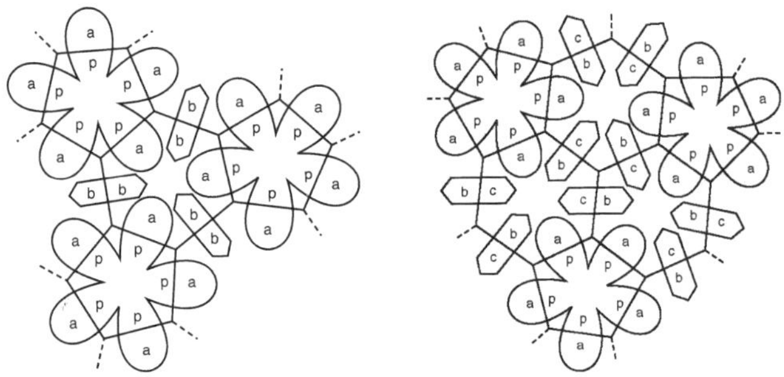

Figure 3. The $T = 3$ and $T = 4$ basic triangles composed by dimers and trimers

These examples suggest that strict association rules may exist providing precise agglomeration pathways for each kind of icosahedral capsid. Let us analyze these rules in more detail.

If the viruses were using undifferentiated hexamers with all their sides equivalent, then there would be no reason for not creating any kind of arbitrary local structures and the final yield of the "proper" ones would be very low (at best like in the fullerenes, less than 10%),

Whereas with differentiated hexamers of the $(ababab)-$ type simple selection rules excluding the $(p - b)$ and (ab) associations while letting the creation of $(p - a)$ and of $(b - b)$ links, we have seen that the issue becomes determined with practically 100% certainty, as it follows from the Fig. 3. The next case presents itself when one uses the next hexamer type, with a two-fold symmetry : $(abcabc)$. Again, supposing that only a-sides can stick to pentamers' sides p, there is no other choice but the one presented in Fig. 3. These sticking rules can be summarized up in a table that we shall call the *"affinity matrix"* , displayed in the Table 1 below:

Table 1. Affinity matrices for the $T = 3$ and $T = 4$ capsid construction

	p	a	b
p	0	1	0
a	1	0	0
b	0	0	1

	p	a	b	c
p	0	1	0	0
a	1	0	0	0
b	0	0	0	1
c	0	0	1	0

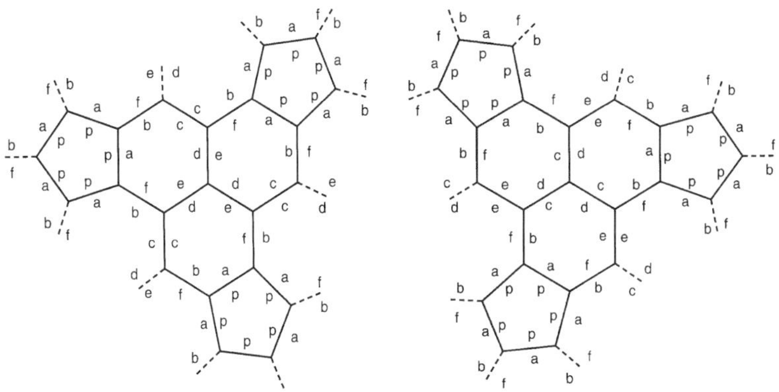

Figure 4. The building formula for the T=7 capsids, left and right

Here a "0" is put at the crossing of two symbols whose agglomeration is forbidden, and a "1" when the agglomeration is allowed. By construction, a "1" can occur only once any line or in any column. The next case presents itself when one uses the next hexamer type, with a two-fold symmetry: $(abcabc)$. Again, supposing that only a-sides can stick to pentamers' sides p, there is no other choice but the one presented in Fig. 3. The corresponding affinity matrix is as follows:

Finally, let us use the most highly differentiated hexamers of the $(abcdef)$-type. Starting with pentamers surrounded by the hexamers sticking via the $(p - a)$-pairing, we discover that now two choices are possible, leading to left- and right-hand sided versions, as shown in the following Fig. 4.

The corresponding affinity matrices can be easily deduced from the figure; they can be found in [14] and [16].

In order to grow capsids with T-numbers greater than 7, one has to introduce new types of hexamers that would never stick to pentamers, but being able to associate themselves with certain sides of the former maximally differentiated hexamers. The result is shown in the Fig. 5 below, where one can see the building scheme for $T = 9$ and $T = 12$ capsids:

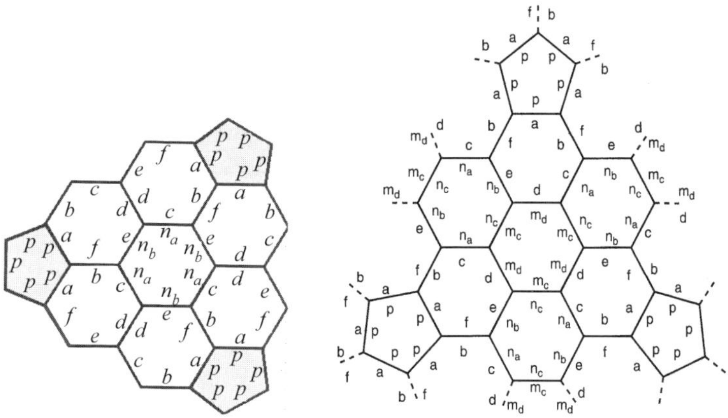

Figure 5. The building formulae for the T=9 and T=12 capsids

The corresponding affinity tables can be found in the book [16] and in the paper [14].

For bigger capsids, in which the rate of pentamers is lower, one can not obtain proper probabilities unless more than one type of hexamers is present, out of which only one is allowed to agglomerate with pentamers. In the case of two different hexamer types one obtains either the $T = 9$ capsid, or, with more exclusive sticking rules, the $T = 12$ capsid.

Finally, in order to get the $T = 25$ adenovirus capsid, one must introduce no less than *four* hexamer types, out of which only one type can agglomerate with pentamers.

3. Four classes of icosahedral capsid symmetries

Now we can organize all these results in a single table that follows. To each value of triangular number T corresponds a unique partition into $1+(T-1)$, where the "1" represents the unique pentamer type and $(T-1)$ is partitioned into a sum of certain number of different hexamer types, according to the formula

$$(T - 1) = 6\,\alpha + 3\,\beta + 2\,\gamma \tag{2}$$

with non-negative integers α, β and γ.

Table 2. Classification of icosahedral capsids. The last column gives the number and type of hexamers needed for the construction

Type (p,q)	$T = p^2 + pq + q^2$	$N_6 = 10(T-1)$	T decomposition
(1,1)	3	20	$1+2$
(2,0)	4	30	$1+3$
(2,1)	7	60	$1+6$
(3,0)	9	80	$1+6+2$
(2,2)	12	110	$1+6+2+3$
(3,1)	13	120	$1+6+6$
(4,0)	16	150	$1+6+6+3$
(3,2)	19	180	$1+6+6+6$
(4,1)	21	200	$1+6+6+6+2$
(5,0)	25	240	$1+(4\times 6)$
(3,3)	27	260	$1+(4\times 6)+2$
(4,2)	28	270	$1+(4\times 6)+3$
(5,1)	31	300	$1+(5\times 6)$
(6,0)	36	350	$1+(5\times 6)+2+3$
(4,3)	37	360	$1+(6\times 6)$
(5,2)	39	380	$1+(7\times 6)+2$
(6,1)	43	420	$1+(7\times 6)$
(4,4)	48	470	$1+(7\times 6)+2+3$
(7,0)	49	480	$1+(8\times 6)$
(5,3)	49	480	$1+(8\times 6)$
(6,2)	52	510	$1+(8\times 6)$
(7,1)	57	560	$1+(9\times 6)+2$

Table 3. Periodic Table of icosahedral capsids. The four types, A, B, C and D are put in separate columns.

T, (p,q)	**A**(5 + 6)	**B**(5 + 6 + 2)	**C**(5 + 6 + 3)	**D**(5 + 6 + 2 + 3)
1, (1, 0)	**(1, 0, 0, 0)**			
3, (1, 1)		**(1, 0, 1, 0)**		
4, (2, 0)			**(1, 0, 0, 1)**	
6	–	–	–	(1, 0, 1, 1)
7, (2, 1)	**(1, 1, 0, 0)**			
9, (3, 0)		**(1, 1, 1, 0)**		
10	–	–	(1, 1, 0, 1)	–
12, (2, 2)				**(1, 1, 1, 1)**
13, (3, 1)	**(1, 2, 0, 0)**			
15	–	(1, 2, 1, 0)	–	–
16, (4, 0)			**(1, 2, 0, 1)**	
18	–	–	–	(1, 2, 1, 1)
19, (3, 2)	**(1, 3, 0, 0)**			
21, (4, 1)		**(1, 3, 1, 0)**		
22	–	–	(1, 3, 0, 1)	–
24	–	–	–	(1, 3, 1, 1)
25, (5, 0)	**(1, 4, 0, 0)**			
27, (3, 3)		**(1, 4, 1, 0)**		
28, (4, 2)			**(1, 4, 0, 1)**	
30	–	–	–	(1, 4, 1, 1)
31, (5, 1)	**(1, 5, 0, 0)**			
33	–	(1, 5, 1, 0)	–	–
34	–	–	(1, 5, 0, 1)	–
36, (6, 0)				**(1, 5, 1, 1)**

Let us continue to organize all icosahedral capsids into different classes following the type of symmetry of hexamers involved in their construction.

We saw already that to each value of triangular number T corresponds a unique partition into $1 + (T - 1)$, where the "1" represents the unique pentamer type and $(T - 1)$ is partitioned into a sum of certain numbers of different hexamer types according to the formula $(T-1) = 6\,\alpha + 3\,\beta + 2\,\gamma$ (2) with non-negative integers α, β and γ, the numbers β and γ taking on exclusively the values 0 or 1; this results from the fact that the corresponding hexamers are centered on a *three-fold* or a *two-fold symmetry axis*, so that the first type must be found at the center of icosahedron's triangular face, whereas the second type must be found in the center of an edge between elementary triangles. The number α of maximally differentiated hexamers follows then from the corresponding partition of a given triangular number, as shown in the table.

It follows that all icosahedral capsids can be divided into four separate groups according to their internal symmetry, dictated by the presence of three- and two-fold centers inside the elementary triangles or on their edges. The result looks like a periodic table of capsids, arranged in four columns, according to the composition of constitutive hexamers: pure $(abcdef)$ hexamers exclusively in the first column; the $(abcdef)$-type hexamers with one $(ababab)$-type hexamer in the second column; with one $(abcabc)$-type hexamer in thye third column, and finally with both, $(ababab)$ and $(abcabc)$-type in the fourth column. Symmetric icosahedral capsids (marked in gold) correspond to specific values of T with $(p, 0)$ or (p, p); the chiral icosahedral capsids, existing in left- and right-handed versions, are marked in magenta.

It is easy to see what is the arithmetic nature of triangular numbers of each group. The first column (type A) the triangular number T must be either a prime number, or a square of a prime number. The T-numbers appearing in the second column (type B) are divisible by 3; the T-numbers appearing in the third column (type C) are multiples of 4; finally, the fourth column (type D) contains triangular numbers divisible both by 3 and by 4, i.e. the T-numbers which are multiples of 12.

4. Prolate and tubular capsids

Prolate and tubular capsids constructed with the same coat proteins that the icosahedral ones are quite common in nature. The smallest prolate capsid, quite well studied and known, is the phage Phi-29, endowed with a powerful molecular motor at one of its ends, propulsing the long DNA chain inside the shell when its construction is completed. Another example of prolate capsid is provided by the $T4$ phage ([4]) (here $T4$ does not mean the triangular number, which in this case is equal to 13). Tubular capsids can reach quite large dimensions, like e.g. the giant Mason-Pfitzer Monkey Virus ([5]). In all these cases, the icosahedral character of the cap can be very well observed. The two prolate capids, $Ph29$ and $T4$, are shown in the following Figure 6.

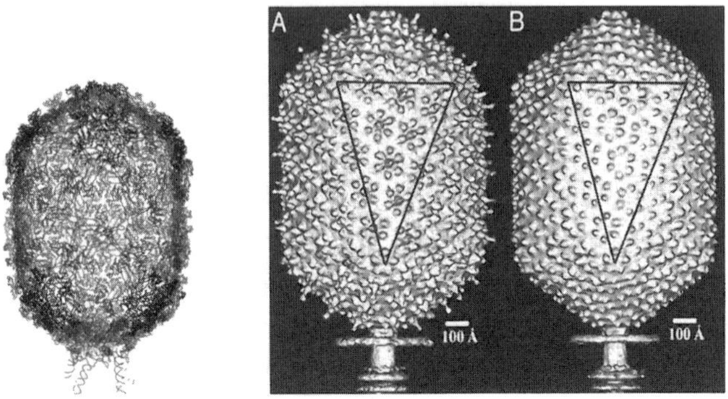

Figure 6. The phage T4 capsid (after A. Fokine et al., PNAS, Vol. 101 (16), (20034)

As we have already shown in the Introduction, adjoining extra hexagons to a tri-coordinated lattice tiling the sphere does not alter the Euler topological formula. Nevertheless, their number is not at all arbitrary if one wants to produce a prolate, but strictly axially-symmetric structure by inserting extra hexagon belts into a given icosahedral capsid.

The structure of icosahedral capsids could be easily analyzed on the example of an *elementary triangle*, because each icosahedron could be composed out of twenty identical triangles. In the case of capped tubular capsids the symmetry is lower, but still, each such structure displays a five-fold symmetry axis, and can be divided into five egual slices. The best way to represent a prolate or tubular capsid on a plane is given by Mercator's projection.

The Mercator projection is often used to represent our globe on a plane. The polar regions are then very strongly extended (and the region in the closest vicinity of North and South Poles are cut out), but the equatorial and tropical regions are mapped quite faithfully.

Mercator's projection is even better suited for representing prolate or tubular structures, containing a large cylindrical part. Below one can see a prolate capsid structure and its schematic view on a plane. There are twelve pentagons, as usual, of which ten are contained in the lateral cylindrical face, and the remaining two are represented as Northern and Southern poles, largely extended.

In the case of icosahedral capsids, each of them composed of twenty identical triangular faces full information can be obtained by considering just one of such faces. In the case of prolate and tubular capsids their axial symmetry suggests another kind of partition, namely, in five identical vertical slices.

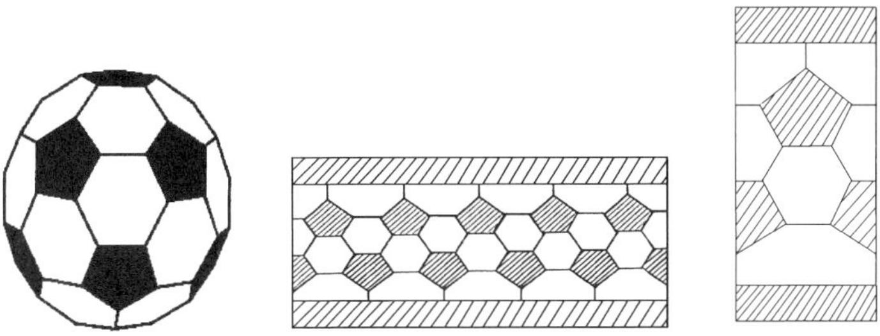

Figure 7. The smallest $T = 3$ icosahedral capsid, its full Mercator projection and the one-fifth vertical slice.

In what follows, we shall adopt the convention for labeling the prolate and tubular capsids following the type of the spherical caps, always supposed identical, and the number of extra hexagon belts that need to be added to the initial icosahedral capsid in order to form the corresponding tubular one. For the caps we shall keep the triangular number T, and the number of extra hexagon belts being denoted by P, so that an icosahedral capsid with triangular number T is noted $(T = 3, P = 0)$, the first prolate capsid (with geometrical structure similar to the prolate fullerene C_{70}, with 30 hexagons and 12 pentagons) will be denoted by $(T = 3, P = 1)$, etc.

Longer structures are also possible, although we are not sure whether they are realized in nature:

In a similar way, on the basis of larger icosahedral capsids characterized by higher triangular numbers $(T = 4, 7, 9...)$ one can built corresponding prolate capsids by adding one or more extra hexamer-filled belts. In the figure (10) we show the Mercator projection of slices of the icosahedral $T = 4, P = 0$ capsid, followed by its prolate extensions $T = 4, P = 1$ and $T = 4, P = 2$. Note the two *isomers* in tha last case $T = 4, P = 2$ prolate structure, due to the fact that the extra belt contains 10 hexamers, and the two icosahedral caps (the Northern and Southern "hemispheres") can be put in two different positions with respect to each other, either straight

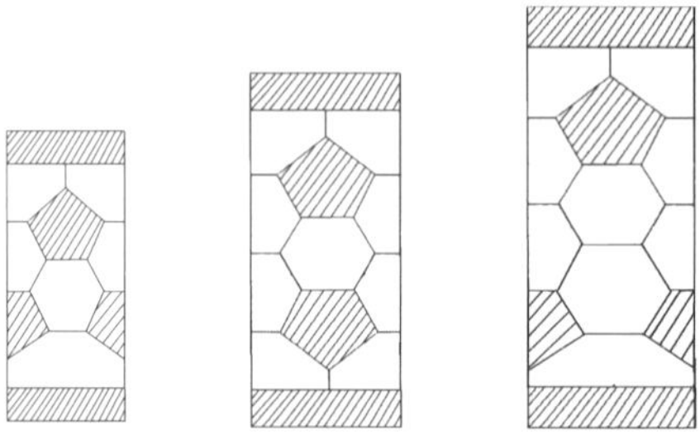

Figure 8. Three examples of tubular capsids $T = 3, P = 0$, $T = 3, P = 1$ and $T = 3, P = 2$. The last one corresponds to the $Ph29$ phage.

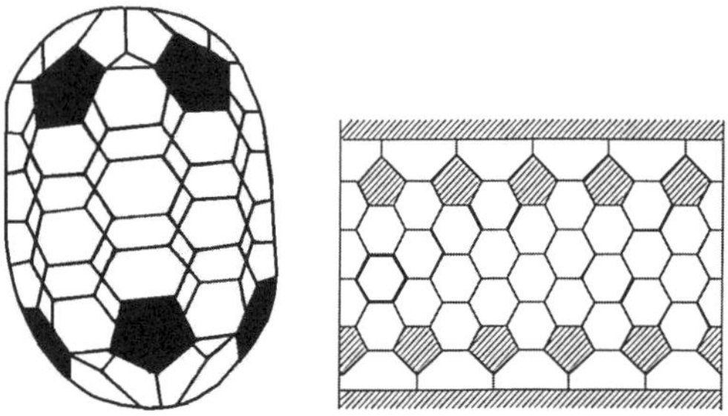

Figure 9. The prolate capsid ($T = 3, P = 3$) and its planar representation in Mercator's projection.

on, or with a 36° angular shift.

Figure 10. Slices of a $T = 4$ icosahedral capsid and its prolate extensions, $T = 4, P = 1$ and $T = 4, P = 2$

It can be easily observed that each slice contains *two* entire pentagons, plus two fifths at the poles; altogether this represents $2 + 2/5 = 12/5$ pentagons, which after multiplication by 5 gives twelve pentagons per capsid, as it should be.

Next T-number is 7; the corresponding icosahedral capsid and its tubular extensions with one ($P = 1$) and two ($P = 2$) extra hexamer belts, each containing 15 more hexamers, are represented in the following Figure 11. Note the skew character of the scheme - the upper and lower caps are turned with respect to each other by 36°.

Figure 11. Slices of a $T = 7$ icosahedral capsid and its prolate extensions, $T = 7, P = 1$ and $T = 7, P = 2$

The next triangular number is $T = 9$.

Here are the slices giving the structure of the newt two icosahedral capsids with triangular numbers $T = 12$, $T = 13$ and $T = 16$. The $T = 13$ is found as the cap ot the prolate phage $T4$ represented in the Figure 6.

All icosahedral capsids can be extended to prolate and tubular forms by adding an appropriate number of extra hexamer belts. The total number of hexamers needed for a new belt formation depends on the T-number of corresponding icosahedral cap, and is always a multiple of 5. Thus, 10 new hexamers are needed to form an extra belt in the case of $T = 3$ and $T = 4$, 15 new hexamers are needed for a new belt in $T = 7$ and $T = 9$ capsids, 20 new hexamers per extra belt for the $T = 12$ and $T = 13$, and so on.

With these building principles we can easily find the total number of hexamers and the total number of coat proteins needed for the construction of the entire capsid of any type. This gives also the possibility to evaluate the volume of each capsid, which in turn gives an approximate information

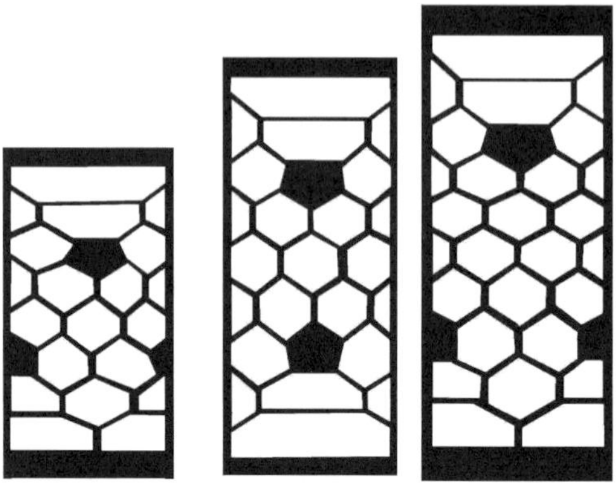

Figure 12. Slices of a $T = 9$ icosahedral capsid and its prolate extensions, $T = 9P = 1$ and $T = 9, P = 2$

Figure 13. The elementary slices of $T = 12$, $T = 12$ and $T = 16$ icosahedral capsids.

of the length of the corresponding DNA chain tthat is to be packed inside the given capsid type.

5. Dimensions of icosahedral and prolate capsids

It is interesting to compare the dimensions, especially volumes, of various viral capsids. Their dimensions are intimately related with the length of the DNA that they host, because it is very densely packed inside, leaving almost no free volume. This gives a hint concerning the evolutionary trends of prolate viruses, supposing that the mutations leading from an icosahedral towards elongated forms are accompanying the growth of length of the DNA that has to be packed inside the capsid. It is also interesting to know how many new protein types have to be produced in order to pass from a given icosahedral assembly scheme to a bigger one, or alternatively, to a prolate form.

With a very good approximation one can assume that all capsomers, pentagonal and hexagonal, have the same side length, close to 8.5 nanometers. We shall denote this commn length by a. The surfaces of a perfect pentagon and a perfect hexagon with side a are given by:

$$S_5 = 1.773\, a^2, \qquad S_6 = 2.6\, a^2. \tag{3}$$

Smaller icosahedral capsids' shapes are very close to spherical, so we can evaluate the radius of such sphere from its total surface. The smallest $T = 3$ icosahedral capsid is composed of 12 pentagons and 20 hexagons; its total surface is then

$$S_{T3} = 12 \times 1.773 + 20 \times 2.6 = 73.28\, a^2$$

The radius R is obtained by expressing the surface as $4\pi R^2$, giving $R = 2.415\, a$. As easily seen from Fig. 8, the circumference of this capsid is made of five hexagons, which gives the total length $15\, a$. Comparing this value with $2\pi R$, we get another radius estimation, $R = 2.387\, a$; the error with respect to the previous estimate is about 1%, which shows how close the capsid is to a perfect spherical shape.

Similar calculus can be made for the newt icosahedral capsid with triangular number $T = 4$. Its surface contains 30 hexagons and 12 pentagons, so its total area is easily found equal to $S_{T4} = 99.28\, a^2$, giving the average radius $R = 2.81\, a$. If we compute the same radius from the length of the equatorial belt, which in this case is composed by six hexagons and two three pentagons, disposed in different ways, we get the result $R = 2.76\, a$, with less than 2% difference.

Finally, the $T = 7$ capsid has 60 hexagons and 12 pentagons, and its average radius is $R = 3.76\, a$, as derived from the surface of the sphere formula; the circumference formula is a bit less obvious in this case due to capsomers' inclined position, but looking at the Mercator projection scheme, we see that the circumference is five times the width of the elementary slice, which is equal to $3a + \sqrt{3}a$. Multiplying this width by 5, we get the circumference equal to $23.66a$, which is practically equal to $2\pi \times 3.76 = 23.62$.

The general formula deriving average radius R_T from the triangular number T of an icosahedral capsid is obtained as follows. The total area of capsids' surface consists of N_6 hexagons and 12 pentagons; this gives

$$S_T = N_6 \times S_6 + 12 \times S_5,$$

with $S_6 = 2.6\, a^2$ and $S_5 = 1.773\, a^2$. Because $N_6 = 10(T-1)$, and $4\pi R_T^2 = S_T$, we get

$$R_T = \sqrt{\frac{10(T-1) \times 2.6\, a^2 + 12 \times 1.773\, a^2}{4\pi}}. \tag{4}$$

Spherical approximation gives less satisfactory results in the case of bigger T-numbers, because the capsids' shapes are closer to an icosahedron than to a sphere. In this case the best way to evaluate the circumference is to use the corresponding Mercator's projection and count the haxagons forming the belt.

The *exterior* volumes of the first three icosahedral capsids are related as cubes of their radii, which means that

$$\frac{V_{T4}}{V_{T3}} = 1.6, \qquad \frac{V_{T7}}{V_{T4}} = 2.4,$$

so the volumes grow quite fast indeed. The useful volumes inside the capsids, serving for ADN storage, are smaller because one has to remove the width of the shell from the total volume; but this effect is not very strong, the capsids being quite thin as compared to their overall dimensions.

It is easy to find the formula for the volume of prolate capsids with two caps with triangular number T and the number of extra cylindrical belts H. We assume that the radius of the cylindrical belt is equal to the radius of the corresponding icosahedral capsid R_T; if its height is h, then the extra volume of the cylinder that should be added to the volume of the icosahedral capsid is given by $\pi h R_T^2$. In the case of $T = 3, H = 1, 2, 3, ...$ we have

$$R = \frac{15a}{2\pi}, \qquad h = \frac{\sqrt{3}}{2} a,$$

and the tubular capsid volume is:

$$V_{(3,H)} = \frac{4}{3}\pi R^3 + \pi h R^2 = \left(\frac{15}{2\pi}\right)^2 \left[10 + \frac{\pi\sqrt{3}}{2} H\right] a^3, \tag{5}$$

or in the simplified form,

$$V_{(3,H)} \left[57 + 15,5\, H\right] a^3.$$

Type	(T, P)	Radius	Surface	Volume
B	(3, 0)	$2.4a$	$73a^2$	$59a^3$
B	(3,1)	$2.4a$	$86a^2$	$75a^3$
C	(4,0)	$2.8a$	$101a^2$	$92a^3$
C	(4,1)	$2.8a$	$127a^2$	$166a^3$
A	(7,0)	$3.8a$	$177a^2$	$229a^3$
A	(7,1)	$3.8a$	$218a^2$	$376a^3$
B	(9,0)	$4.2a$	$229a^2$	$326a^3$
B	(9,1)	$4.2a$	$268a^2$	$365a^3$
D	(12,0)	$4.9a$	$308a^2$	$493a^3$
D	(12,1)	$4.9a$	$360a^2$	$624a^3$
A	(13,0)	$5.2a$	$334a^2$	$573a^3$

6. Internal symmetries of prolate capsids

The internal symmetries in the prolate capsids are inherited from the symmetries of the corresponding icosahedral capsids that provide their two identical halves to serve as two caps of the tubular structure. As in the icosahedral case, there are four distinct classes with triangular numbers which are either prime numbers or products of prime numbers (class A), or divisible by 3 (class B), then multiples of 4 (class C) and finally, multiples of 12 (class D).

The prolate structures with two caps belonging to one of these classes fall into a subsequent classification according to the number of extra hexamer belts, labeled by the second integer P. The larger the capsid, the more isomers are possible due to the different angles between the two polar caps. This is the theory; what is observed in reality in known prolate capsids, is the preference given to skew configurations, with relative angle being either 36° or 72°.

7. Discussion and conclusion

The details of internal structure and symetries of various icosahedral and prolate capsids should provide some hints concerning their evolutionary pattern. There are several conjectures that seem reasonable in the light of geometrical data exposed here:

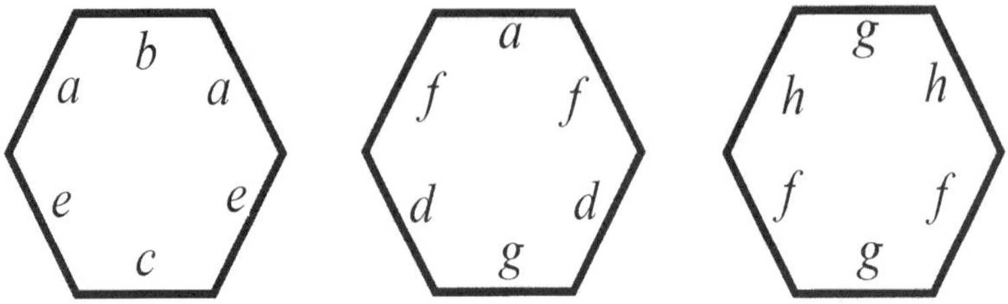

Figure 14. The three types of hexamers with axial symmetry.

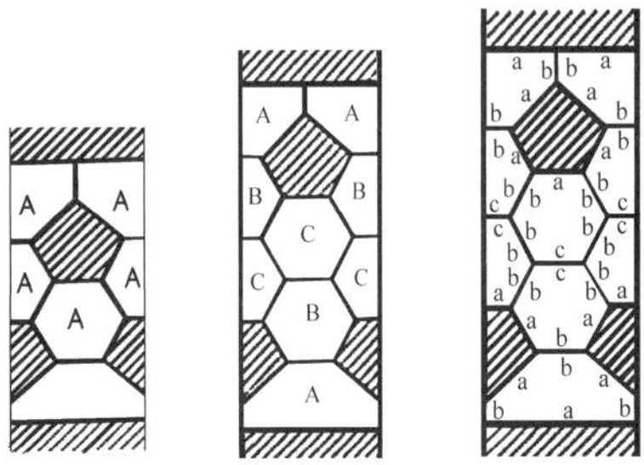

Figure 15. The building formula for the T=3 capsid and its first two prolate extensions

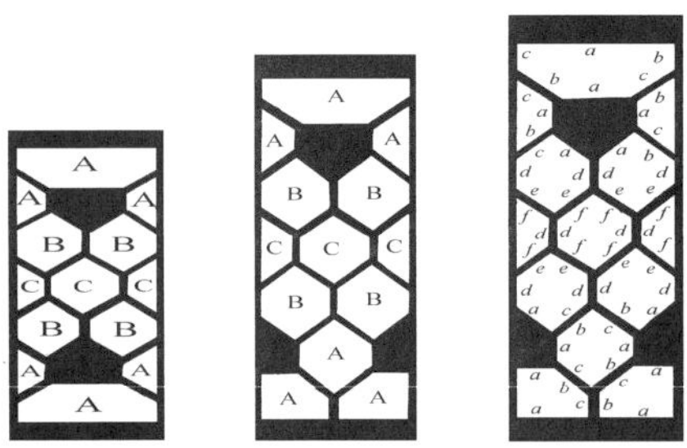

Figure 16. The building schemes for the $T = 4$ prolate capsids.

1) Among the icosaherdal capsid viruses the easiest mutations shouls occur *inside* each column corresponding to one of the four symmetries. Thus one can expect genetic affinities between viruses with triangular numbers $T = 7$, $T = 13$, $T = 17$, $T = 19$, (A-type) and another group with $T = 3$, $T = 9$, $T = 21$, $T = 27$ (B-type). The next two groups, with $T = 4$, $T = 16$, $T = 28$,..., (C-type) and the last one, $T = 12$, $T = 36$, $T = 48$ (D-type), are much less populated, and their genetic relationship is less obvious.

2) Prolate and tubular capsids closing with spherical caps identical with halves of icosahedral capsids fall into the same classification scheme, but display an extra substructures resulting from their axial symmetry. They can be obtained progressively from the icosahedral ones via addition of new hexamer belts. The number of extra hexagons is always multiple of 5.

References

1. M.C.M Coxeter, *Regular Polytopes*, Methuen & Co., London (1948).
2. D.L.D. Caspar and A. Klug, *Cold Spring Harbor Symp. Quant. Biology*, **27** (1), (1962).
3. N.A. Kislev and A. Klug, *Journal of Molecular Biology*, **40** (155) (1969) *Phys. Rev. Lett.* **86**, 4492 (2001).
4. A. Fokine *et al.*, *P.N.A.S.* **101** (16), pp. 6003-6008 (2004) 2072 (1995).
5. M. V. Nermut *et al.*, *Journal of Virology* **76** (9), pp. 4321-4330 (2002) *Nucl. Phys.* **A632**, 287 (1998).
6. W.M. Gelbart and C.M. Knobler, *Physics Today*, Jan. 2008, pp. 42-47.
7. H. Kroto, J.R. Heath, S.C. O'Brien, R.F. Curl and R.E. Smalley, *Nature*, **318**, 162 (1985).
8. A. Zlotnick, *J. Mol. Biology* **241**, pp. 59-67 (1994).
9. "The Role of Topology in Growth and Agglomeration", chapter in the book *Topology in Condensed Matter*, ed. M.I. Monastyrski, Springer Series in Solid-State Sciences Volume 150, pp. 61-91 (2006).
10. H.R. Hill, N.J. Stonehouse, S.A. Fonseca, and P.G. Stockley, *J. Mol. Biol.* **266**, pp. 1-7. (1997).
11. B. Buckley, S. Silva and S. Singh, *Virus Research*, **30**, pp. 335-349 (1993).
12. R. Twarock, *Journal of Theoretical Biology*, **21 226** (4), pp. 477-482 (2004) *Phys. Rev. Lett.* **86**, 4492 (2001).
13. T. Keef, A. Taormina and R. Twarock, *Journal of Physics: Condensed Matter* **18** (14) S375 (2006).
14. R. Kerner *Journal Computational and Mathematical Methods in Medicine*, **9**, Issue 3 & 4, pp. 175-181 (2008). 2072 (1995).
15. R. Kerner *Journal Computational and Mathematical Methods in Medicine*, Volume 6, Issue 2, pp. 95-97 (2007).
16. R. Kerner, *Models of Agglomeration and Glass Transition*, *Imperial College Press* (2007).

DISCLINATION PRODUCTION AND THE ASSEMBLY OF SPHERICAL SHELLS*

JONATHAN P. REUTER
Undergraduate Program in Bioengineering, University of California, San Diego La Jolla, CA 92093-0412.

ROBIJN F. BRUINSMA
Department of Physics and Astronomy, University of California, Los Angeles. Los Angeles, CA 90095.

WILIAM S. KLUG
Department of Mechanical and Aerospace Engineering, University of California, Los Angeles, CA 90095.

We present a simple Landau theory for the growth kinetics of solid spherical shells, as a model for the assembly of large viral capsids. The equations of motion are solved by the Finite Element Method. As a partial shell grows, compressional stresses build up along the perimeter. At a critical size the elastic stress suddenly condenses into a limited number of Volterra Disclinations. These disclinations migrate inwards, as they relieve the compressional stress. Our results indicate that defect-free assembly of solid spherical shells requires substantial plastic flow generated by the disclination migration.

1. Introduction

Crystals grow by the repetitive addition of identical building blocks to an exposed surface. For low growth rates, crystals often can be practically free of assembly defects. The assembly of ordered *spherical shells* composed of identical building blocks is of interest to structural virology - as discussed by the other contributors in this symposium - and also in materials science. Flat 2D crystalline layers can grow defect free by the (slow) addition of units to the growth perimeter, but numerous numerical simulations show that growing a defect free *curved* crystal like a viral shell is not so simple

*This work was supported by the nsf under dmr grant 04-04507

[1]. If, for example, one packs discs on the surface of a sphere with a radius large compared to that of the discs, then the preferred *local* packing arrangement would be hexagonal, as for a flat 2D plane. However, it is not possible to fully cover or tile the surface of a sphere with a hexagonal layer. Euler's Theorem requires that twelve sites with five-fold symmetry have to be introduced, for example of the form shown in Figure 1.

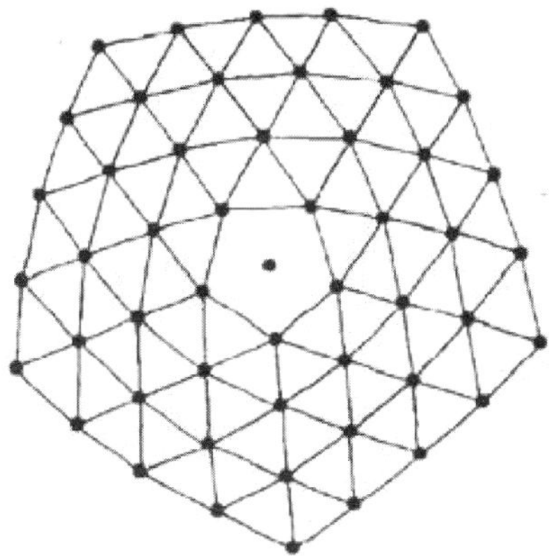

Figure 1. Five-fold symmetry site imbedded in a hexagonal lattice (from Ref.5)

The canonical Caspar-Klug icosahedral shells of structural virology [2] are composed of twelve pentamers located at equidistant sites on the icosahedral vertices with an additional 10 (T-1) hexamers - with T = 1,3,4,7,... - located in between the pentamers. Generating the twelve five-fold sites during the assembly of a spherical shell and placing them at their proper minimum free energy locations is clearly a challenge.

In this contribution, we examine the assembly kinetics of spherical shells in the limiting case that the size of the building blocks is small compared to the radius of the sphere to which the building blocks are confined. For example for the case of the assembly of the *Herpes* viral capsids, the sphere could represent the assembly scaffold. In the limit of large shell sizes, the minimum free energy state of a shell is believed to have icosahedral symmetry, with the twelve five-fold symmetry sites located at the equidistant vertices of an icosahedron. Long defect lines ("scars") can emanate from the twelve five-fold sites [3]. The five-fold symmetry site in a hexagonal array shown in Fig. 1 is an example of a *disclination*, topologically stable structures surrounded by deformation fields [4]. In continuum elasticity theory, a disclination of a solid sheet is constructed by removing a wedge-shaped section from the sheet and then pasting together the exposed edges [5], known as a *Volterra Disclination* (for the five-fold disclination in a hexagonal sheet shown in Fig. 1, the apex angle of the wedge equals $2\pi/6$). Disclinations of the same strength mutually repel each other so the elastic energy of twelve five-fold disclinations distributed over a spherical surface is indeed minimized by placing them on the vertices of an icosahedron.

In continuum theory, the physical mechanism that drives the appearance of disclinations during the growth of a shell works as follows. Assume that - initially - only a small part of the sphere is covered by the elastic sheet, say in the form of a spherical cap (in the unstrained state, the sheet would be flat). As the size of the cap grows, the perimeter section of the sheet becomes increasingly compressed while the central part of the cap is increasingly stretched [6]. The introduction of a disclination along the perimeter reduces the compression elastic energy because it squeezes out a wedge of material from the perimeter region.

The continuum equations of elasticity can be solved numerically by computing the elastic displacement field, typically by the Finite Element Method (FEM). For the present problem, numerical solution is complicated by the fact that we must allow for the appearance of lines of *mathematical singularities* because the removal of a wedge corresponds to the introduction of a discontinuity, or "cut", in the displacement field along the cut proportional to the distance from the apex. Physical quantities like the stress tensor of course remain continuous along the cut. In this paper we show how that this problem can be resolved by combining FEM with the Landau theory of phase transitions. The computation is carried out for a simple model system and the implications of the results for capsid assembly are discussed in the conclusion.

2. Time-dependent Landau Equations

In this section we will develop the time-dependent Landau equations for the growth of an elastic layer confined to a sphere surface of radius R. The part of the sphere not covered by the layer will be assumed to be occupied by a disordered assembly of units in the form of a 2D gas. The assembly process will be described as the progression of a *solidification front* associated with a first-order phase transition across the surface of the sphere. The area density difference η between the gas and solid phases plays here the role of a scalar order parameter [a].

If one allows for exchange of assembly units between the surrounding 3D volume and the sphere surface, then this order-parameter must be treated as a *non-conserved variable.* The order parameter field $\eta(\vec{r}, t)$ is defined as a function the 3D position vector field $\vec{r}$ that gives the location of the units on the surface of the sphere. The Landau variational free energy associated with this order parameter is

$$G_\eta = \int d^2S \left[\frac{\Gamma}{2}(\nabla\eta)^2 + f(\eta) - \mu\eta\right] \tag{1}$$

with d^2S an element of the sphere surface. The free energy per unit area at the phase-coexistence point between the two phases is taken to be $f(\eta) = 16 \cdot \Delta f \cdot (1-\eta)^2 \cdot \eta^2$, a double-well function of the order parameter. The minimum at $\eta = 0$ represents the gas phase and the minimum at $\eta = 1$ the solid phase.Δf represents the free energy cost for units located in the interface between the solid and liquid phases and $\mu\eta$ represents the deviation from phase-coexistence point, with μ proportional to the chemical potential difference between units in the ordered and disordered phases in the absence of elastic stress. The term $\frac{\Gamma}{2}(\nabla\eta)^2$ measures the *cooperativity* of the phase transition. Large values of the prefactor Γ correspond to narrow solid/gas interfaces. Specifically, the width ξ of the interface equals $\xi = \sqrt{\frac{\Gamma}{32 \cdot \Delta f}}$ while the interface line energy equals $\tau = \sqrt{32 \cdot \Delta f \cdot \Gamma}$. If the perimeter of the shell is sufficiently smooth, we can approximate $G_\eta \approx -\mu A + \tau P$, with A the area of the shell and P the perimeter length of the interface (consistent with Gibbs thermodynamics). For given A, G_η is minimized if the shell has the shape of a *spherical cap.* The dependence of G_η on the number of units in the solid phase exhibits the classical nucleation-and-growth form. If the radius of a circular interface is less than a critical radius $\frac{\tau}{2\mu}$, then the cap will shrink and disappear while it will continue to grow if the radius exceeds $\frac{\tau}{2\mu}$. Note that at phase coexistence (η =0), a free energy barrier $\Delta G = 2\pi R\tau$ separates the ordered and disordered phases.

The elastic free energy of the sheet will be assumed to be that of a 2D layer with hexagonal symmetry. Let $\overrightarrow{X}(s_1, s_2)$ be the position vector of

---

[a] In actuality we are using a version of the Landau theory for the *liquid-gas* transition (see Chaikin and Lubensky, Pinciples of Condensed Matter Physics, Cambridge, 1995, pg. 159. The true order parameter for solidification is a set of complex numbers (*ibid*, Ch. 4.7), but this leads to mathematical complexities.

the assembly units in the "reference state": a flat undeformed, defect free hexagonal layer resting on a flat surface (see Fig. 2). The surface is chosen so it cuts the sphere into two equal parts.

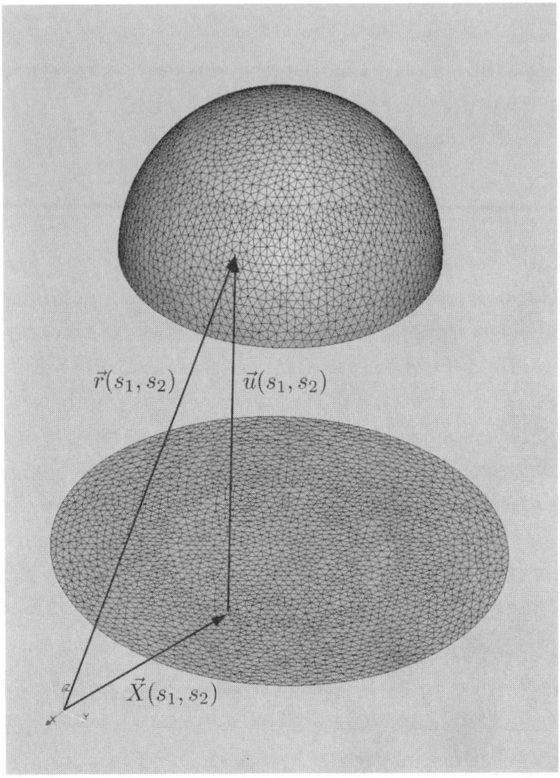

Figure 2. Geometry used in the finite element calculations. The locations of nodes in the undeformed reference state are indicated by the vector field $\vec{X}(s_1, s_2)$ with s_1, s_2 a coordinate system in the plane. The elastic displacement field of the deformed state is indicated by $\vec{u}(s_1, s_2)$. The final locations of the nodes on the assembly sphere are indicated by $\vec{r}(s_1, s_2) = \vec{X}(s_1, s_2) + \vec{u}(s_1, s_2)$.

Here, s_1, s_2 is a two-component coordinate system along the surface. In the reference state, the elastic energy is of course zero. The location vector of the units on the sphere surface is $\vec{r}(s_1, s_2) = \vec{X}(s_1, s_2) + \vec{u}(s_1, s_2)$, with $\vec{u}(s_1, s_2)$ the displacement vector away from the strain free reference state. This displacement is the sum of a vertical displacement that brings a unit from the flat surface to the sphere surface plus a displacement to a different location on the sphere surface. Figure 2 shows a schematic of the displacement vectors for the case that the displacement is only along the vertical direction. The strain tensor is defined as

$$\varepsilon_{ij} = \frac{1}{2} \cdot \left(\frac{\partial u_i}{\partial X_j} + \frac{\partial u_j}{\partial X_i} + \frac{\partial u_k}{\partial X_i} \cdot \frac{\partial u_k}{\partial X_j} \right) \tag{2}$$

where $\vec{u}\,(s_1, s_2)$ must be expressed as a function of the location $\vec{X}(s_1, s_2)$ of the undeformed system. The elastic free energy of the isotropic layer equals

$$F_\varepsilon = \int d^2S \cdot \left[\frac{1}{2} \lambda_0\,(\eta)\, \varepsilon_{ii}^2 + \mu_0\,(\eta)\, \varepsilon_{ij}^2 \right] \tag{3}$$

with $\lambda_\square$ and $\mu_\square$ the Lamé coefficients [b]. We will explore a simple model where both Lame coefficients are proportional to the order parameter η . The free energy is thus zero in the disordered phase with η =0. We will express our results in terms of the 2D Young's Modulus Y and ν Poisson's Ratio, using the relations $\lambda_0 = \frac{Y\nu}{(1-\nu)^2}$ and $\mu_0 = \frac{Y}{2(1+\nu)}$. The Young's Modulus is again proportional to the order parameter while Poisson's Ratio is independent of the order parameter. The static elastic free energy of a spherical cap of radius ρ_0 with no disclinations on a sphere of radius R can be obtained analytically by classical methods of elasticity theory [6]:

$$F_\varepsilon = \begin{cases} \dfrac{\pi Y}{384} \dfrac{\rho_0^6}{R^4} & \Gamma \ll 1 \\[2ex] \dfrac{\pi(1-\nu)}{Y} \tau^2 & \Gamma \ll 1 \end{cases} \tag{4}$$

with $\Gamma = \frac{\tau}{Y\rho_0}$ the dimensionless line tension. As mentioned, this elastic energy is due to compression along the perimeter of the cap and stretching in the center section. Note that it grows rapidly, as the sixth power of the radius of the cap. When compared with the order parameter part of

---

[b] By defining the strain energy using the "large-strain" tensor, we recover a 2D version of the St. Venant-Kirchhoff model for large-deformation elasticity commonly employed in finite-deformation solid mechanics. Even though strains in the shell itself are typically rather small, it is important to retain the nonlinear terms in the strain tensor to properly account for finite rotations of points on the shell surface as it is deformed from the initially flat geometry to the hemisphere.

the free energy with no elasticity,$G_\eta = -\mu A + \tau P$, it follows that in the absence of dislocation nucleation, the growth of a cap will be "suffocated" by the elastic energy when the cap radius is of the order of $\frac{\rho_0}{R} : \left(\frac{384\mu}{Y}\right)^{1/4}$ or less (depending on τ).

The kinetic equations will be taken to have the simplest possible dissipative form, as appropriate for non-conserved hydrodynamic variables:

$$C_\eta \frac{\partial \eta}{\partial t} = -\frac{\delta \left(G_\eta + F_\varepsilon\right)}{\delta \eta}$$
$$C_\mu \frac{\partial \mu_i}{\partial t} = -\frac{\delta \left(G_\eta + F_\varepsilon\right)}{\delta \mu_i} \tag{5}$$

The kinetic coefficients C_h and C_u are effective friction coefficients for order parameter and strain tensor relaxation respectively. The terms on the right hand side of Eq. (5) are functional derivatives. Note that these kinetic equations might be viewed as providing "local rules" for the growth but that the long-range strain field couples different parts of the shell and the perimeter.

3. Finite Element Method

In order to integrate the coupled equations of motion, we applied the FEM in the form of a "weak" boundary value problem. We will briefly sketch the method with the details presented elsewhere. We discretized the position function $\vec{X}(s_1, s_2)$ of the undeformed sheet as a triangular mesh (the triangles are the "finite elements") covering the reference state (see Fig. 2). The function $\vec{X}(s_1, s_2)$ is expressed as a piecewise interpolation between the nodal points $\vec{X}_\alpha$ of the mesh:

$$\vec{X}(s_1, s_2) = \sum_\alpha \vec{X}_\alpha N_\alpha \left(s_1, s_2\right) \tag{6}$$

where the $N_\alpha(s_1, s_2)$ are piecewise polynomial functions that combine to form a conforming position mesh. Similar interpolations can be constructed for the order-parameter and displacement fields. In the deformed state, the nodes of the triangular mesh are located at $\vec{X}_\alpha + \vec{\mu}(\vec{X}_\alpha)$, as shown in Fig. 2. Importantly, the displacement and order parameter fields are piecewise continuous *and have no discontinuities along the triangular mesh.*

A semi-discrete set of "nodal equations" can be obtained from the equations of motion so the displacement field can be monitored by following the movement in time of the nodes of the mesh. We constructed a sequence of time steps and used forward and backward Euler integration schemes to update the location of the nodal points and the fields defined on the nodal points. We started from a small circular seed where the order parameter was equal to one with the order parameter equal to zero on the remaining part of the sphere. We choose parameter values reasonable for viral assembly in terms of a characteristic interaction energy $\Delta\varepsilon$ between assembly units and a characteristic assembly unit size a. The chemical potential per unit area μ and the Young's Modulus can be expressed in units of $\Delta\varepsilon\Box a^2$, the line energy τ in units of $\Delta\varepsilon\Box a$ and the width ξ of the interface of in units of a. In these dimensionless units, μ and τ varied between 0.1 and 10 while ξ was equal to 2.

During the initial growth stage, the sheet maintained circular symmetry as shown in Figure 3.

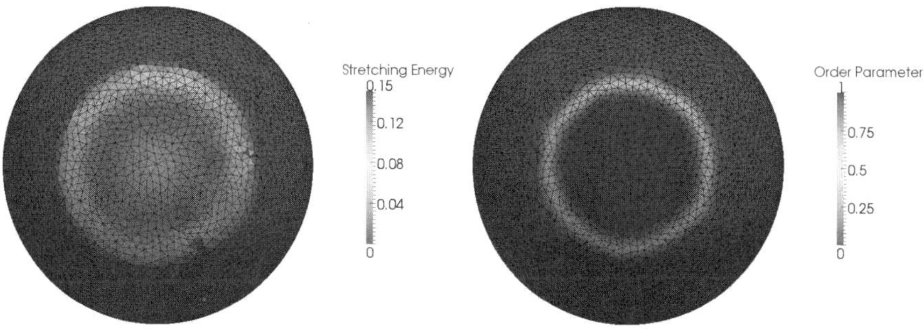

Figure 3. Figure 3A and 3B, distribution of the elastic energy (Fig. 3A, dimensionless units) and order parameter (Fig. 3B) during the initial growth stage, obtained by the finite element method. Note that the elastic energy is not quite rotationally invariant

The perimeter is approximately circular. The elastic energy is concentrated along the perimeter Note that the energy distribution along the perimeter is not quite uniform. At earlier stages of the growth, when the ordered region was smaller, the elastic energy distribution was more closely rotationally symmetric. For the order parameter (right panel), the boundary line between the ordered (red) and disordered (blue) region shows up as a white line where the order parameter is close to one half.

Shortly after the snapshot shown in Fig. 3, the non-uniformity of the elastic energy distribution increased quite suddenly. The elastic energy was nearly entirely focused on a number of small regions distributed at roughly equal distances along the perimeter (see Fig. 4).

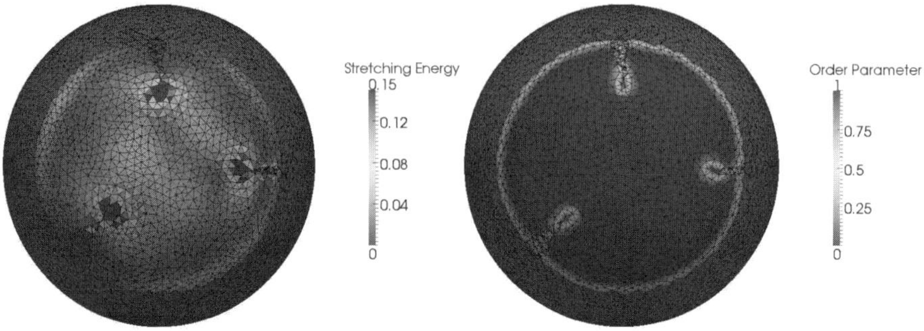

Figure 4. Figures 4A and 4B, the elastic energy condenses around the rim in a few isolated sites, which then migrate inwards from the perimeter (Fig. 4A). They leave behind a trail where the order parameter is zero (Fig. 4B). For two of the three trails, the elastic energy is only slightly higher than background along the section of the trail near the perimeter.

The energy foci then moved inwards towards the center, leaving behind a trail of deformed FEM mesh. Note that the three singular regions in Fig. 4 are, approximately, located on the vertices of an equilateral triangle. A magnified view of one of the trails, and a schematic of the mesh, is shown in Figure 5.

The mesh is folded up into a conical shape with apex at the center of the elastic energy concentration. Inside the fold, the order parameter is zero. Because the free energy is zero when the order parameter is zero, the section inside the fold does not contribute to the energy of the system. *The displacement field, but not the order parameter field, suffers a discontinuity proportional to the distance from the apex.* Effectively, the FEM created a cut and carried out a Volterra Disclination construction! The combination of the Landau description with the FEM is what makes this possible. Suppose that the order parameter of a section of the triangular mesh is set equal to zero. Since the elastic moduli of this section are zero, the section can be twisted and bend with no energy penalty. Effectively, the FEM of the elastic sheet proceeds in the presence of a "reservoir" of nodes with zero chemical potential. This allows the removal of wedge-shaped regions from the perimeter and the generation of Volterra Disclinations. The FEM

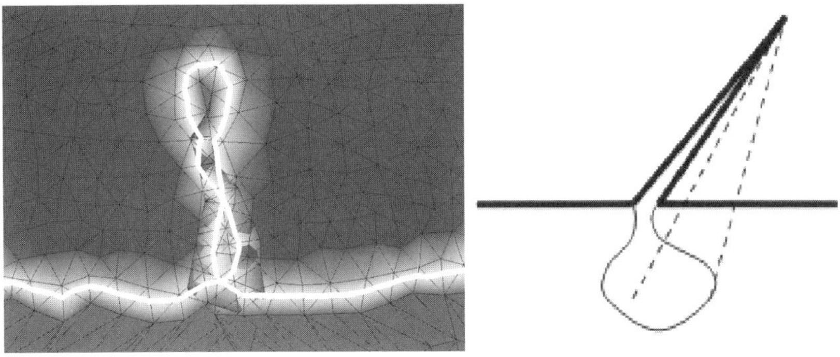

Figure 5. Close up of a fold. The white line indicates the perimeter where the order equals 1/2. The line drawing shows a possible structure of a fold. The perimeter, where the order parameter drops to zero is shown as a solid line. The order parameter is zero inside the fold.

representation of a Volterra Disclination is not perfect. The edges of the trail are *stress free*, apart from the line energy, and this is in general not the case for a Volterra Dislocation. In other words, the cut did not heal. This is not a serious problem for the present case, since disclination production practically completely drains away perimeter elastic stress, but this defect would have to be corrected if one wanted to apply our method to other elasticity problems involving the generation of topological defects.

4. Conclusion

In summary, we have presented a simple numerical method for the study of the growth of an elastic layer confined to a spherical surface allowing for the generation of disclinations. The elastic energy distribution of the growing layer induced by the curvature initially had circular symmetry. At a critical size, the elastic energy "condensed" into a few sites along the rim, which then migrated inward away from the perimeter. Lines extended from the condensation sites to the perimeter of the sheet along which the displacement field suffers a discontinuity. The sites can be identified as approximate realizations of the Volterra Disclinations of continuum elasticity theory.

The most important implications for capsid assembly are as follows (i) disclinations are generated along the growth perimeter by the process of *elastic stress condensation* and (ii) during the assembly process, disclinations can not appear at the locations they are destined to eventually occupy in the final minimum free energy state. They have to migrate a considerable distance away from the perimeter driven by the perimeter lateral compression that removes excess material in the form of the Volterra Disclination wedge. Note that in *flat* elastic sheets, disclinations are in fact *attracted* to the perimeter (through an elastic image force). Disclination motion in a solid requires large-scale rearrangements of the bonding pattern since a significant amount of material must be transported out of the growing region by a form of *flow* [7] and the emission/absorption of dislocations [5]. This suggests that if five-fold disclination generation during initial assembly is to take place following the scenario described above, then some form of plastic flow must be allowed to take place. Note that if the disclinations were generated at their final locations then the plastic flow would not have been necessary. Plastic flow in a solid material is possible if the bonding strength between the units is not too large compared to the thermal energy so, at least initially, assembly unit association energies must be weak.

References

1. H. D. Nguyen, V. S. Reddy and C. L. Brooks, *J. Am. Chem. Soc.* **131**, 2606 (2009), and references therein.
2. D. L. D. Caspar and A. Klug, *Symp. Quant. Biol.* **27**, 1 (1962).
3. A.R. Bausch, M.J. Bowick, A. Cacciuto, A.D. Dinsmore, M.F. Hsu, D.R. Nelson, M.G. Nikolaides, A. Travesset and D.A. Weitz, *Science* **299**, 1716(2002).
4. A. Kroner and K-H Anthony, *Ann. Rev. Mat. Sc.*, **43**, (1975).
5. M. Kleman and J. Friedel, *Rev. Mod. Phys.*, **80**, 63 (2008).
6. A. Y. Morozov and R. F. Bruinsma, *Phys Rev E.*, **81**, 041925 (2010).
7. E. I. Kats, V. V. Lebedev and S. V. Malinin, *J. Exp. Theor. Phys. (Russia)*, **95**, 1063 (2002).

A PROPOSAL FOR MODELLING THE STRUCTURE OF BIOMACROMOLECULES

R. P. MONDAINI, S. P. VILELA

UFRJ- Federal University of Rio de Janeiro,
Centre of Technology - COPPE, BL. H, Sl. 319
P. O. Box 68511, CEP: 21941-972- Rio de Janeiro, RJ, Brasil.
E-mail: mondaini@cos.ufrj.br

Some considerations about the modelling of biomolecular structure are presented with a special emphasis in an approach used to circumvent problems with the systematic applications of the techniques of Berlin surfaces to macromolecules. A new paradigm is proposed for predicting bond angles in biopolymers.

1. Introduction

In the present work we intend to apply the technique of Berlin diagrams [1, 2, 3] for studying forces in molecular structures, to biopolymers. These are the forces among nuclei and the surrounding electronic cloud [4]. In section 2 we introduce the Hellmann-Feynman theory and some elementary concepts of Density Functional Theory [5] as a useful guide for the derivations of following sections. In section 3, we introduce the Berlin diagrams and we show the behaviour of the diagrams on centre of mass and geometric centre systems for diatomic molecules. In section 4, we stress that in despite of the dependence of Berlin diagrams on the adopted form for the total representative forces on polyatomic molecules, there is adequacy of the diagrams as applied to a biopolymer. In this section we also introduce an alternative way of treating the local equilibrium around nuclei of the main backbone sequence in biomolecules. In section 5 we comment on the influence of the dipolar terms to be introduced with the same alternative treatment of section 4. Some general remarks are also made to promote future research on force patterns with the help of the Density Functional Theory [5].

2. The Hellmann-Feynman Theorem and Applications in Molecular Structure

For a stationary molecule, the force on each nucleus due to other nuclei and the surrounding electronic cloud is given by

$$\overrightarrow{F_k} = e^2 Z_k \sum_{j \neq k} Z_j \frac{\left(\overrightarrow{R_k} - \overrightarrow{R_j}\right)}{\left\| \overrightarrow{R_k} - \overrightarrow{R_j} \right\|^3} - e^2 Z_k \int \rho\left(\vec{r}\right) \frac{\left(\overrightarrow{R_k} - \vec{r}\right)}{\left\| \overrightarrow{R_k} - \vec{r} \right\|^3} d^3\vec{r}, \qquad (1)$$

where $e\, Z_k\, \rho(\overrightarrow{r})\, d^3\overrightarrow{r}$ is the total electronic charge at point $\overrightarrow{r}$. Nuclei are supposed to be clamped at their positions $\overrightarrow{R}_k$, according the Born-Oppenheimer approximation [6]. The time-independent Schroedinger equation is then solved to obtain the electronic wave function ψ_e and the corresponding energy eigenvalue. The effective potential which is used to study the subsequent nuclear motion is the sum of this eigenvalue and the supposed form of the interaction potential energy, or

$$\hat{H} \left|\psi_e\right\rangle = \left(\hat{T}_e + \hat{V}_e + \hat{V}_N\right) \left|\psi_e\right\rangle = U \left|\psi_e\right\rangle, \qquad (2)$$

where

$$\hat{T}_e = \frac{1}{2m_e} \sum_{j=1}^{N} {p_j}^2, \qquad (3)$$

$$\hat{V}_e = -\sum_{k=1}^{N} \sum_{j=1}^{N} \frac{Z_k e^2}{\left\| \vec{R}_k - \vec{r}_j \right\|} + \sum_{j=1}^{N} \sum_{i>j}^{N} \frac{Z_k e^2}{\left\| \vec{r}_i - \vec{r}_j \right\|}, \qquad (4)$$

$$\hat{V}_N = \sum_{k=1}^{N} \sum_{l>k}^{N} \frac{Z_k Z_l e^2}{\left\| \vec{R}_k - \vec{R}_l \right\|}, \qquad (5)$$

with Z_k standing for atomic number of kth nucleus and e, m_e, the electronic charge and electronic mass. Since the molecule is in a stationary state, this force on a nucleus of position vector $\overrightarrow{R_k}$ is given by the virtual work principle as

$$\vec{F}_k . \overrightarrow{\Delta R_k} + \Delta U = 0,$$

or

$$\vec{F}_k = -\frac{\partial U}{\partial \vec{R}_k} . \qquad (6)$$

From Hellmann-Feynman theorem, after choosing $\overrightarrow{R_k}$ as the parameter, we have

$$\vec{F_k} = -\frac{\partial U}{\partial \vec{R}_k} = -\langle \psi_e | \frac{\partial \hat{H}}{\partial \vec{R}_k} | \psi_e \rangle , \tag{7}$$

we can also write,

$$\vec{F_k} = \langle \psi_e | \hat{F}_k | \psi_e \rangle = \frac{i}{\hbar} \langle \psi_e | \left[\hat{H}, \hat{P}_k \right] | \psi_e \rangle , \tag{8}$$

where $\overrightarrow{P_k}$ is the conjugate momentum operator satisfying

$$\frac{d\vec{P_k}}{dt} = \vec{F_k} = \frac{i}{\hbar} \left[\hat{H}, \hat{P}_k \right] = - \left[\hat{H}, \frac{\partial}{\partial \vec{R}_k} \right] . \tag{9}$$

From eqs. (8), (9), we get eq. (7). After using eqs. (2)-(5), we can write eq. (1). In this derivation the following comments are in order:

For a cloud of n electrons, an expected value like $\langle \psi_e | \left\| \vec{R}_k - \vec{r}_j \right\|^{-1} | \psi_e \rangle$ means to calculate the integral over a 3n-dimensional volume (3n variables),

$$\langle \psi_e | \left\| \vec{R}_k - \vec{r}_j \right\|^{-1} | \psi_e \rangle = \int\limits_{V^{3n}} \frac{\psi^*{}_e \, \psi_e}{\left\| \vec{R}_k - \vec{r}_j \right\|} \, d^{3n} \vec{r}. \tag{10}$$

The summation over all spin states is also included in the integral symbol. The right hand side can be also written as

$$\int\limits_{V^j} \int\limits_{V^{3n-3}} \frac{\psi^*{}_e \, \psi_e}{\left\| \vec{R}_k - \vec{r}_j \right\|} \, d^{3n-3} \vec{r} \, d^3 \vec{r}_j \, .$$

The probability of finding the *j-th* electron in the volume element $d^3\vec{r}_j$ does not depend on the position of the other electrons. This probability will be the same for any other electrons and it will be given by

$$d^3 \vec{r}_j \int\limits_{V^{3n-3}} \psi_e^* \, \psi_e \, d^{3n-3} \vec{r} \, .$$

Since electrons are indistinguishable, the probability of finding an electron in a 3-dimensional volume is

$$\rho \, d^3 \vec{r} = n \, d^3 \vec{r} \int\limits_{V^{3n-3}} \psi_e^* \, \psi_e \, d^{3n-3} \vec{r} \, ,$$

where ρ is the electron probability density of the n-electron molecule. We then have, from eq. (10):

$$\sum_{j=1}^{n} \langle\psi_e| \left\|\vec{R}_k - \vec{r}_j\right\|^{-1} |\psi_e\rangle = n \int \frac{\rho}{n} \frac{1}{\left\|\vec{R}_k - \vec{r}\right\|} d^{\,3}\vec{r}$$
$$= \int \frac{\rho\left(\vec{r}\right)}{\left\|\vec{R}_k - \vec{r}\right\|} d^{\,3}\vec{r}, \tag{11}$$

where the fact that electrons are indistinguishable has been used again.

3. Berlin Surfaces

For a polyatomic molecule, the moment of the nuclei in the centre of mass and the geometric coordinates are given by

$$\vec{P}_{k\;CM} = \vec{P}_k - \frac{m_k}{M} \sum_{l=1}^{N} \vec{P}_l, \qquad M = \sum_{k=1}^{N} m_k, \tag{12}$$

$$\vec{P}_{k\;GC} = \vec{P}_k - \frac{m_k}{M} \sum_{l=1}^{N} \frac{\vec{P}_l}{m_l}, \tag{13}$$

respectively.

The forces on these nuclei can be written,

$$\vec{F}_{kCM} = \frac{i}{\hbar} \langle\psi_e| [\hat{H}, \vec{P}_{k\;CM}] |\psi_e\rangle = \vec{F}_k - \frac{m_k}{M} \sum_{l=1}^{N} \vec{F}_l, \tag{14}$$

$$\vec{F}_{k\;GC} = \frac{i}{\hbar} \langle\psi_e| [\hat{H}, \vec{P}_{k\;GC}] |\psi_e\rangle = \vec{F}_k - m_k \sum_{l=1}^{N} \frac{\vec{F}_l}{m_l}. \tag{15}$$

From eq. (1), we can also write:

$$\begin{aligned}\vec{F}_{k\;CM} = {} & e^2 Z_k \sum_{l\neq k} Z_l \frac{\left(\vec{R}_k - \vec{R}_l\right)}{\left\|\vec{R}_k - \vec{R}_l\right\|^3} \\ & -e^2 \int \rho \left(Z_k \frac{\left(\vec{R}_k - \vec{r}\right)}{\left\|\vec{R}_k - \vec{r}\right\|^3} - \frac{m_k}{M} \sum_{s=1}^{n} Z_s \frac{\left(\vec{R}_s - \vec{r}\right)}{\left\|\vec{R}_s - \vec{r}\right\|^3} \right) d^3\vec{r},\end{aligned} \tag{16}$$

$$\begin{aligned}\vec{F}_{k\;GC} = {} & e^2 Z_k \sum_{l\neq k} Z_l \frac{\left(\vec{R}_k - \vec{R}_l\right)}{\left\|\vec{R}_k - \vec{R}_l\right\|^3} - e^2 \frac{m_k}{n} \sum_{s=1}^{n} \sum_{l>s} \frac{Z_s Z_l \left(\vec{R}_s - \vec{R}_l\right)}{m_s \left\|\vec{R}_s - \vec{R}_l\right\|^3} \\ & -e^2 \int \rho \left(Z_k \frac{\left(\vec{R}_k - \vec{r}\right)}{\left\|\vec{R}_k - \vec{r}\right\|^3} - \frac{m_k}{n} \sum_{s=1}^{n} \frac{Z_s \left(\vec{R}_s - \vec{r}\right)}{m_s \left\|\vec{R}_s - \vec{r}\right\|^3} \right) d^3\vec{r}.\end{aligned} \tag{17}$$

From eq. (16) it follows the trivial relation.

$$\sum_{k=1}^{n} \vec{F}_{k_{CM}} = 0 \tag{18}$$

In the case of a diatomic molecule, we choose the relative vector as a special combination. We thus have form eqs. (16) and (17) some proposal for characterizing the internuclear forces on nuclei:

$$\vec{F}_{1_{CM}} - \vec{F}_{2_{CM}} = -\frac{e^2 Z_1 Z_2 \left(\vec{R}_2 - \vec{R}_1\right)}{\left\|\vec{R}_1 - \vec{R}_2\right\|^3} - \frac{2e^2 m_1 m_2}{M} \int \rho\left(\vec{r}\right) f\left(\vec{r}\right) d^3\vec{r}, \tag{19}$$

$$\vec{F}_{1_{GC}} + \vec{F}_{2_{GC}} = -\frac{e^2\left(m_1 - m_2\right)}{2} \int \rho\left(\vec{r}\right) f\left(\vec{r}\right) d^3\vec{r}, \tag{20}$$

$$\vec{F}_{1_{GC}} - \vec{F}_{2_{GC}} = -\frac{e^2\left(m_1 + m_2\right)}{4m_1 m_2} \cdot \frac{Z_1 Z_2 \left(\vec{R}_2 - \vec{R}_1\right)}{\left\|\vec{R}_1 - \vec{R}_2\right\|^3} - \frac{e^2(m_1 + m_2)}{2} \int \rho\left(\vec{r}\right) f\left(\vec{r}\right) d^3\vec{r}, \tag{21}$$

where

$$f(\vec{r}) = \frac{Z_1}{m_1} \frac{\left(\vec{R}_1 - \vec{r}\right)}{\left\|\vec{R}_1 - \vec{r}\right\|^3} - \frac{Z_2}{m_2} \frac{\left(\vec{R}_2 - \vec{r}\right)}{\left\|\vec{R}_2 - \vec{r}\right\|^3}. \tag{22}$$

The Berlin diagrams are obtained by the projection of eq. (22) in a specified direction, say Ox, and we have:

$$f_x = \frac{Z_1}{m_1} \frac{\left(X_1 - x\right)}{\left\|\vec{R}_1 - \vec{r}\right\|^3} - \frac{Z_2}{m_2} \frac{\left(X_2 - x\right)}{\left\|\vec{R}_2 - \vec{r}\right\|^3}. \tag{23}$$

The Berlin surfaces $f_x = 0$ separate the regions of binding ($f_x > 0$) and antibinding ($f_x < 0$) forces.

If we consider a potential energy in which the law

$$\left\| \vec{R}_k - \vec{r} \right\|^{-(m-2)},$$

is the predominant one, the equation $f_x = 0$ can be written as

$$y^2 + z^2 = \frac{(x - X_1)^{2/m}(x - X_2)^{2/m}\left[\eta^{2/m}(x - X_2)^{\frac{2(m-1)}{m}} - (x - X_1)^{\frac{2(m-1)}{m}}\right]}{(x - X_2)^{2/m} - \eta^{2/m}(x - X_1)^{2/m}}, \tag{24}$$

where $\eta = \frac{Z_1/m_1}{Z_2/m_2}$

For $\eta = 1$, $m > 2$, we have,

$$y^2 + z^2 = \sum_{j=0}^{m-2} (x - X_1)^{\frac{2(j-1)}{m}} (x - X_2)^{\frac{2(m-j-1)}{m}}. \tag{25}$$

These equations correspond to open surfaces and the generation of the cones

$$y^2 + z^2 = (m - 1)\left(x^2 - \frac{(X_1 + X_2)}{2}\right)^2, \tag{26}$$

are their asymptotes, or

$$y = \pm (m - 1)^{1/2}\left(x - \frac{(X_1 + X_2)}{2}\right). \tag{27}$$

Some useful notes are now in order. In the case $m = 3$ (Coulombian interaction), the angle between asymptotes is $2\, arctan\, 2^{1/2} = arccos\left(-\frac{1}{3}\right) = 109^0\, 28'$ and this is the tetrahedron angle[9]. With the predominance of a term $\left\| \vec{R}_k - \vec{r} \right\|^{-2}$ which can be introduced by considering a dipolar interaction, corresponding to $m = 4$, we get $2\, arctan\, 3^{1/2} = arccos\left(-\frac{1}{2}\right) = 120^0$. This is the angle between edges of the usual Steiner problem. In the general case, we can assume the predominance of a term $\left\| \vec{R}_k - \vec{r} \right\|^{-(m-2)}$ and the angle between asymptotes will be also given by [9].

$$\hat{R}_{k\,r} \,.\, \hat{R}_{l\,r} = -\frac{1}{a - 1}\,(1 - a\,\delta_{k\,l}), \tag{28}$$

where $a = \frac{2(m-1)}{(m-2)}$ and $\hat{R}_{kr}$ is the unit vector

$$\hat{R}_{k\,r} = \frac{\vec{R}_k - \vec{r}}{\left\| \vec{R}_k - \vec{r} \right\|}. \tag{29}$$

4. The Adequacy of Berlin Surfaces for the Classification of Binding and Non-binding regions in Macromolecules

For a triatomic heteronuclear molecule, some feasible combinations of forces can be written as

$$\vec{F}_{1\,CM} - \vec{F}_{2\,CM} + \vec{F}_{3\,CM} = -e^2 Z_2 \left(Z_1 \vec{R}^3_{12} - Z_3 \vec{R}^3_{23} \right)$$
$$-\frac{2e^2 m_2 \left(m_1 + m_3\right)}{M} \int \rho\left(\vec{r}\right) \left[\frac{Z_1 \vec{R}^3_{1r} + Z_3 \vec{R}^3_{3r}}{m_1 + m_3} - \frac{Z_2 \vec{R}^3_{2r}}{m_2} \right] d^3\vec{r}, \tag{30}$$

where $\vec{R}^{\,3}_{k\,r} = \frac{\vec{R}_k - \vec{r}}{\left\|\vec{R}_k - \vec{r}\right\|^3}$ and $\vec{R}^{\,3}_{k\,s} = \frac{\vec{R}_k - \vec{R}_s}{\left\|\vec{R}_k - \vec{R}_s\right\|^3}$.

$$\vec{F}_{1GC} + \vec{F}_{2GC} + \vec{F}_{3GC} = e^2 \frac{M}{6} \left[\frac{Z_1 Z_2}{m_1 m_2} \left(m_1 - m_2\right) \vec{R}^3_{12} \right.$$
$$+ \frac{Z_1 Z_3}{m_1 m_3} \left(m_1 - m_3\right) \vec{R}^3_{13} + \frac{Z_2 Z_3}{m_2 m_3} \left(m_2 - m_3\right) \vec{R}^3_{23}$$
$$-\frac{2e^2}{3} \int \rho\left(\vec{r}\right) \left[\left(1 - \frac{m_2 + m_3}{2m_1}\right) Z_1 \vec{R}^3_{1r} + \left(1 - \frac{m_1 + m_3}{2m_2}\right) Z_2 \vec{R}^3_{2r} \right.$$
$$\left. + \left(1 - \frac{m_1 + m_2}{2m_3}\right) Z_3 \vec{R}^3_{3\,r} \right] d^3\vec{r}. \tag{31}$$

$$\vec{F}_{1\,GC} - \vec{F}_{2\,GC} + \vec{F}_{3\,GC} = -e^2 Z_2 \left(Z_1 \vec{R}^3_{12} - Z_3 \vec{R}^3_{23} \right)$$
$$+ \frac{e^2}{6} \left(m_1 - m_2 + m_3\right) \left(\frac{Z_1 Z_2 (m_1 - m_2) \vec{R}^{\,3}_{12}}{m_1 m_2} + \frac{Z_1 Z_3 (m_1 - m_3) \vec{R}^{\,3}_{13}}{m_1 m_3} \right.$$
$$\left. + \frac{Z_2 Z_3 (m_2 - m_3) \vec{R}^{\,3}_{23}}{m_2 m_3} \right) - \frac{2e^2}{3} \int \rho\left(\vec{r}\right) \left[\left(1 - \frac{m_3 - m_2}{m_1}\right) Z_1 \vec{R}^{\,3}_{1\,r} \right.$$
$$\left. - \left(1 + \frac{m_1 + m_3}{2m_2}\right) Z_2 \vec{R}^{\,3}_{2\,r} + \left(1 - \frac{m_1 - m_2}{2m_3}\right) Z_3 \vec{R}^{\,3}_{3\,r} \right] d^3\vec{r} \tag{32}$$

Actually, for n nuclei, there are $(2^{n-1} - 1)$ non-trivial possibilities of proposals for generalized Berlin surfaces for a heteronuclear molecule in the centre of mass systems and 2^{n-1} candidates in the geometric center systems. These are given by the number of combinations of signs $\pm$ after disregarding combinations corresponding to a global multiplication by (-1). For a homonuclear molecule there are $(2^{n-1} - 1)$ possibilities for the two systems, CM and GC. All these possibilities and the complexity of the resulting diagrams (surfaces) are enough for convince us of the difficulty of working with the representation of binding and non-binding regions in the analysis of Berlin surfaces [5]. However, there is an alternative procedure which has been proved to be applicable to the modelling of polypeptides, since the perturbation in the angular parameters of these structures, like bonding angles [9] and dihedral angles have a good agreement with values reported in the literature [7, 8] and obtained from an intensive statistical analysis. This has been done with dipeptides only and as far as we know, there are no statistical results for large polypeptides like proteins. The new approach is able to treat a protein structure and the present work should be considered as an introduction. The work reported in ref [9].

In order to introduce the new paradigm on modelling of macromolecular structure, we take the example of Coulombian interaction given in eqs. (16) and (17). We start form the usual concept of Hellmann-Feynman force of eq. (1). We then consider the equilibrium of the entire molecule as obtained from local equilibrium of backbone nuclei and their nearest neighbours. This is done by assuming u equality of force intensities between a backbone nucleus and its neighbors after getting rid of non-neighbour terms.

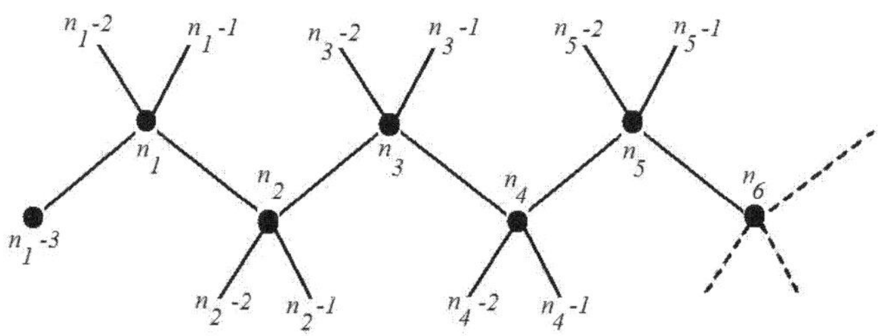

Figure 1. A diagram for modelling a macromolecule with four neighbours to each backbone nucleus.

We have from eq. (1):

$$\vec{F}_{n_1} \approx e^2 Z_{n_1} \sum_{l=1}^{n_1-1} Z_l \frac{\left(\vec{R}_{n_1} - \vec{R}_l\right)}{\left\|\vec{R}_{n_1} - \vec{R}_l\right\|^3} + e^2 Z_{n_1} Z_{n_2} \frac{\left(\vec{R}_{n_1} - \vec{R}_{n_2}\right)}{\left\|\vec{R}_{n_1} - \vec{R}_{n_2}\right\|^3}$$
$$-e^2 \int \rho(\vec{r})\, Z_{n_1} \frac{\left(\vec{R}_{n_1} - \vec{r}\right)}{\left\|\vec{R}_{n_1} - \vec{r}\right\|^3} d^3\vec{r}, \tag{33}$$

where the assumption of deformation of the electronic cloud and subsequent concentration of the same over the nucleus n' due to the inspection of this nucleus n_1, is taken into account.

From the equality of force intensities, or

$$\frac{e^2 Z_{n_1} Z_{n_2}}{\left\|\vec{R}_{n_1} - \vec{R}_{n_2}\right\|^2} = \frac{e^2 Z_{n_1} Z_l}{\left\|\vec{R}_{n_1} - \vec{R}_l\right\|^2} = \frac{e^2 Z_{n_1} Z_k}{\left\|\vec{R}_{n_1} - \vec{R}_k\right\|^2} = F_{n_1}, \tag{34}$$

where $l,\ k = n_1 - 3,\ n_1 - 2,\ n_1 - 1$

we have, from eq.(33),

$$\vec{F}_{n_1} \approx F_n \left(\sum_{l=n_1-3}^{n_1-1} \hat{R}_{n_1 l} + \hat{R}_{n_1 n_2} \right) - e^2 \int \rho(\vec{r})\, Z_{n_1} \frac{\left(\vec{R}_{n_1} - \vec{r}\right)}{\left\|\vec{R}_{n_1} - \vec{r}\right\|^3} d^3\vec{r}, \tag{35}$$

where $\hat{R}_{n_1,s}$ is the unit vector

$$\hat{R}_{n_1,s} = \frac{\left(\vec{R}_{n_1} - \vec{R}_s\right)}{\left\|\vec{R}_{n_1} - \vec{R}_s\right\|}. \tag{36}$$

According to the prescription for local equilibrium as derived from a Fermat Problem the parenthesis above vanishes and four n_1 vectors satisfy a relation of the form given by eq. (28), with $a = 4$.

The same argument can be used again for another backbone nucleus Z_{n_2} and its neighbours $Z_{n_1}, Z_{n_2-2}, Z_{n_2-1}, Z_{n_3}$. We can then write for a natural candidate for total force in two nuclei:

$$\vec{F}_{n\,1} + \vec{F}_{n\,2} \approx -e^2 \int \rho\left(\vec{r}\right) \left(Z_{n\,1} \frac{\left(\vec{R}_{n\,1} - \vec{r}\right)}{\left\|\vec{R}_{n\,1} - \vec{r}\right\|^3} + Z_{n\,2} \frac{\left(\vec{R}_{n\,2} - \vec{r}\right)}{\left\|\vec{R}_{n\,2} - \vec{r}\right\|^3} \right) d^3\vec{r}. \tag{37}$$

The representation of forces according a Berlin diagram is then done by considering the forces applied to backbone nuclei of the macromolecule, only. We also notice that the contributions to the forces on each backbone nucleus in CM and GC systems can be disregarded.
These contributions are given by

$$\Delta\vec{F}_{n\,1_{CM}} = e^2 \frac{m_{n\,1}}{M} \int \rho\left(\vec{r}\right) \sum_{s=n_1-3}^{n_1-1} Z_s \frac{\left(\vec{R}_s - \vec{r}\right)}{\left\|\vec{R}_s - \vec{r}\right\|^3} d^3\vec{r}, \tag{38}$$

$$\begin{aligned}\Delta\vec{F}_{n\,1_{GC}} &= -e^2 \frac{m_{n\,1}}{2n} \sum_{s=n_1-3}^{n_1-1} \sum_{l\neq s} \frac{Z_s Z_l}{m_s} \frac{\left(\vec{R}_s - \vec{R}_l\right)}{\left\|\vec{R}_s - \vec{R}_l\right\|^3} \\ &+ e^2 \frac{m_{n\,1}}{n} \int \rho\left(\vec{r}\right) \sum_{s=n_1-3}^{n_1-1} \frac{Z_s}{m_s} \frac{\left(\vec{R}_s - \vec{r}\right)}{\left\|\vec{R}_s - \vec{r}\right\|^3} d^3\vec{r},\end{aligned} \tag{39}$$

where we have already disregarded the terms with $s \geq n_1 - 1$ in the sums above due to the large values of the corresponding norms $\left\|\vec{R}_s - \vec{R}_l\right\|$, $\left\|\vec{R}_s - \vec{r}\right\|$. Actually, we have $\Delta\vec{F}_{n_{1\,CM}} \approx 0$ and $\Delta\vec{F}_{n_{1\,GC}} \approx 0$ for a macromolecule, respectively.

We then see that we can extend the Berlin formulation for describing regions of force the macromolecules and this treatment will be free from difficulties found with small molecules or molecular clusters.

5. Contribution of Dipolar Terms

Some ideas could be reported on the predominance of terms with dependence $\left\|\vec{R}_s - \vec{r}\right\|$ in the potential energy. These will correspond to the interaction of a protons of a backbone nuclei with a dipole which could be formed by the probability of occurrence of an electron of the electronic cloud near a proton of another backbone nuclei. These terms will predominate over the Coulombian terms of eq. (37) in the binding forces among backbone nuclei, and they are given by

$$\vec{F}_{n_1} + \vec{F}_{n_2} \approx \int \rho(\vec{r}) \left(D_{n_1} \frac{\left(\vec{R}_{n_1} - \vec{r}\right)}{\left\|\vec{R}_{n_1} - \vec{r}\right\|^4} + D_{n_2} \frac{\left(\vec{R}_{n_2} - \vec{r}\right)}{\left\|\vec{R}_{n_2} - \vec{r}\right\|^4} \right) d^3\vec{r}, \quad (40)$$

where D_{n_1}, D_{n_2} are the dipole parameters of the two backbone nuclei n_1, n_2.

If a Berlin diagram could be used for an experimental analysis of the force pattern in a macromolecule, the dominance of the pattern by terms like those given in eq. (40) lead to a number of three nearest neighbours to each backbone nuclei instead of four as in the case of eq. (37). In a protein architecture these values 3 and 4 correspond to the number of neighbours around nitrogen or carbonyl and α-carbon, respectively.

Eq. (34) should be written in this case as

$$\frac{eZ_{n_1} D_{n_2}}{\left\|\vec{R}_{n_1} - \vec{R}_{n_2}\right\|^3} = \frac{eZ_{n_1} D_l}{\left\|\vec{R}_{n_1} - \vec{R}_l\right\|^3} = \frac{eZ_{n_1} D_k}{\left\|\vec{R}_{n_1} - \vec{R}_k\right\|^3} = F_{n_1}, \quad (41)$$

where $l, k = n_1 - 3,\ n_1 - 2,\ n_1 - 1$.

The requirements given in eqs. (34) and (41) have an interesting consequence for the case of a homonuclear region of a macromolecule. If we assume the validity of relations (34) and (41) for a homonuclear sequence of consecutive nuclei $n_2 - n_1 = n_3 - n_2 = ... = 1$, we have,

$$\left\|\vec{R}_{n_1} - \vec{R}_{n_2}\right\| = \left\|\vec{R}_{n_2} - \vec{R}_{n_3}\right\| = \ldots, \quad (42)$$

and this corresponds to a sequence of evenly spaced backbone nuclei according the Euclidean distance.

Actually, in the proposal for modelling which we have been using in recent works [9], we assume the validity of relations (42) even for a heteronuclear sequence. Eqs. (42) can be satisfied by choosing the coordinates of the position vectors $\vec{R}_{n_1}$ as

$$\vec{R}_{n_1} = \left(r(\varphi)\cos n_1\varphi,\ r(\varphi)\sin n_1\varphi,\ n_1 h(\varphi)\right) \quad (43)$$

where $r(\varphi), h(\varphi)$ are arbitrary.

$$\left\|\vec{R}_{n_1} - \vec{R}_{n_2}\right\| = \left(r^2(\varphi) + h^2(\varphi)\right)^{1/2}.$$

After using eq. (28) for the associated unit vectors we get

$$h^2 = \frac{2}{a-2} r^2 \left(1 - \cos\ \varphi\right)\left(1 - (a-1)\cos\ \varphi\right).$$

The representation of binding and non-binding regions from eqs. (30)-(32) is an unpleasant task, due to the exponential number of terms when $n \gg 1$. Fortunately, the assumption of local equilibrium and the generalized Fermat problem lead to an alternative way of considering the representation of Berlin diagrams for extended macromolecules like biopolymers. Some theoretical development including the analysis of Berlin diagrams for selected regions of biopolymers are now in progress and will be published elsewhere.

As a final remark [10], let us notice that an Ansatz like that given in eq. (43) which means evenly spaced consecutive atom sites, satisfies eq. (28) only for $a = 3, 4$, corresponding to the values $m = 4, 3$ in the potential interaction energy term $\left\| \vec{R}_{n\,1} - \vec{R}_{n\,2} \right\|^{-(m-2)}$ and this corresponds to the structures of bond angles in biomacromolecules, where the angles between bond edges are given by $arccos\ \left(-\frac{1}{a-1}\right) = arccos\ \left(-\frac{1}{2}\right),\ arccos\ \left(-\frac{1}{3}\right)$, respectively.

References

1. I. N. Levine, *Quantum Chemistry*, 4th Edition, Prentice-Hall Inc. 1999.
2. T. Berlin, Binding Regions in Diatomic Molecules, *Journal Chem. Phys.* **19(1)** (1951) 208.
3. B. M. Deb, The Force concept in Chemistry, *Rev. Mod. Phys.* **45(1)** (1973) 22.
4. R. P. Feynman, Forces in Molecules, *Phys. Rev.* **56** (1939) 340.
5. T. Koga, H. Nakatsuji, T. Yonezawa, Generalized Berlin diagram for Polyatomic Molecules, *J. Amer. Chem. Soc.* **100**.
6. M. Born, J. R. Oppenheimer, Zur Quantentheorie der Molekeln (On the Quantum Theory of Molecules) *Ann. Phys.* **84** (1927) 457.
7. C. Dale Keefe, J. K. Pearson, Ab Initio Investigations of Dipeptide Structures, *J. Mol. Structure (Theochem)* **679** (2004).
8. A. S. Edison, Linus Pauling and the Planar Peptide Bond, *Nature Structural & Molecular Biology* **8(3)** (2001) 201-202.
9. R. P. Mondaini, A Correlation between Atom sites and Amide Planes in Protein Structures in BIOMAT 2009, *Int. Symp. Math. Comp. Biology, World Scientific* (2010).
10. R. P. Mondaini, The Steiner Tree Problem and its Application to the Modelling of Biomolecular Structures, in *Mathematical Modelling of Biosystems*, Applied Optimization Series, Vol. 102, Spring Verlag, 2008.

ON THE USE OF MECHANISTIC AND DATA-DRIVEN MODELS IN POPULATION DYNAMICS: THE CASE OF TUBERCULOSIS IN THE US OVER THAT PAST TWO CENTURIES

JUAN PABLO APARICIO

Instituto de Investigación en Energías no Convencionales, CONICET-UNSa
Universidad Nacional de Salta, Argentina.
E-mail: juan.p.aparicio@gmail.com

CARLOS CASTILLO-CHAVEZ

Arizona State University, Tempe, 85287
Santa Fe Institute, 1399 Hyde Park Road
Santa Fe, NM, 87501
E-mail: cchavez@asu.edu

A model is built and used to explore the effects of variations in transmission and/or progression on the decline of tuberculosis rates during the twentieth century. This study also makes use of available data generated over two significantly distinct spatial scales: one (global) involving the United States and the second (local) utilizing the tuberculosis data generated by the state of Massachusetts.

1. Introduction

The population dynamics theory most often addresses biological questions through its use of simple mechanistic deterministic or stochastic models. In the context of specific applications, however, the use of models tailored to fit the case under study is essential. Most frequently the use of statistical models of the data are the preferred approach regularly bypassing the advantages provided by mechanistic dynamical models parameterized with population-relevant parameters. In this manuscript we show how data may be used to parameterize what is often referred as a hybrid model. That is, a mechanistic data-driven, parameter-scarce, dynamical systems model is fitted to data with time-dependent population-level parameters. The case of the tuberculosis epidemic in United States over long-time scales is used to illustrate this approach.

The assumption that no changes in progression rates from latent to active tuberculosis have taken place leads to conclusion that the maximum rate of variation of tuberculosis transmission must have taken place before any major medical intervention were implemented (including the use of antibiotics). Next, under the assumption of constant transmission rates, it is shown that the maximum rate of variation in tuberculosis progression rates must have occurred around the beginning of the twentieth century, that is, when the maximum variation was recorded for the mortality rates. These results, among others discussed in this manuscript, support the view that improved living conditions must have been the main force driving the tuberculosis decline over the past two centuries in the United States.

Deterministic and stochastic mathematical models have played and continue to play a fundamental role in the development of population dynamics theory. The key to their success in advancing biological theory can be traced directly to the fact that they are built under specific (albeit often oversimplified) mechanistic assumption and therefore naturally incorporating population-level parameters. These models capture our aggregated knowledge of or perspectives on the fundamental processes that drive the dynamics of a population. The first-order goal of population dynamics is to *explain* or tie in observed population patterns and the mechanisms that might have generated them, is met in this setting.

However, like with any other approach, the limitations or shortcomings that come from the use of simplified models must be addressed. The issue of structural stability and model resilience come to mind. In fact, model *structure* depends, perhaps too much, on *the* modeler as quite often implicit assumptions are made that lead to potentially non-robust conclusions (see for example the paper by Wood [1]).

The classic Lotka-Volterra predator-prey model provides but one albeit often rehashed example. The L-K model predicts sustained oscillations with their amplitude a function of the initial conditions, results that critically depend on the model structure. The use of alternative mechanistic population-level assumptions (Holling [2,3,4]) lead, for example, to modified models that also predicts sustained oscillations (stable limit cycles) via a radically different mechanism where, for example, the amplitude of the oscillations does not depend on initial conditions. Alternative mechanisms have been identified. For example, deterministic predator-prey models that robustly support damped oscillations, give rise to sustainable oscillations when the always present, demographic stochasticity, is taken into account[5]. The use of refuges that mediate predator-prey interactions not only can support the type of transitions driven by demographic stochasticity but enrich even further the set of possibilities [8]. In the context of epidemiology, the use of mathematical models has had a powerful impact ([9,7,10,11,12,13] and references there in). In the case of tuberculosis, for example, the role of theoretical models may be seen directly from our study of the role of re-infection on the generation of bi-stability ([37,19,21,20,17] and references there in)

Dealing with applications in the context of specific cases, typically begins with observational data. Most often statistical explanations are desired and consequently, statistical, data-driven models, like linear or non-linear regression models are widely used. On some situations the use of detailed simulations involving complex models is the preferred procedure. Detailed models are parameter rich and consequently, hardly ever we have enough reliable data to validate these approaches. Here, we use a hybrid model with potentially relevant predictive power. Hence, we combine a mechanistic simple model to build a data-driven models via the introduction of time-dependent parameters (modeled from data). The model together with demographic and tuberculosis epidemiological data from the United States, is used to shed light on the probable causes behind the historical decline of tuberculosis observed over the past two centuries in the US.

2. A mechanistic model for tuberculosis dynamics over long-temporal scales

Tuberculosis (TB) is an infectious disease with unusual characteristics. The overwhelming majority of the infected individuals *now* remain non-symptomatic for life while the small fraction that develops the clinical disease (active-TB) may do it years after infection. Risk of progression from latent infection to active-TB decreases (almost exponentially) with the age of infection. Most of the new cases of active-TB arise within few years after infection[14,15]. This particular way of progression is captured well using a model with two latent classes: a high risk latent class E and a low risk latent class L [16,17].

After acquiring infection, previously uninfected individuals (U) are moved to the high-risk latent class (E). High-risk latent-individuals progress to active-TB at the per capita rate k. However, the proportion q of new active-TB cases is assumed to be constituted by individuals with pulmonary cases (infectious class A_p), while the remaining proportion $1-q$ is constituted by individuals who have developed extra-pulmonary, non-infectious, active TB (class A_e).

Individuals in the high risk class E not progressing to the active-TB class, enter the low risk latent class L at the per-capita rate α. L-individuals may develop active TB at the rate k_L but since previous infectious only provide partial immunity, latent individuals, these individuals may become re-infected. We assume only that low risk latent individuals may become reinfected and these cases are moved to a new high risk class E^* where either they progress to the active-TB class (at the per-capita rate k^*) or "regress" to the low risk latent class L (at the rate α). Infectious cases die from the disease at the per capita rate d or recover at per capita rate r (d_e and r_e denote the death and recovery rates for non-infectious TB-cases, respectively). Active-TB cases may recover either naturally or as a result of treatment at the per-capita rate r (r_e for extra-pulmonary cases). Recovered individuals develop active TB at the rate k_{Rp} (TB relapse).

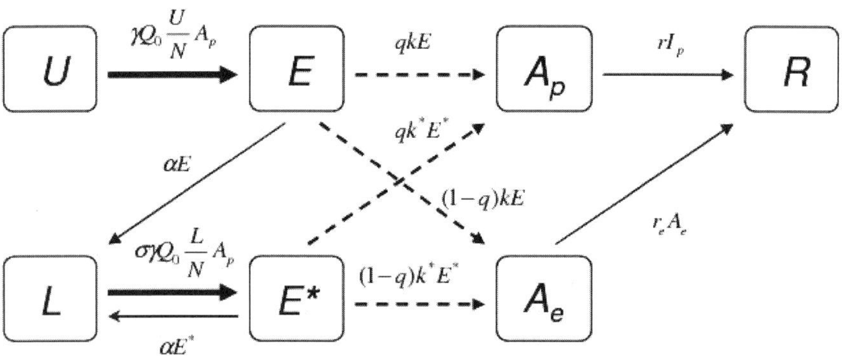

Figure 1. Transfer Diagram. Individuals are recruited into the uninfected class U and moved, after infection, into the high risk latent class E. Individuals in the E. class may progress to active TB. In fact, it is assumed that the fraction q develops pulmonary TB (A_p) while the fraction 1-q develops extra-pulmonary TB (A_e). However, most individuals move to the low-risk latent class L where they may become reinfected moving to the high-risk class E^* where they may either develop active-TB or return to the low-risk class L. Active cases may recover moving to the R class. Transmission is represented with thick arrows progression with dashed arrows. Progression from the low-risk latent class L is not shown. Transfer rates per unit of time are shown.

Transmission is modeled as follow. A "typical" pulmonary case placed in an completely uninfected population produces Q_0 secondary infections on the average (Q_0 is called the *contact number*). This number it is not the basic reproduction number because only a (generally small) fraction of infected individuals will become active (infectious) cases. The basic reproduction number is here is defined as the number of secondary active infectious cases produced by an average infectious case placed in a fully susceptible population. When only a fraction U/N of the contacts of an average pulmonary case is uninfected then the number of first infections becomes Q_0U/N. Therefore the per-capita rate of new infections (that is, per infectious individual) is obtained by dividing Q_0U/N by the infectious period $1/\gamma$. Similarly, the reinfection rates are modeled as $\sigma\gamma Q_0 L/N$ where $\sigma \leq 1$ denotes the partial protection conferred by previous infections if any. The compartmental deterministic model becomes (see Fig. 1 for a transfer diagram):

$$\frac{dU}{dt} = B - \mu U - \gamma Q_0 \frac{U}{N} A_p, \tag{1}$$

$$\frac{dE}{dt} = \gamma Q_0 \frac{U}{N} A_p - (k + \mu + \alpha)E, \tag{2}$$

$$\frac{dL}{dt} = \alpha(E + E^*) - (\mu + k_L)L - \sigma\gamma Q_0 \frac{L}{N} A_p, \tag{3}$$

$$\frac{dE^*}{dt} = \sigma\gamma Q_0 \frac{L}{N} A_p - (k^* + \mu + \alpha)E^*, \tag{4}$$

$$\frac{dA_p}{dt} = q(kE + k^*E^* + k_L L + k_{Rp} R) - \gamma A_p, \tag{5}$$

$$\frac{dA_e}{dt} = (1 - q)(kE + k^*E^* + k_L L + k_{Rp} R) - \gamma_e A_e, \tag{6}$$

$$\frac{dR}{dt} = rA_p + r_e A_e - (\mu + k_{Rp})R, \tag{7}$$

where $\gamma = \mu + d + r$ and $\gamma_e = \mu + d_e + r_e$.

2.1. *Basic reproduction number*

The basic reproduction number (R_0) is defined as the average number of secondary cases produced by a typical infectious individual in a fully susceptible population. The clustering of contacts play a significant role in shaping the transmission tuberculosis patterns[19,18].

The number of secondary infections produced by a source case under the above conditions is the contact number here denoted by Q_0. From this number only the fraction f will develop active-TB, and only the fraction q of them will result in a pulmonary, infectious cases. Therefore, the basic reproduction number by be estimated as follows:

$$R_0 = qQ_0 f. \tag{8}$$

Numerical simulation shows that expression (8) is an excellent estimate of the epidemic threshold for model (1-7).

In the above formulation the basic reproduction number does not depend *explicitly* on γ. However, the contact number Q_0 is in general a non-linear function of the mean infectious period $1/\gamma$. For example, if the mean infectious period is assumed to be exponentially distributed then Q_0 would be given by [19,20,17]

$$Q_0(\gamma) = \frac{\beta n}{\beta + \gamma} + m\frac{\beta}{\gamma} \tag{9}$$

where n is the mean size of the network of close and frequent contacts (*generalized household*) of an infectious individual (household and workplace contacts, for example); β is the per susceptible risk of infection in generalized households; and m is a measure of the size of the network of casual contacts (contacts outside the generalized household) of the average infectious individual [19]. The more realistic approximation, using an infectious period of fixed length, results in $Q_0 = 1 - exp(-\beta/\gamma)$ [22].

3. Data series, data-driven models and model parameterization

The demographic and epidemiological data from United States and from the state of Massachusetts (USA) are incorporated explicitly in the modeling framework. The dynamics of tuberculosis at the population level is extremely slow with epidemics spanning centuries[24,25,26]. Therefore, we cannot ignore the demographic changes, that is, we must take into account the historical variation in the values of parameters like the mortality and

Table 1. Parameter values used in the models. For details see Appendix A.

	Parameter description	Value	Source and comments
B	Recruitment rate	variable	From data, US Bureau of the census
μ	non-TB mortality	variable	From data, US Bureau of the census
d	Pulmonary-TB mortality	r	From data, Styblo 1991
d_e	Extrapulmonary-TB mortality	μ	The case $d_e = d$ was also considered
k	Progression rate from high risk latent class E	variable	From eq. $f = k/(k + \mu + \alpha)$
k^*	Progression rate from high risk latent class E^*	variable	$k^* = \sigma_p k$
α	High-risk to low-risk transfer rate	1.5/year	Aparicio et al 2002a using data from Styblo 1991, Sutherland 1968
k_{Rp}	TB-relapse rate	0.0008/year	From data, CDC 1999
q	proportion of pulmonary cases	0.7	CDC 1999 $q(t)$ variable was also considered
γ	removal rate	2/year	Exact value does not play a significant role
Q_0	Secondary infections produced by one primary case	variable	In the cases where Q_0=constant we considered $Q_0 = 10$ and 20. Otherwise see Appendix B.
r	Pulmonary-TB recovery rate	d	From data, Styblo 1991
r_e	Extra-pulmonary recovery rate	$\gamma - \mu - d_e$	From model construction

birth rates. In order to account for these demographic changes, historical data series used include: population sizes, proportion of urban populations, mortality rates, TB related deaths, and new cases of active-TB per year. For a more detailed discussion about model parameterization see ref [17].

3.1. *Mortality rates*

The total per-capita mortality rates decreased from about 0.02 yr^{-1} in 1850 to the present average value of 0.0087 yr^{-1} [27,28]. Before 1900 the data presents fluctuations. The pattern was captured by the smooth sigmoid function of the calendar year t (see Fig. 2),

$$\mu_{TOT}(t) = \mu_f - \frac{\mu_f - \mu_i}{1 + exp[(t - t_\mu)/\Delta_\mu]}. \tag{10}$$

The parameter values $\mu_i = 0.021yr^{-1}$, $\mu_f = 0.00887yr^{-1}$, $t_\mu = 1910yr$, and $\Delta_\mu = 16yr$ were obtained by standard least squares fit. Non TB-related mortality μ used in the model was estimated from total mortality μ_{TOT} by subtracting TB's contribution. Deaths, mostly attributable to tuberculosis have been recorded in the state of Massachusetts since around 1850 (see section 3.4 below).

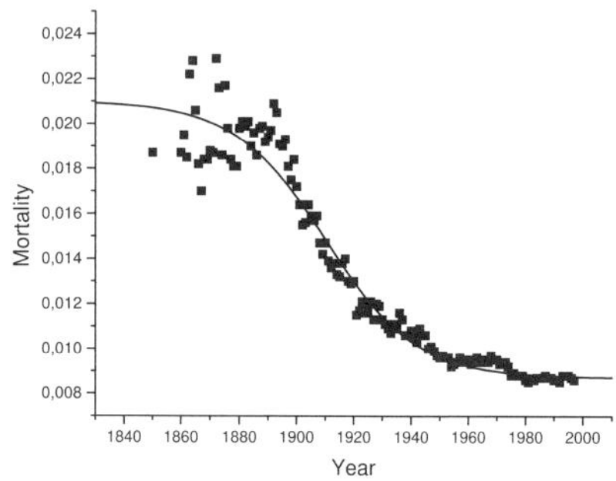

Figure 2. Observed total mortality rates (squares) and their approximations given by the expression 10 (continuous line).

3.2. *Population sizes*

Tuberculosis is primarily an urban disease and therefore we have disregarded the contributions to the TB transmission-dynamics by rural populations as *first* approximation. Census data on total population and the proportion of of individuals living in urban centers has been available since 1850 for United States (as well as for the state of Massachusetts) [27,28], see Fig. 3). From the first 50 years of data (which show an exponential growth), we back-extrapolate population values going back to 1700. The proportion of urban populations were modeled as

$$P_U(t) = P_f - \frac{P_f - P_i}{1 + exp[(t - t_{PU})/\Delta_{PU}]} \tag{11}$$

using least squares fit (see Fig. 3 for parameter values). Values for the urban populations were obtained using census data for the total populations and expression (11).

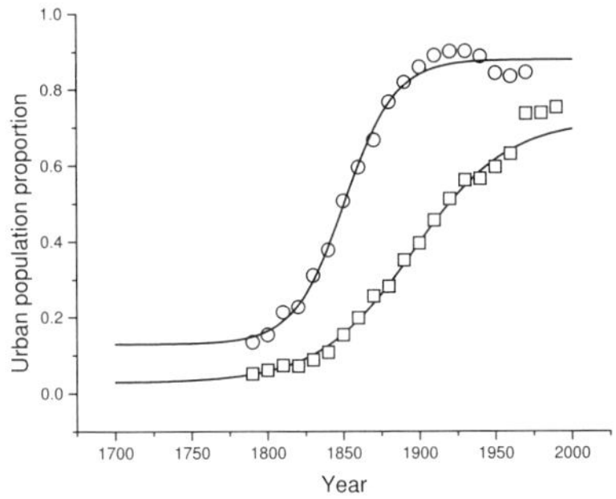

Figure 3. Proportion of urban populations for United States (squares) and the state of Massachusetts (circles). In both cases we fit data to the Boltzman function (11). Parameter values obtained from best least squares fit are: $P_i = 0.028236$, $P_f = 0.71835$, $t_{PU} = 1895.6$, $\Delta_{PU} = 30.998$ for US, and $P_i = 0.12965$, $P_f = 0.87949$, $t_{PU} = 1851.0$, $\Delta_{PU} = 16.817$ for Massachusetts.

3.3. *Recruitment rates*

The recruitment of uninfected individuals $B(t)$ was estimated by the difference between the observed net (urban) population growth, (which include births and immigration) and the number of total deaths per unit of time dt (from expression 10); that is $B(t) = [N(t + dt) - N(t)]/dt + \mu_{tot}N(t)$. The values for the urban population $N(t)$ between successive census data were generated by linear interpolation. Because immigration to cities was largely the result of the movement of people free of TB infection, the assumption of recruitment only in the uninfected class may be a reasonable approximation.

3.4. *Incidence of active-TB*

The incidence of active-TB for the United States and the state of Massachusetts have been recorded since 1953. However, there exist records of pulmonary TB mortality since 1850. Hence, before 1953 we assume that the incidence of active-TB is proportional to pulmonary TB-mortality. Before the antibiotic treatment era, pulmonary TB mortality was around 50%. Therefore a relation of 2:1 between the incidence of pulmonary active-TB and mortality may be reasonable. Here, we use the value of 2.875 as the proportional constant between TB incidence and mortality under the assumption that pulmonary TB represented the 70% of the total cases. Estimated and observed incidence were log transformed and the results were fit to polynomials functions.

USA data. We used the polynomial of grade two that produced the best fit (in the least squares sense) to the natural logarithm of 2.875 times the mortality data form 1850 and 1944 and 1944 and 1979. For $t > 1979$ we used a polynomial of grade one. The fitted function $Inc(t)$ obtained is

$$Inc(t) = \begin{cases} exp(-1599.38143 + 1.70781t - 0.000454057t^2) & 1840 < t < 1944 \\ exp(2667.46 - 2.65571t + 0.0006615t^2) & 1944 < t < 1979 \\ exp(101.683531 - 0.0501t) & 1979 < t < 2000 \end{cases}$$

Massachusetts data. Between 1850 and 1953, and after 1953 we approximated the incidence by polynomials of grade two. These two curves do not intersect but were connected with a horizontal line. The function $Inc(t)$ obtained is:

$$Inc(t) = \begin{cases} exp(-1214.31417 + 1.31483t - 0.000353917t^2) & 1840 < t < 1950 \\ exp(3.834937) & 1950 < t < 1952.325 \\ exp(2327.06362 - 2.30263t + 0.00056991t^2) & 1952.325 < t < 2000 \end{cases}$$

The simulated incidence (per 10^5 population, per year) of active-TB was estimated from the model (1-7) as

$$Simulated\ incidence \ = (kE + k^*E^* + k_L L + k_{Rp} R)10^5/N_{tot}, \qquad (12)$$

where N_{tot} is the total population (note that in the model N represents the urban population).

The simulated incidence depends on the progression rates (k, k^*, k_L, k_{Rp}) and the sizes of the latent populations (E, E^* L, R) which in turn depend on the transmission processes controlled by the value of Q_0. Here we take the actual observed value for k_{Rp} constant over time. As first approximation we disregard progression from low risk latent class, that is, $k_L = 0$ (but see Aparicio and Castillo-Chavez[17] for a results using $k_L > 0$). It is also assumed that the partial protection conferred by previous infections decreased k^* in a factor $\sigma_p = 0.7$, that is, we considered $k^* = \sigma_p k$. These observations left only two parameters to be determined from the incidence historical data series: k and Q_0.

We considered the non-dimensional lifetime risk of progression to active-TB of latent individuals (Instead of the progression rate k) which in our framework may be estimated as

$$f = \frac{k}{k + \alpha + \mu} \qquad (13)$$

when reinfection and progression from the low risk latent class is disregarded. Values of k used in the model are obtained from expression (13).

From the incidence's time series the values of Q_0 and f cannot be obtained simultaneously. We estimate one of them independently and used the data series to estimate the other. We start the simulations in the year 1700 to avoid the influence of the (on the other hand unknown) initial conditions. $P(t)$ denotes the parameter being estimated from the incidence time series ($P(t)$ is either $Q_0(t)$ or $f(t)$). From 1700 to 1850 $P(t)$ is kept constant and future values must be chosen in such a way that the simulated incidence for the year 1850 matches (within some small error) the value $Inc(1850)$ obtained above. After 1850 new $P(t)$ values are obtained as follows: After each time step dt of the numerical integration scheme we computed the relative error

$$\varepsilon = \frac{Simulated\ incidence(t) - Inc(t)}{Inc(t)}$$

where *Simulated incidence* is given by (12) and $Inc(t)$, for the cases of US or Massachusetts, is given by the functions defined in 3.4. The parameter P is actualized as

$$P(t + dt) = (1 - \varepsilon)P(t)$$

after each step of the numerical integration. This estimation of $P(t)$ may be improved in several ways but in this case, the above simple scheme was enough to find values for $Q_0(t)$ or $f(t)$ for which simulated incidence accurately reproduce the observed trends of TB incidence (see Fig. 4).

4. Application: A study of the causes of the secular decline of tuberculosis

Tuberculosis transmission and progression rates have varied over the past two centuries but what has been the pattern? Here we considered two possible extreme explanations:

a) TB decline is the result of (solely) reductions in transmission rates (f=constant) and

b) TB decline is the result of (solely) reduction in progression rates (Q_0=constant).

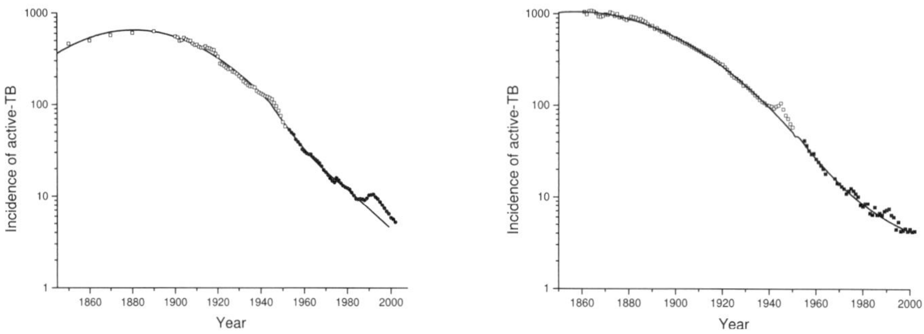

Figure 4. Observed incidence of active-TB (all forms, solid squares, per year and per 100,000 population), estimated incidence of active-TB (estimated as 2.875 times TB mortality rates, open squares), for United States (left) and Massachusetts state (right). Simulated incidence obtained from model (1-7) are shown in continuous lines.

Variation in the transmission per case are reflected in the variability in Q_0 while variations in progression are reflected in variations of f. Most of the parameters used were determined from data (see Table 1). The possible influence of the uncertainty on parameter values is discussed in the appendix. In each case we analyzed the trends of $Q_0(t)$ or $f(t)$ obtained fitting model solutions to the data.

In order to compare variations in transmission (or progression) per case at different times we normalized the absolute variations per unit of time $[P(t+\Delta t) - P(t)]/\delta t$ dividing it by the value of $P(t)$. For δt small enough, the relative rate of variation is given by

$$\theta \equiv \frac{1}{P}\frac{dP}{dt} = \frac{d(\ln P)}{dt}. \tag{14}$$

Since the values of $\ln Q_0(t)$ or $\ln f(t)$ were obtained at discrete times, numerical derivatives exhibit large, meaningless, fluctuations. We considered the best least square fit to smooth functions. For the USA data the fit was excellent with the use of Lorentz functions $[y(t) = y_0 + \frac{2A}{\pi}\frac{w}{4(t-t_0)^2+w^2}$, $y \equiv \ln Q_0$ or $\ln f$, see Fig. 5]. Parameter values obtained from the best fit in each case were:

y	P	A	t_0	w	y_0
$\ln Q_0$	$f = 0.1$	$299,48$	1859	$94,444$	$2,4280$
$\ln Q_0$	$f = 0.2$	$338,23$	$1861,4$	$124,27$	$1,6099$
$\ln f$	$Q_0 = 10$	$235,86$	$1864,9$	$149,83$	$2,4280$
$\ln f$	$Q_0 = 20$	$320,43$	$1861,9$	$153,50$	$1,0021$

For Massachusetts data we used polynomials of order four $[y(t) = \sum_{i=0}^{4} a_i t^i]$(see Fig. 6). Parameter values obtained from the best fit in each case were:

y	f or Q_0	a_0	a_1	a_2	a_3	a_3
$\ln Q_0$	$f = 0.1$	$-22634,869$	$47,19016$	$-0,03683$	$1,27522E^{-5}$	$-1,65362E^{-9}$
$\ln Q_0$	$f = 0.2$	$-26954,7165$	$55,6619$	$-0,04306$	$1,479E^{-5}$	$-1,90354E^{-9}$
$\ln f$	$Q_0 = 10$	$-17990,489$	$36,6019$	$9,4296966E^{-6}$	$1,27522E^{-5}$	$-1,19411E^{-9}$
$\ln f$	$Q_0 = 20$	$-19301,075$	$39,2871$	$-0,02994$	$1,01245E^{-5}$	$-1,28188E^{-9}$

The smooth functions obtained were constructed in order to quantify the relative rate of variation $\theta = \frac{1}{P}\frac{dP}{dt}$. These relative rates of variation were then used to determine when reductions in transmission or in progression were more significant. In particular, major changes in transmission or progression per case should had left its signature in the trends and the relative rates of variation should allow us to find them.

4.1. *Reduction in transmission (f constant)*

Influenced by prior European experiences the first TB sanatoria in the United States opened at the end of the eighteenth century. Yet, by 1915 only about 4% of the active TB cases were isolated in sanatoria[29]. This percentage increased to 25% in 1934 [30] and by 1954 almost 50% of the active TB cases were confined to TB hospitals[29]. The era of effective antibiotic treatment began in 1946 when *streptomycin* was introduced.

The isolation or effective treatment of infectious individuals reduces the average length of the effective infectious period. That is, these two measures would tend to reduce TB transmission. This reduction is modeled via a reduction in the number of secondary infections caused by an infectious individual, that is, as a reduction in the value of Q_0.

We used our model to obtain functions of the time $Q_0(t)$ for which simulated incidence reproduced the observed incidence of active-TB (see appendix B) under the assumption that the risk of progression to active-TB (f) remained constant. Actual estimates of the risk f for the United States are in the range 0.05 to 0.10. We considered two scenarios: $f = 0.1$ and $f = 0.2$ (an overestimate) and the results are summarized in Table 2 and Figs. 5 and 6.

The figures obtained for the USA and Massachusetts are similar (see Figs. 5 and 6). For the USA data (with $f = 0.1$) an extremely high value

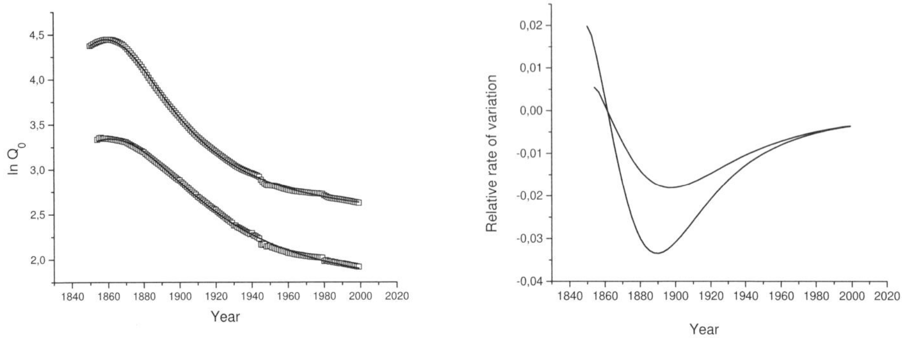

Figure 5. Time evolution of lnQ_0 (open squares, left figures). When these values of Q_0 are used in model 1-7 numerical solutions reproduce the observed incidence of active-TB when f is assumed constant (with values 0.1 and 0.2). Continuous lines correspond to the best fit to a Lorentztian function (see section 4). In the right column we show the relative rates of variation θ. USA data.

of $Q_{0MAX} = 80$ was obtained for 1860. For the Massachusetts data this number drops to 40 for 1850 but the trends indicate that substantial higher numbers could have been obtained for prior times. Because today values of f are likely lower than ten percent, those values of Q_{0MAX}, are underestimates under the assumption of constant f.

Maximum basic reproduction numbers, $R_{0MAX} = qQ_{0MAX}f$ are in the range 1.89 to 5.6, while minimum values ($R_{0MIN} = qQ_{0MIN}f$) are close to 1 (see Table 2).

In the both cases considered ($f = 0.1$ and $f = 0.2$, see Figs. 5 and 6) the maximum absolute value of the relative rate of variation is reached before 1900, that is before any major epidemiological intervention (reclusion of active cases or treatment) was in practice. No effect of antibiotic treatment after its introduction in the fifties is observed in incidence trends although its impact in TB-mortality trends is evident.

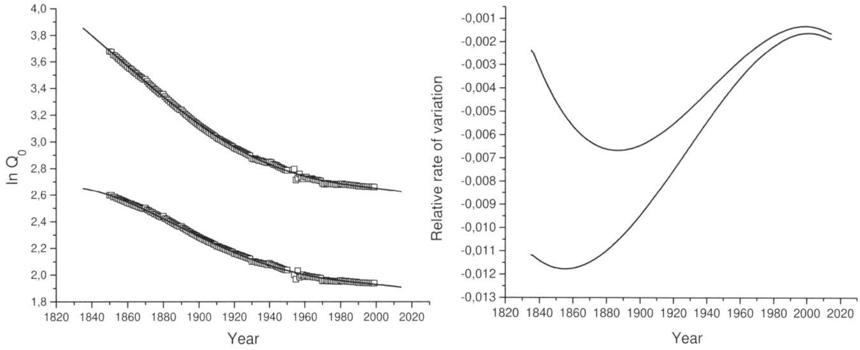

Figure 6. Time evolution of lnQ_0 (open squares, left figures). When these values of Q_0 are used in model 1-7 numerical solutions reproduce the observed incidence of active-TB under the assumption that f is constant (with values 0.1 and 0.2). Continuous lines correspond to the best polynomial (order four) fit (see section 4). In the right column we show the relative rate of variation θ. Massachusetts data.

4.2. *Reduction in progression (Q_0 constant)*

Currently, only a relatively small fraction of the latently-infected individuals develops active TB over their entire life-span. We use this proportion (f) as a non-dimensional measure of the risk of progression to active-TB. Improved living conditions (particularly the massive reduction in malnutrition), may had reduced progression rates by enhancing individuals's immune system responses[31,32,33,34,35].

Improved living conditions is a broad concept that include amelioration of poverty, improved water sanitation, housing quality, or reduced crowding. These, among other factors, may affect both, the transmission and the progression processes, while in general also improve immune function response (for example, just by reducing co-infections with other diseases).

Reduction in progression may (partially) had been also the result of host-parasite coevolution. However, no matter what caused the reductions in progression rates, they are reflected in reductions in the fraction f.

As in the previous case we used the model to obtain functions of the time $f(t)$ for which simulated incidence reproduced the observed incidence of active-TB when Q_0 is assumed to be constant. Results are remarkably independent of the populations studied in this work (see Table 3). We considered two scenarios: $Q_0 = 10$ and $Q_0 = 20$. For $Q_0 = 10$ the maximum value obtained for f was around 0.29 (for 1840) while the minimum around 0.14 which is above the actual upper estimate of 0.1. For $Q_0 = 20$ these values drop to 0.18 and 0.07 respectively. In all cases a slightly more than a twofold reduction in the risk f was obtained (Figs. 7 and 8).

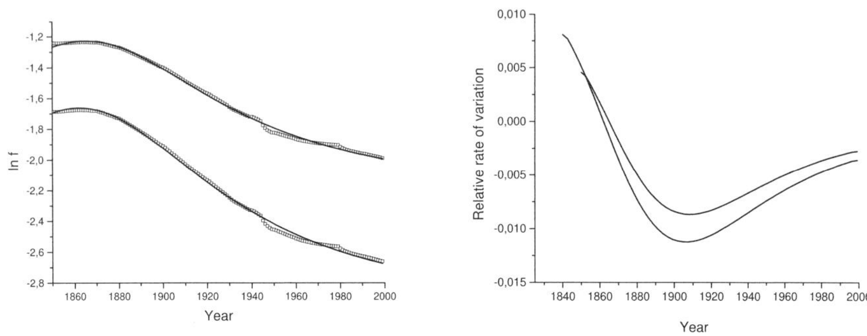

Figure 7. Time evolution of $lnf(t)$ (open squares, left figures) for USA data. When these values of $f(t)$ are used in Model 1-7 numerical solutions reproduce the observed incidence of active-TB when Q_0 is assumed constant (with values 10 and 20 respectively). Continuous lines correspond to the best fit to a Lorentz functional form (see section 4). In the right column we show the relative rate of variation θ.

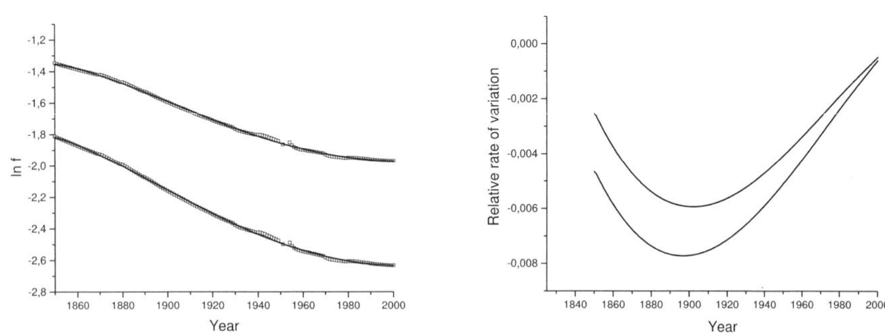

Figure 8. Time evolution of $lnf(t)$ (open squares, left figures) for Massachusetts data. When these values of $f(t)$ are used in model 1-7 numerical solutions reproduce the observed incidence of active-TB when Q_0 is assumed constant (with values 10 and 20 respectively). Continuous lines correspond to the best fourth order polynomial fit (see section 4). In the right column we show the relative rates of variation θ.

Maximum basic reproduction numbers, $R_{0\,MAX} = qQ_0 f_{MAX}$, are between 2 to 2.6, while minimum values ($R_{0\,MIN} = qQ_0 f_{MIN}$) are close to 1 (see Table 3 and compare with Table 2). Minimum values of the risk of progression to active-TB, f_{MIN}, are well within current estimates for $Q_0 = 20$ but resulted in values slightly above the estimated upper value of 0.1 for $Q_0 = 10$.

Table 2. Maximum and minimum values obtained for Q_0 when different constant values of f were used. Maximum values correspond to 1850 (except for the USA and f=0.1, where the maximum are reached at 1860) while minimum values correspond to the year 2000. The corresponding reproductive basic numbers are also included.

Population	f	Q_{0MAX}	R_{0MAX}	Q_{0MIN}	R_{0MIN}
USA	0.10	80	5.6	13.85	0.97
USA	0.20	30	2.1	6.8	0.952
Mass.	0.10	40	2.8	14.3	1.001
Mass.	0.20	13.5	1.89	6.965	0.975

Table 3. Maximum and minimum values obtained for $f(t)$ when different constant values of Q_0 were used . Maximum values correspond to 1840 while minimum values correspond to the year 2000. The corresponding reproductive basic numbers are also included.

Population	Q_0	f_{MAX}	R_{0MAX}	f_{0MIN}	R_{0MIN}
USA	10	0.29	2.03	0.136	0.952
USA	20	0.187	2.618	0.07	0.98
Mass.	10	0.28	1.96	0.14	0.98
Mass.	20	0.18	1.89	0.07	0.98

5. Discussion

The lack of general biological laws is one of the reasons behind the fact that biological theory follows a program radically different than physics, for example. However, ecological theory was build, in large extent, upon simple mechanistic models with constant parameters (or periodically varying parameters). Because its nature, mechanistic models give us a qualitative understanding of the population dynamics that result from basic processes like, births and deaths. Analytical, general qualitative results, are the outcome generated from such a program. However, these models (often "simple") cannot be directly applied in specific studies. Statistical data-driven models' aim is not tied in to the identification of mechanisms that help connect cause with effect, important factors population dynamics. Instead, statistical methods focus (simplistically speaking) in describing the correlations between variables and/or parameters. Although these models can provide excellent descriptions of the data set, they do not provide us with insights on what could happen under different circumstances. Neither type of models can or should be used to generate long-term projections.

In this work we combine both approaches. We use statistical models to capture (parametrically) the time evolution of several parameters from time series data with some caveats.

Mortality data present some large fluctuations between 1850 and 1900 likely due to poor recording. This problem is solved using smooth functions (sigmoid Boltzman functions in this case) fitting the observed data. The same procedure was applied to the data on the proportion of urban populations. These functions also provide interpolated values between observations. Furthermore, the sigmoid shape form of the functions used provide in these cases reasonable extrapolations. We used back-extrapolation to estimate mortalities and population values between 1700 and 1850. But extrapolation may not be used to produce long-term predictions [23]. An additional advantage is that functions are easier to use than the data sets when implementing a numerical integration scheme for the model (1-7).

The fitting used for the parameter values series' $f(t)$ and $Q_0(t)$ are of a different nature. First we found smooth functions ($Inc(t)$) which fitted the data corresponding to the observed (or estimated) incidence of active TB. Then we used the dynamical model (1-7) to obtain values for the simulated incidence close to the $Inc(t)$ values. This was achieved using a simple adaptive approach where the system 'learns' from the previous error $\varepsilon = [Simulated\ incidence(t) - Inc(t)]/Inc(t)$ as it generates the new parameter value as $P(t+dt) = (1-\varepsilon)P(t)$. Therefore, results are *model dependent* and all conclusions obtained from these results should be exercised with care. The mechanistic, dynamical model, becomes at this point, a data-driven model in the sense that future values of the parameters are unknown and are determined as the model is solved by comparing model solutions to data. Finally, in order to study the generated series $P(t)$, we fit the series $lnP(t)$ to smooth functions which were differentiated in order to find the rate of variation $\theta(t)$.

Dynamical model are used to examine the potential influence of reductions in transmission or progression on the course of the tuberculosis epidemics in the United States. The results show that under the assumption of constant rates of progression (constant f), the maximum relative rates of variations for Q_0 must occur before 1900, that is, before any major medical or public health interventions were in place. We did not find noticeable decreases in the value of the relative rate of variation of Q_0 as a consequence, for example, of the massive use of antibiotics beginning in the fifties. Our results suggest that reductions in the mean infectious period had also minimal effects in reducing TB transmission[31,32,30,29,36] and therefore, observed declining trends have been most likely due *primary* to other causes. Setting f at its upper actual estimated level ($f = 0.10$) leads to values for Q_{0MAX} that are too high (around 80 for USA and 40 for Massachusetts). *These results support the view that progression rates must have been declining.*

Reductions in the risk of progression to active-TB may have resulted from several causes including *improved living conditions* a concept that includes improved socioeconomic conditions and the use of broad medical interventions including, for example, water sanitation.

The assumption of constant transmission rates (Q_0 constant over time) led to a twofold decrease in the risk f and this was enough to explain the historical trends. In other words, a moderate reduction in progression rates may have accounted for much of the observed historical declines in active tuberculosis.

The maximum rates of variation for $f(t)$ were placed in the range 1900-1910 in all the cases considered while the maximum rate of variation for the mortality 10 rate was reached about 1910. That is, about the same time when progression decreased the most. Mortality or life expectancy are good surrogate candidates for variables that measure 'living conditions'. The fact that changes in mortality (or life expectancy) mapped fairly well to changes progression rates suggest the possibility that both decreases may have the same underlying factors[16].

Acknowledgments

JPA is a member of the CONICET.This project have been partially supported by grants from the National Science Foundation (NSF - Grant DMS - 0502349), the National Security Agency (NSA - Grant H98230- 06-1-0097), the Alfred T. Sloan Foundation and the Office of the Provost of Arizona State University.

Apendix. Parameter estimation and the impact of omitted factors

Tuberculosis mortalities and recovery rates. Average case fatality of untreated pulmonary cases is about 50% [15]. This value is compatible with the relation between mortality and incidence observed in Massachusetts before the chemotherapy era [30]. Therefore, we have set $d = r$. Because we fix γ and μ is estimated from demographic data, we set $d = r = 0.5(\gamma - \mu)$. Since $d \gg \mu$, $\gamma = d + r + \mu \simeq d + r$. Extrapulmonary TB mortality rate was set equal to non-TB related mortality, $d_e = \mu$. This assumption underestimate its impact but, fortunately, it does not play a significant role in the final analysis. We also considered the case $d_e = d$, which overestimate extrapulmonary TB mortality. For the USA case, in example, when $Q_0 = 10$ the value of f_{MAX} decreased from 0.29 to 0.27; while for $Q_0 = 20$ this value change to 0.174 from 0.187. We set $\gamma_e = \mu + d_e + r_e = 1yr^{-1}$ from where we obtained the value of r_e.

Transfer rate from high-risk to low-risk classes. The value of α was set to reproduce the observed pattern of progression. Its value is almost independent of the exact value of progression rates because $\alpha \gg k$. The value used in this work ($\alpha = 1.5yr^{-1}$) produced an excellent fit between model solutions and data on active-TB progression[16,17].

Partial immunity. Individuals previously infected may develop some degree of protection against reinfection. Reduced susceptibility is modelled through $\sigma\gamma Q_0 \frac{L}{N} A_p$ while reductions in progression rates are modelled through $k^* = \sigma_p k$. Both, σ and σ_p are smaller or equal to one. Here, we have set $\sigma = 1$ and considered only reductions in progression rates. We used a default value of $\sigma_p = 0.7$, but qualitatively similar results were obtained for the full range of values of σ_p in [0-1][16].

Rate of TB-relapse. During the last few years TB relapse accounted for about a 5% of the total incidence of active-TB (Centers for Disease Control, 1999). The rate of TB relapse (k_{Rp}) was set at $0.0008yr^{-1}$. Using this value the model produce a contribution of TB relapse which results 5% of the total simulated incidence of active-TB.

Proportion of pulmonary cases. The observed proportion of pulmonary cases in Massachusetts at present is $q \simeq 0.7$ [38]. This proportion was probably not higher in the past since extrapulmonary tuberculosis occurs mostly in children. Past higher risk of infection leads to an overall smaller *average age of first infection* what increase the proportion of extra pulmonary cases (and therefore reducing the proportion of pulmonary cases). The role associated with its variation should arise naturally in an age structured model but here, we have assumed that q is constant through time. This assumption does not alter the nature of our conclusions. We have considered $q(t)$ varying between an initial value of 0.5 to the actual value of 0.7. For USA data, for example, f_{MAX} increased from 0.29 (when $Q_0 = 10$ is assumed constant through the time) to 0.37. Conversely, when f is kept constant through the time, maximum values of Q_{0MAX} increased from 80 (when $q = 0.7$ constant) to 130 (when q is variable between 0.5 and 0.7). Rates of variation of Q_0 obtained are vertically shifted, but the location of the maximum remains essentially unchanged. The expected increase in q already turn the hypothesis of reductions in transmission as the sole explanation of the trends more unlikely. At this respect our assumption of constant q is conservative.

Pulmonary cases may be sputum-positive, sputum-negative, and culture-negative. Infectiousness varies among pulmonary cases with higher values for sputum-positive cases. Here we do not distinguished among these classes and therefore we considered the infectiousness (Q_0) of an average pulmonary case.

Deterministic approximation. For each equation of Model (1-7) there is always at least one of its right side rates (for example μU in eq. 1) for which the product $\mu U dt$, where dt is the time step used in the numerical integration, is much greater than one. Therefore, quasi-deterministic approximation holds [6]. Stochastic simulations (not presented here) converge to deterministic solutions in few simulated years[17].

References

1. Wood, S.N.*Aspects of Applied Biology,* **53**, 41-49, (1999).
2. Holling, C., *Can. Entomol.* **91**, 385-398,(1959).
3. Tanner, J.*Ecology* **56**, 1835-1841,(1975).
4. Renshaw E., *Modelling Biological POpulations in Space and Time*, Cambridge University Press, (1991).
5. Juan P. Aparicio & Hernán G. Solari, *Mathematical Biosciences*, **169**, 15-25, (2001).
6. Aparicio, J.P., Solari, H.G. *Phys. Rev. Lett.* **86**, 4183-4186, (2001).
7. Anderson, R.M. and R.M. May., *Infectious Diseases of Humans*, Oxford Science Publications, (1991).
8. Berezovskaya, F., B Song and C Castillo-Chavez, *SIAM Journal of Applied Mathematics* (in press, 2010).
9. Brauer Fred and Carlos Castillo-Chavez, *Texts in Applied Mathematics*, **40**. Springer-Verlag, (2001).
10. Castillo-Chavez C., S. Blower, P. van den Driessche, D. Kirschner, and A-

A Yakubu (eds.), *Mathematical Approaches for Emerging and Reemerging Infectious Diseases: An Introduction*, Volume **125**, Springer-Verlag, Berlin-Heidelberg-New York, 368 pages, (2002).
11. Castillo-Chavez C., S. Blower, P. van den Driessche, D. Kirschner, and A-A Yakubu (eds.), *Mathematical Approaches for Emerging and Reemerging Infectious Diseases: Models, Methods, and Theory*, Volume **126**, Springer-Verlag, Berlin-Heidelberg-New York, (2002).
12. Chowell G., J.M. Hyman, L.M.A. Bettencourt, C. Castillo-Chavez (Eds.), *Mathematical and Statistical Estimation Approaches in Epidemiology*, Springer. 2009, Approx. 430 p., Hardcover, ISBN: 978-90-481-2312-4.
13. Gumel A., Castillo-Chavez, C., Clemence, D.P. and R.E. Mickens, *American Mathematical Society*, Volume 410, 389 pages, (2006).
14. Sutherland, I., *TSRU Progress Report.* KNVC, The Hague, Netherlands (1968).
15. Styblo, K., *Epidemiology of Tuberculosis. Selected Papers*, **24**,(1991).
16. Aparicio, J.P., Capurro, A.F., Castillo-Chavez, C., *J. Theor. Biol*, **215**, 227-237. (2002).
17. Juan. P. Aparicio & Carlos Castillo-Chavez, *Mathematical Biosciences and Engineering*, **6** **(2)**, 209–237,(2009).
18. Aparicio, J.P., Pacual, M., *Proc. Roy. Soc. B*, **274**, 505-512, (2007).
19. Aparicio, J.P., Capurro, A.F., Castillo-Chavez, C., *J. theor. Biol.* **206**, 327-341, (2000).
20. Song, B., Castillo-Chavez, C., Aparicio, J.P., *Math. Biosci* **180**, 187-205, (2002).
21. Castillo-Chavez, C. and B. Song, *Journal of Mathematical Biosciences and Engineering*,**1(2)**: 361-404, (2004).
22. Keeling, M.J., Grenfell, B.T., *J. theor. Biol.* **203**, 51-61, (2000).
23. Aparicio, J.P., Capurro, A.F., Castillo-Chavez, C., *Mathematical Approaches for Emerging and Reemerging Infectious Diseases: An Introduction*, Springer-Verlag; pp. 351-360, (2002).
24. Grigg, E.R.N., *Am. Rev. Tuberculosis and Pulmonary Diseases* **78**, 151-172, (1958).
25. Grigg, E.R.N., *Am. Rev. Tuberculosis and Pulmonary Diseases* **78**, 426-453, (1958).
26. Grigg, E.R.N., *Am. Rev. Tuberculosis and Pulmonary Diseases* **78**, 583-608 (1958).
27. U. S. Bureau of the Census, *Historical statistics of the United States: colonial times to 1970.* Washington DC: Government Printing Office, (1975).
28. U. S. Bureau of the Census, *Statistical Abstracts of the United States 1996.* See also, http://www.census.gov/population.
29. Davis, A.L., *History of the sanatorium movement. In: Rom WN, Garay SM*, (eds). Tuberculosis. Little, Brown and co, (1996).
30. Drolet, G., *Am. Rev. of Tuberc.* **37**, 125-151, (1938).
31. Barnes, D.S.,*The making of a social disease: tuberculosis in the nineteenth-century France.* University of California Press, pp 5-13, (1995).
32. McKeown, T., Record, R.G., *Population Studies* **16**, 94-122, (1962).

33. McKeown T., *The origins of human disease. Basil Blackwell*, (1987).
34. Raloff, J., *Science* **150**, 374, (1996).
35. Chan, J., Tian, Y., Tanaka, E., Tsang, M.S., Yu, K., Salgame, P., Carroll, D., Kress, Y., Teitelbaum, R., Bloom, B.R., *Proc. Natl. Acad. Sci. USA* **93**, 14857-14861, (1996).
36. Pesanti, E.L., *A history of tuberculosis. In: Lutwick, L.I., editor. Tuberculosis. A clinical handbook.*, Chapman & Hall Medical, (1994).
37. Wang X., Feng, Z., Aparicio, J. and C. Castillo-Chavez, *BIOMAT 2009, International Symposium on Mathematical and Computational Biology*, Ed. Rubem P Mondaini, World Scientific, (2009).
38. Centers for Disease Control, *Reported Tuberculosis in the United States, 1999*, order: (99-6583), (1999).

A TWO-PATCHES POPULATION AFFECTED BY A SIS TYPE DISEASE. INFECTION IN THE SOURCE. CONSEQUENCES IN THE SINK

F. CÓRDOVA-LEPE* and R. DEL-VALLE
Universidad Católica del Maule,
3605 San Miguel Avenue,
Talca, Chile
E-mail: fcordova@ucm.cl & rvalle@ucm.cl

J HUINCAHUE-ARCOS
Pontificia Universidad Católica de Valparaíso,
2950 Brasil Avenue,
Valparaíso, Chile
E-mail: jaime.huincahue.a@mail.ucv.cl

We consider a mathematical model of a source-sink metapopulation, where it is assumed a SIS disease affecting the source although not the sink. The vital process and the infection process have a continuous type representation, but migration is considered impulsive. A basic reproductive number R_0 is determined in the source. Our main objective is to know, under an endemic situation ($R_0 > 1$) in the source, what are the consequences in the sink. It was proved the existence of a non trivial globally attractive periodic trajectory for the infective population in the sink.

1. Introduction

It is known that some infectious diseases require the concurrence of certain specific environmental conditions for its spread. On the other hand, it is a fact that some populations are spatially structured in fragmented environments. These fragments or patches may have different quality enviroments for the spread of a disease. The problem of understanding the behavior of a SIS disease in the framework of a metapopulation is not new. The work of Allen et al.[1] (continuous time and deterministic) is a good study for unde-

*Work partially supported by grant 15/2009 of the Universidad Metropolitana de Ciencias de la Educación.

standing, through the determination of the basic reproductive number R_0, how "spatial heterogeneity, habitat connectivity, and rates of movement can have large impacts on the persistence and extinction" of a SIS disease with several patches. There, it is shown that if the disease–free trajectory is unstable, then there exists a unique endemic-equilibrium, some derivations about low–risk patches and hight–risk patches are made. Remarkable work is one of Arino et al.[2], where a continuous-time and deterministic SEIR model with p–patches and $4p$ ordinary differential equations is studied. In this work, R_0 is calculated and it is proved that "for $R_0 > 1$, solutions tend to an equilibrium with disease presents in each patch". The work of Foley et al.[5], incorporating stochasticity, predicts that disease persistence would be finite.

The mentionded works above are very generalistic. To the best of our knowledge, if we reduce the scope to a SIS disease in a two–patches metapopulation, more precisely, to a source–sink metapopulation, see Pulliam[6] and Córdova-Lepe[4], there not exist many works in the literature exploring this particular case.

In this work, with the use of a system of impulsive differential equations, a metapopulation with two patches is considered, a source and a sink. The source is affected by a SIS disease (Britton[3] & Thieme[7]), although there is not conditions in the sink for its propagation. We suposse, by seasonal motives, that emigration from source to sink occurs every τ units of time, *i.e.*, in a periodic sequence $\{k\tau\}$, $k \geq 0$. The susceptible group and the infectious group in the source (resp. sink) are represented by S_1 and I_1 (resp. S_2 and I_2) respectively.

A diagram of the proposed model is given in Fig.1, where it can be observed the lines of flux between the different groups. So that, the model is described by the impulsive system of differential equations:

$$
\left\{
\begin{array}{ll}
\left.
\begin{array}{ll}
\dot{S}_1(t) & = (\nu_1 - \mu_1)S_1(t) + (\nu_1 + \gamma_1)I_1(t) - \beta S_1(t)I_1(t), \\
\dot{I}_1(t) & = \beta S_1(t)I_1(t) - (\gamma_1 + \mu_1)I_1(t), \\
\dot{S}_2(t) & = -(\mu_2 - \nu_2)S_2(t) + (\nu_2 + \gamma_2)I_2(t), \\
\dot{I}_2(t) & = -(\gamma_2 + \mu_2)I_2(t),
\end{array}
\right\} & t \neq k\tau, \\
\left.
\begin{array}{ll}
S_1(t^+) & = (1-m)S_1(t), \\
I_1(t^+) & = (1-m)I_1(t), \\
S_2(t^+) & = S_2(t) + mS_1(t), \\
I_2(t^+) & = I_2(t) + mI_1(t),
\end{array}
\right\} & t = k\tau.
\end{array}
\right.
\tag{1}
$$

This model considers vital dynamics. The per capita birth rate and the per capita death rate in source (resp. sink) are ν_1 and μ_1 (resp. ν_2 and μ_2)

respectively. Notice that it is assumed $\nu_1 > \mu_1$ (resp. $\nu_2 < \mu_2$) because the first (resp. second) patch is a source (sink). The transition rate between I_1 and S_1 will be the standard βI_1, with $\beta > 0$. The recovery rate in the source is γ_1 and in the sink is γ_2.

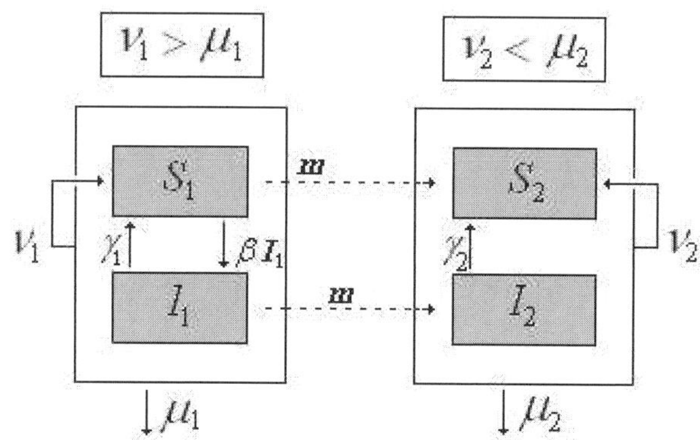

Figure 1. The SIS disease in the metapopulation represented diagrammatically. Segmented lines symbolize the impulsive migration.

The novelty of this work is to put in a very simple (mathematically) metapopulation model the determination of a SIS disease behavior in sink by its development in source.

2. Assumptions on the Source

2.1. *Source is in a periodic equilibrium*

It will be assumed that the source is a patch in equilibrium in an impulsive framework, this means that $P_1 = S_1 + I_1$ has a periodic trajectory, for instance of period τ. Note that the total subpopulation in the first patch satisfies the impulsive subsystem of (1) defined by

$$\begin{cases} P_1'(t) = (\nu_1 - \mu_1)P_1(t), \ t \neq k\tau, \\ P_1(t^+) = (1-m)P_1(t), \ t = k\tau. \end{cases} \tag{2}$$

By (2) it is obtained the general relation that follows

$$P_1((k+1)\tau) = (1-m)e^{r_1\tau}P_1(k\tau), \tag{3}$$

where $r_1 = \nu_1 - \mu_1 > 0$, for any $k \geq 0$.

It is observed that (3) determines the behavior for the subpopulation source in times $\{k\tau\}$. For (2), piecewise continuous solutions that at times $\{k\tau\}$ form a geometric progression are obtained. In order to have an equilibrium in the source, we will assume that $(1-m)e^{r_1\tau} = 1$, that is

$$m = 1 - e^{-r_1\tau}. \tag{4}$$

It will be considered that $P_1(0) = (S_1 + I_1)(0) = 1$, which means that at the begining of the process the source patch is on its maximal capacity. So that, the assumed τ-periodic trajectory in the source is given by

$$P_1(t) = e^{r_1(t-(k+1)\tau)} \quad \text{if} \quad k\tau < t \leq (k+1)\tau. \tag{5}$$

2.2. *Condition for an endemic case in source*

By using (5) combined with (1), it follows that the size of the infective group in source is described by

$$\begin{cases} \dot{I}_1(t) = \alpha_k(t)I_1(t) - \beta I_1^2(t), \; t \neq k\tau, \\ I_1(t^+) = e^{-r_1\tau}I_1(t), \qquad\qquad t = k\tau, \end{cases} \tag{6}$$

where $\alpha_k(t) = \beta e^{r_1(t-(k+1)\tau)} - (\gamma_1 + \mu_1)$, if $k\tau < t \leq (k+1)\tau$, which is a τ-periodic function.

Theorem 2.1. *Let us define*

$$R_0 = \left(\frac{\beta}{\nu_1 + \gamma_1}\right)\frac{1 - e^{-r_1\tau}}{r_1\tau}. \tag{7}$$

i) If $R_0 < 1$, then it follows that any solution $I_1(\cdot)$ of (6) verifies $I_1(t) \to 0$ as $t \to \infty$.

ii) If $R_0 > 1$, then (6) has a non trivial τ-periodic solution $\hat{I}_1(\cdot)$, such that any solution $I_1(\cdot)$, $I_1(0) \neq 0$ converges to $\hat{I}_1(t)$ as $t \to \infty$.

Proof: Equation (6) can be solved directly between time of impulses. So that, the solution is given by

$$I_1(t) = \frac{I_1(k\tau)\exp\left(\int_0^{t-k\tau}\alpha_0(v)dv\right)}{e^{r_1\tau} + I_1(t_k)\beta\int_0^{t-k\tau}\exp\left(\int_0^v \alpha_0(u)du\right)dv}, \quad t \in (k\tau, (k+1)\tau], \;\; k \geq 0. \tag{8}$$

Observe that (8) determines a relation $I_1((k+1)\tau) = F(I_1(k\tau))$, where $F : [0,\infty) \to [0,\infty)$ is such that

$$F(I) = \frac{I \cdot a}{b + I \cdot c}, \quad \text{with } I \geq 0, \tag{9}$$

this is a one dimensional dynamical system, where

$$a = \exp\left(\int_0^\tau \alpha_0(u)du\right), \quad b = e^{r_1\tau} \text{ and } c = \beta\int_0^\tau \exp\left(\int_0^v \alpha_0(u)du\right)dv.$$

Note that if $F'(0) > 1$, then there exists a unique and globally attractive fixed point of (9) on $(0,1]$, which determines a globally attractive periodic trajectory of (6), *i.e.*, of the infected population in source.

Considering that $F'(I) = ab/(b+cI)^2$, $I \geq 0$, then $F'(0) = a/b$. Since condition $a/b > 1$, this is

$$\int_0^\tau \alpha_0(u)du > r_1\tau,$$

is equivalent to $R_0 > 1$, it is follows item (ii) of the theorem.

If $R_0 < 1$, *i.e.*, $F'(0) < 1$, then system (9) has zero as a globally attractive fixed point. Hence, it is not difficult to prove that the disease free solution is a global attractor. ◇

Notice that when the time between migrations τ tends to zero, then R_0 tends to $\beta/(\nu_1 + g)$, *i.e.*, we obtained the basic reproductive number of the continuous SIS model with vital dynamics.

2.3. *The endemic trajectory*

If $R_0 > 1$, there exists $\hat{I}_1 \neq 0$ a fixed point of F, this is, $\hat{I}_1(b + c\hat{I}_1) = a\hat{I}_1$. Hence,

$$\hat{I}_1 = \frac{a-b}{c}. \tag{10}$$

Note that, $a > b$ is equivalent to $R_0 > 1$.

The τ–periodic global attractive trajectory representing the endemic level is given by

$$\hat{I}_1(t) = \frac{\hat{I}_1 \exp(\int_0^s \alpha_0)}{b + \hat{I}_1\beta\int_0^s \exp\left(\int_0^u \alpha_0\right)du}, \quad s = t - k\tau, \ \ t \in (k\tau, (k+1)\tau], \ k \geq 0. \tag{11}$$

By (5), it follows that the respective τ–periodic trajectory of the susceptibles in the source is

$$\hat{S}_1(t) = e^{r_1(s-\tau)} - \hat{I}_1(t),, \quad s = t - k\tau, \ \ t \in (k\tau, (k+1)\tau], \ k \geq 0. \tag{12}$$

Theorem 2.2. *The arithmetic mean of the τ-periodic trajectories defined in (11) and (12) are given by:*

$$\frac{M[(\hat{S}_1(\cdot), \hat{I}_1(\cdot))]}{M[P_1(\cdot)]} = \left(\frac{1}{R_0}, 1 - \frac{1}{R_0}\right). \tag{13}$$

Proof: Notice that,

$$M[\hat{I}_1(\cdot)] = \frac{1}{\beta\tau}\int_{w(0)}^{w(\tau)} \frac{dw}{w} = \frac{1}{\beta\tau}\ln(\frac{a}{b}),$$

where $w(s)$, $s > 0$, is the denominator of the right side of (11), and a and b are defined in the proof of Theorem 2.1.

So that, it is have

$$M[\hat{I}_1(\cdot)] = \frac{1}{\beta\tau}\left[\frac{\beta}{r_1}(1 - e^{-r_1\tau}) - (\gamma_1 + \mu_1)\tau - r_1\tau\right] = \frac{1 - e^{-r_1\tau}}{r_1\tau}\left(1 - \frac{1}{R_0}\right).$$

The mean of the susceptibles is obtained by straigh forward computations. From (5) is clear that $M[P_1(\cdot)] = (1 - e^{-r_1\tau})/(r_1\tau)$, then (13) follows. $\diamond$

3. Results

3.1. *About total population in sink*

By (5) follows that $P_1(k\tau) = (S_1 + I_1)(k\tau) = 1$, $k \geq 0$, then the subpopulation sink $P_2 = S_2 + I_2$ satisfies

$$\begin{cases} P_2'(t) &= -r_2 P_2(t), \qquad t \neq k\tau, \\ P_2(t^+) &= P_2(t) + (1 - e^{-r_1\tau}), \; t = k\tau, \end{cases} \tag{14}$$

where $r_2 = \mu_2 - \nu_2 > 0$.

Theorem 3.1. *System (14) has a τ–periodic trajectory $\hat{P}_2(\cdot)$, such that any solution $P_2(\cdot)$ converges to $\hat{P}_2(t)$ as $t \to \infty$. Moreover,*

$$\frac{M[\hat{P}_2(\cdot)]}{M[P_1(\cdot)]} = \frac{r_1}{r_2}. \tag{15}$$

Remark: If $r_1 > r_2$, this is, $\nu_1 + \nu_2 > \mu_1 + \mu_2$, then the mean population in sink can be greater than mean population in source.
Proof: If $t \in (k\tau, (k+1)\tau]$, $k \geq 0$, then

$$P_2(t) = e^{-r_2(t-k\tau)}[P_2(k\tau) + (1 - e^{-r_1\tau})]. \tag{16}$$

Then, at times of impulse $\{k\tau\}$, we have $P_2((k+1)\tau) = G(P_2(k\tau))$, with $G : [0, \infty) \to [0, \infty)$ such that $G(P) = e^{-r_2\tau}[P + (1 - e^{-r_1\tau})]$, a dynamical system with non trivial global stable fixed point

$$\hat{P}_2 = \frac{1 - e^{-r_1\tau}}{e^{r_2\tau} - 1}.$$

Notice that $\hat{P}_2$ determines a τ–periodic attractive trajectory in sink given by

$$\hat{P}_2(t) = e^{-r_2(s-\tau)}\hat{P}_2, \quad s = t - k\tau, \quad t \in (k\tau, (k+1)\tau], \; k \geq 0. \tag{17}$$

Moreover, the mean value of the τ–periodic solution (17) is

$$M(\hat{P}_2(\cdot)) = \frac{\hat{P}_2}{\tau}\int_0^\tau e^{-r_2(t-\tau)}dt = \frac{1 - e^{-r_1\tau}}{r_2\tau}. \tag{18}$$

Hence, it is clear that (15) follows. ⋄

3.2. *About infectious group in sink*

From (1), the dynamics of infective group in sink is given by

$$\begin{cases} \dot{I}_2(t) = -(\gamma_2 + \mu_2)I_2(t), \, t \neq k\tau, \\ I_2(t^+) = I_2(t) + mI_1(t), \quad t = k\tau. \end{cases} \tag{19}$$

So, the solution of (19) are determined by initial values $I_2(0)$ and $I_1(0)$.

Theorem 3.2. *Any solution of (19) tends to a τ–periodic trajectory defined by*

$$\hat{I}_2(t) = e^{-(\gamma_2+\mu_2)s}[\hat{I}_2 + m\hat{I}_1], \quad s = t - k\tau, \quad t \in (k\tau, (k+1)\tau], \quad k \geq 0, \tag{20}$$

with

$$\hat{I}_2 = m\hat{I}_1/[e^{(\gamma_2+\mu_2)\tau} - 1], \tag{21}$$

where m and $\hat{I}_1$ are respectively given by (4) and (10). Moreover,

$$\frac{M[\hat{I}_2(\cdot)]}{M[\hat{P}_2(\cdot)]} = \frac{\mu_2 - \nu_2}{\mu_2 + \gamma_2}\hat{I}_1. \tag{22}$$

Before the proof, we will introduce a technical results.

Lemma 3.1. *Let us consider the recursive equation*

$$x_{k+1} = r(x_k + \omega_k), \quad x \in \mathbb{R}, \; k \geq 0, \tag{23}$$

where $0 < r < 1$ *and* $\omega_k \to \omega_\infty$ *as* $k \to \infty$. *Then any solution* $\{x_k\}$ *satisfies that*

$$\lim_{k\to\infty} x_k = \frac{r}{1-r}\omega_\infty.$$

Proof: We compare the solutions of (23) with

$$y_{k+1} = r(y_k + \omega_\infty), \quad y \in \mathbb{R},\ k \geq 0. \tag{24}$$

From an initial values x_0 and y_0, by direct calculus it is obtained that $\{x_k - y_k\}$ satisfies $z_{k+1} = r(z_k + (\omega_k - \omega_\infty))$. Since

$$z_k = r^k z_0 + \sum_{j=0}^{k-1} r^{k-j}(\omega_j - \omega_\infty),\ k \geq 0, \tag{25}$$

the sum in right side of (25) is less than

$$r^{k-[\frac{k}{2}]}\left(\left[\frac{k}{2}\right]+1\right)\max_{j\in J_1}|\omega_j - \omega_\infty| + \left(\frac{1}{1-r}\right)\max_{j\in J_2}|\omega_j - \omega_\infty|,$$

where $J_1 = \{0, \cdots, [k/2]\}$ and $J_2 = \{[k/2]+1, \cdots, k\}$. In this context $[\cdot]$ denotes the integer part of a real number. In the last expression the first maximun is bounded and given $\varepsilon > 0$, the second one is less than ε for k big enough. So that, $z_k \to 0$ as $k \to \infty$. Hence, the sequence $\{x_k\}$ tends to the global attractor $r\omega_\infty/(1-r)$ of (24). ◇

Proof of Theorem 3.2.

Notice that

$$I_2((k+1)\tau)) = e^{-(\gamma_2+\mu_2)\tau}[I_2(k\tau) + mI_1(k\tau)],\ k \geq 0, \tag{26}$$

will be compared with

$$J_{k+1} = e^{-(\gamma_2+\mu_2)\tau}[J_k + m\hat{I}],\ k \geq 0. \tag{27}$$

So that, we are in the conditions of the above Lemma. Therefore, it is obtain that $I_2(k\tau) \to \hat{I}_2$ as $t \to \infty$, where $\hat{I}_2$ is given by (21). Note that $\hat{I}_2$ determines the τ–periodic solution defined in (20).

The mean value of the infectives in the sink is

$$M(\hat{I}_2) = \frac{1}{\tau}[\hat{I}_2 + m\hat{I}_1]\int_0^\tau e^{-(\gamma_2+\mu_2)t}dt = \frac{1-e^{-r_1\tau}}{(\gamma_2+\mu_2)\tau}\hat{I}_1. \tag{28}$$

Hence, by (18) it is follows (22). ◇

4. Numerical Simulations

We illustrate our previous results by some simulations. Let us consider the system (1), with $\tau = 1$ and where:

Source				Sink		
ν_1	μ_1	β	γ_1	ν_2	μ_2	γ_2
0.05	0.03	0.32	0.02	0.02	0.03	0.02

In Fig.2, with three different initial conditions for infectives in source, it is shown the convergences towards a τ–periodic globally attractive trajectory, *i.e.*, the "impulsive" endemic–equilibrium.

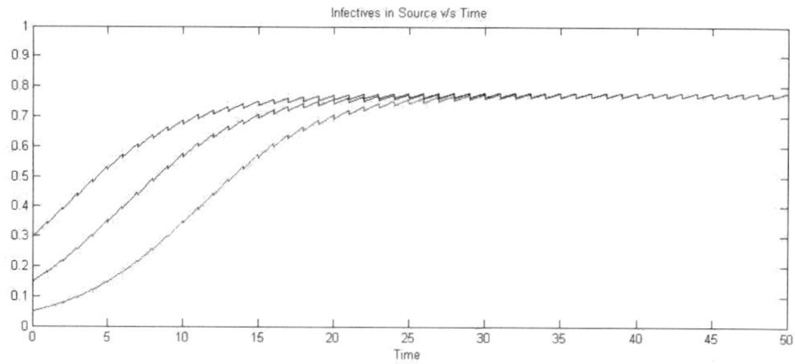

Figure 2. Evolution of infectives in source. Initial conditions $I_1(0)$ with values 0.05, 0.15 & 0.70.

In Fig.3, with three different initial conditions for infectives in sink, it is illustrated the asymptotic behavior towards the "impulsive" endemic–equilibrium. Notice that, in model of this type, but with ordinary differential system, it is obtained only monotonic behavior towards punctual equilibrium.

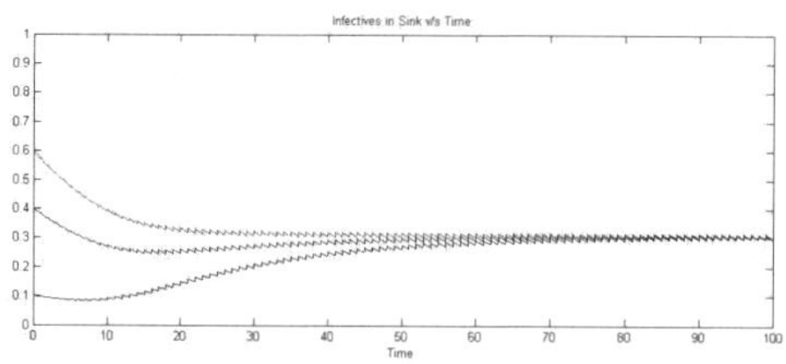

Figure 3. Evolution of infectives in sink. Initial conditions $I_1(0) = 0.05$ and $I_2(0)$ with values 0.10, 0.40 & 0.60.

In Fig.4, can be observed the possibility of having more population in sink than in source what confirms the Pulliam claim.

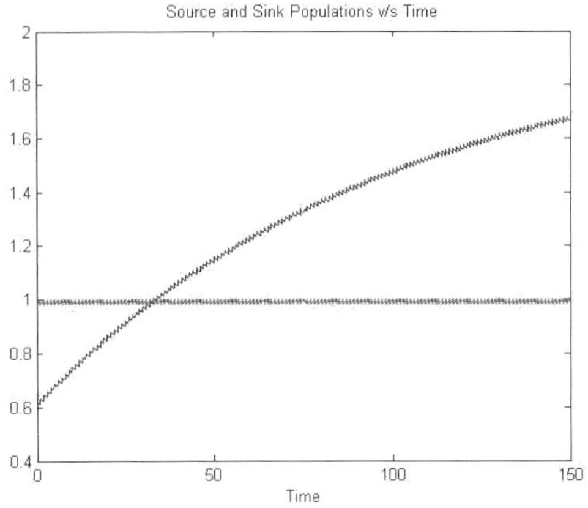

Figure 4. If $\nu_1 + \nu_2 > \mu_1 + \mu_2$, then between patches, the subpopulation in sink can be the biggest.

References

1. L.J.S. Allen, B.M. Bolker, Y. Lou, and A.L. Nevai, *SIAM Journal on Applied Mathematics*, **67**, N 5, pp. 1283–1309 (2007).
2. J. Arino and P. van den Driessche, *Fields Institute Communications.* **48**: 1.12 (2006).
3. N. Britton, *Essential Mathematical Biology*, Springer-Verlag, London, (2003).
4. F. Córdova-Lepe et al., *Journal of Difference Equations and Applications*, accepted.
5. J.E. Foley, P. Foley and N.C. Pedersen, *Journal of Applied Ecology*, **36**, N 4, pp. 555-563 (1999).
6. H. R. Pulliam, *The American Naturalist* **132**, pp. 652-661 (1988).
7. H. Thieme, *Mathematics in Populations Biology*, Princeton University Press, Princeton, (2003).

AGE-STRUCTURED MODELING FOR THE DIRECTLY TRANSMITTED INFECTIONS – I: CHARACTERIZING THE BASIC REPRODUCTION NUMBER*

C. H. DEZOTTI

Depart. Estat. e Inform., UFRPE
Rua Dom Manoel de Medeiros, s/n
CEP: 52171-900, Recife, PE, Brazil
E-mail: diretoria@deinfo.ufrpe.br

H. M. YANG†

Depart. Matem. Aplicada, IMECC, UNICAMP
Praça Sérgio Buarque de Holanda, 651
CEP: 13083-859, Campinas, SP, Brazil
E-mail: hyunyang@ime.unicamp.br

One of the main features of directly transmitted infections is the strong dependency of the risk of infection with age. We propose and analyze a simple mathematical model where the force of infection (per-capita incidence rate) is age-depending. The existence and stability of the non-trivial equilibrium point are determined based on the basic reproduction number. For this reason we deal with a characterization of the basic reproduction number by applying the spectral radius theory.

1. Introduction

Directly transmitted childhood infections, like rubella and measles, have been used as good examples for the application of mathematical models to the study and comprehension of the epidemiology of these diseases. The models are formulated basically by taking into account the force of infection depending on the contact rate, which is related to the pattern of contacts among susceptible and infectious individuals. Therefore, the assumptions on the contact rate lead to quite different approaches when one deals with the models[2].

---

*This work is supported by fapesp – projeto temático

†Work partially supported by grant from cnpq. Corresponding author.

A first assumption, and also the simplest, is to consider a constant contact rate among individuals over all ages and time. Consequently the force of infection becomes constant. The resulting mathematical model is described by a time-depending system of differential equations without age structure. This assumption can generate non-realistic outputs when modeling childhood diseases with a strong age depending pattern. A second and better assumption is, therefore, to take into account the age dependency in the pattern of contacts. A mathematical model with this assumption yields a time- and age-depending system of differential equations (see Dietz[5] who was the first author to apply this formalism to epidemiology with constant contact rate), resulting in the well established concept of the age depending force of infection[8].

When dealing with a constant contact rate modeling, there are classical results related to the basic reproduction number and the lower value (threshold) for the vaccination rate above which the disease can be considered eradicated[1]. With respect to age-structured modeling, results related to the basic reproduction number R_0 and the threshold vaccination rate are more complex. For instance, Greenhalgh[7] and Inaba[10] showed the existence and uniqueness of the non-trivial solution for the Hammerstein equation similar to that presented in Yang[18]. They showed that the bifurcation from the trivial to non-trivial solution of the Hammerstein integral equation occurs when the spectral radius assumes unity value. Furthermore, they related this spectral radius with the basic reproduction number, and stated that whenever $R_0 < 1$ the disease fades out in the community, and when $R_0 > 1$, the disease can be settle at an endemic level. Following the same arguments, they showed the procedure to calculate the threshold vaccination rate.

Two attempts of representing the age-structured contact rate can be found in the literature: a matrix with constant elements and a constant value for different age classes. Anderson and May[1] developed the concept they called *Who-Acquires-Infection-From-Whom* matrix (WAIFW). Briefly, this is a matrix where the elements of rows and columns are the contact rates, constant values, over the discrete age classes of susceptible and infectious individuals. Schenzle[15] developed an age-structured contact pattern where constant values on several age intervals are assigned and, then, structured the dynamics in a coupled differential equations to estimate the contact rate from notified data. Although both methods represent good approaches to modeling the dynamics of direct transmitted diseases, they are applicable to the description of different kinds of data collection:

the WAIFW method is appropriate to analyze seroprevalence data, while Schenzle's method is better applied to incidence records.

The purpose of this paper is to develop a model with age-structured contact rate. However, as pointed by Tudor[17], data on contact rates do not exist, although most parameters related to the disease transmission can be estimated directly. This fundamentally theoretical paper is divided as follows. In section 2 the general model is presented and analyzed. In section 3 we present a characterization of the basic reproduction number. Discussion and conclusion are presented in section 4.

2. The model

Farrington[6] obtained an age depending force of infection from cumulative distribution function of age at infection. Here, the age depending force of infection is obtained from a compartmental model taking into account an age-structured contact rate. Let a closed community be subdivided into four groups $X(a,t)$, $H(a,t)$, $Y(a,t)$ and $Z(a,t)$ which are, respectively, susceptible, exposed, infectious and immune individuals, distributed according to age a at time t. According to the natural history of infection, susceptible individuals are infected at a rate $\lambda(a,t)$, known as force of infection (per-capita incidence rate), and transferred to exposed class. The age-specific force of infection at time t is defined by

$$\lambda(a,t) = \int_0^L \beta(a,a') Y(a',t)\, da', \tag{1}$$

where $\beta(a,a')$ is the age-structured contact rate, that is, the contact among susceptible individuals of age a with infectious individuals of age a', and L is the maximum age attainable by human population. The exposed individuals are moved to the infectious class at a constant rate σ, and enters to the immune class at a rate γ. All individuals are under a constant mortality rate μ. We remark that the additional mortality due to the disease, the loss of immunity and the protective action of maternal antibodies in newborns are not considered in the model.

Based on the above considerations, the dynamics of directly transmitted infectious diseases model considering age-structured contact rate is described by a system of partial differential equations[16],

$$\begin{cases} \frac{\partial}{\partial t}X\left(a,t\right)+\frac{\partial}{\partial a}X\left(a,t\right) = -\left[\lambda\left(a,t\right)+\nu\left(a\right)+\mu\right]X\left(a,t\right) \\ \frac{\partial}{\partial t}H\left(a,t\right)+\frac{\partial}{\partial a}H\left(a,t\right) = \lambda\left(a,t\right)X\left(a,t\right)-\left(\mu+\sigma\right)H\left(a,t\right) \\ \frac{\partial}{\partial t}Y\left(a,t\right)+\frac{\partial}{\partial a}Y\left(a,t\right) = \sigma H\left(a,t\right)-\left(\mu+\gamma\right)Y\left(a,t\right) \\ \frac{\partial}{\partial t}Z\left(a,t\right)+\frac{\partial}{\partial a}Z\left(a,t\right) = \nu\left(a\right)X\left(a,t\right)+\gamma Y\left(a,t\right)-\mu Z\left(a,t\right), \end{cases} \tag{2}$$

where $\nu\left(a\right)$ is the age depending vaccination rate. Note that the last equation is decoupled from the system, hence, hereafter, we will omit the equation for the immune individuals, $Z\left(a,t\right)$. Defining the total population as $N\left(a,t\right)=X\left(a,t\right)+H\left(a,t\right)+Y\left(a,t\right)+Z\left(a,t\right)$, we obtain the equation for the age-distribution of the population irrespective of the disease as

$$\frac{\partial}{\partial a}N\left(a,t\right)+\frac{\partial}{\partial t}N\left(a,t\right)=-\mu N\left(a,t\right).$$

Using this equation, we can obtain the decoupled variable as $Z\left(a,t\right)=N\left(a,t\right)-X\left(a,t\right)-H\left(a,t\right)-Y\left(a,t\right)$.

The boundary conditions of (2) are $X\left(0,t\right)=X_b$, which is the newborn rate, and $H\left(0,t\right)=Y\left(0,t\right)=0$, which come out from the assumptions of the model. Let us assume that a vaccination strategy is introduced at $t=0$ in a non-vaccinated population ($\nu\left(a\right)=0$) in which the disease encounters in the steady state. Then the initial conditions of (2) are the solutions of

$$\begin{cases} \frac{d}{da}X_0(a) = -\left[\lambda_0(a)+\mu\right]X_0(a) \\ \frac{d}{da}H_0(a) = \lambda_0(a)X_0(a)-\left(\sigma+\mu\right)H_0(a) \\ \frac{d}{da}Y_0(a) = \sigma H_0(a)-\left(\gamma+\mu\right)Y_0(a), \end{cases} \tag{3}$$

where $\lambda_0(a)=\int_0^L\beta(a,a')Y_0(a')da'$ is equation (1) in the steady state. From the boundary conditions, the initial conditions are $X_0(0)=X_b$ and $H_0\left(0\right)=Y_0\left(0\right)=0$.

3. A characterization of $\mathbf{R_0}$ in the steady state

Let us characterize the reproduction number R_ν as a spectral radius of a integral operator. The basic reproduction number R_0 is defined as the average number of secondary infections produced by one susceptible individual in a completely homogeneous and susceptible population in the absence of any kind of constraint ($\nu=0$). Hence, R_0 describes the epidemiological situation in a non-vaccinated population. Greenhalgh[7] considered an age-structured contact rate $\beta\left(a,a'\right)$ being a separable function

$$\beta\left(a,a'\right)=\sum_{i=1}^{n}p_i\left(a\right)q_i\left(a'\right),$$

and applied results from the spectral radius of linear operators on Banach spaces in a finite dimension. In our case, we consider $\beta(a,a') \in C[0,L]$ and use results from bifurcation points and positive operators in cones[3] and fixed points[11]. In a companion paper[4], as Inaba[10], we discuss about the stability of the trivial solution and the uniqueness of the non-trivial solution, and provide the estimations for upper and lower bounds of R_0 for special contact rates.

Let us introduce some definitions.

We consider a Banach space X with a solid cone K and an operator $T : X \to X$. A **cone** is a proper convex closed subset (K is proper if $K \cap -K = \emptyset$) such that for all $k > 0$ we have $kK \subset K$. A cone is **solid** if it has $int(K) \neq \emptyset$ (particularly, if a cone is solid it is **reproducing**, that is, $X = K - K$). K establishes on X a partial ordering relation, that is, if $x, y \in X$, we say that $x \leq y$ if $y - x \in K$ and $x < y$ if $y - x \in K$ and $x \neq y$. Particularly, we say that $0 \leq x$ if $x \in K$ and $0 < x$ if $x \in K$ and $x \neq 0$. A cone K is called **normal** if exists a $\delta > 0$ such that $\|x_1 + x_2\| \geq \delta$ for $x_1, x_2 \in K$ and $\|x_1\| = \|x_2\| = 1$. For example, the cone of non-negative continuous real functions in a closed interval with the sup-norm is normal. An operator $T : X \longrightarrow X$ is **positive** if $T(K) \subset K$, and **strongly positive** if for $0 \neq x \in K$, then $T(x) \in int(K)$ (see Deimling[3]).

An operator $T : X \longrightarrow X$ is (strongly) **Fréchet differentiable at the point** $u_0 \in X$ **in the directions of the cone** K if there exist a linear operator $T'(u_0) : X \to X$ and an operator $\omega(u_0, \cdot) : K \longrightarrow X$ so that

$$T(u_0 + h) = T(u_0) + T'(u_0)\, h \mp \omega(u_0, h), \forall h \in K,$$

where $\lim\limits_{\|h\| \to 0} \frac{\|\omega(u_0,h)\|}{\|h\|} = 0$. $T'(u_0)$ is called (strong) **Fréchet derivative with respect to the cone** K **at the point** u. A function $y : t \in \mathbf{R} \to y(t) \in X$ is called **differentiable at infinity** if the ratio $\left(\frac{1}{t}\right) y(t)$ converges to some element $y'(\infty) \in X$ as $t \to \infty$, and it is usual to speaking about **strong** or **weak differentiability at infinity** depending on whether $\left(\frac{1}{t}\right) y(t)$ converges strongly or weakly to $y'(\infty)$. The operator T is called (strongly) **differentiable at infinity in the directions of the cone** K if for all directions $h \in K$, $h \neq 0$, we have the derivative $y'(\infty)$ of $T(th)$ is representable in the form $y'(\infty) = T'(\infty)\, h$, where $T'(\infty)$ is some continuous linear operator, which is the derivative at infinity with respect to the cone K. The operator $T'(\infty)$ is called the **strong asymptotic derivative with respect to the cone** K, and the operator T is called

strongly asymptotically linear with respect to the cone K**,** if

$$\lim_{R\to\infty}\sup_{\|x\|\geq R, x\in K}\frac{\|Tx - T'(\infty)x\|}{\|x\|} = 0$$

(see Krasnosel'skii[11]).

If X is not a complex Banach space, we can consider its complexification $X_{\mathbf{C}}$, the complex Banach space of all pairs (x, y) with $x, y \in X$, where

$$(x_1, y_1) + (x_2, y_2) = (x_1 + x_2, y_1 + y_2)$$

and

$$(\lambda_1 + i\lambda_2)(x, y) = (\lambda_1 x - \lambda_2 y, \lambda_1 y + \lambda_2 x),$$

with norm given by $\|(x, y)\| = \sup_{\theta\in[0,2\pi]} \|x\cos\theta + y\sin\theta\|$. In this case, X is isometrically isomorphic to the real subspace $\widehat{X} = \{(x, 0); x \in X\}$ of $X_{\mathbf{C}}$. If $T : X \to X$ is a linear operator, its complexification $T_{\mathbf{C}} : X_{\mathbf{C}} \to X_{\mathbf{C}}$ is defined by

$$T_{\mathbf{C}}(x, y) = (Tx, Ty),$$

and $\|T_{\mathbf{C}}\| = \|T\|$ (see Deimling[3]).

Let $T : X \longrightarrow X$ be a linear operator on a complex normed space X, and λ is a complex number, $\lambda \in \mathbf{C}$. Then we have associate operators

$$T_\lambda = T - \lambda I$$

and

$$\Re(\lambda) = T_\lambda^{-1}$$

when the inverse operator exists. The application $\Re$ which for $\lambda \in \mathbf{C}$ associates $\Re(\lambda)$, when it is possible, is called the **resolvent operator of** T. $\lambda \in \mathbf{C}$ is called a **regular value of** T if there exists $\Re(\lambda)$, $\Re(\lambda)$ is bounded linear operator and $\overline{Dom(\Re(\lambda))} = X$. The set of all regular values of T is called resolvent set of T, or simply, the **resolvent of** T, which is denoted by $\rho(T)$. $\sigma(T) = \mathbf{C} - \rho(T)$ is called the **spectrum of** T. The set of all $\lambda \in \mathbf{C}$ so that $R(\lambda)$ does not exist is called the **point spectrum of** T, $\sigma_p(T)$, and their elements are called **eigenvalues**, so $\lambda \in \sigma_p(T)$ if and only if there exists $x \in X$, $x \neq 0$, so that $Tx = \lambda x$, and x is called an **eigenvector of** T **associates to eigenvalue** λ, or simply, an eigenvector of T. Following the definition of spectrum of T, $\lambda \in \sigma(T)$ if one of the above conditions is not true, that is, whether T_λ^{-1} does not exist, or if there exists but it is not bounded, or $\overline{Dom\,(R\,(\lambda))} \neq X$.

Let us consider that X is a complex Banach space and T is bounded. Then if $\Re\,(\lambda)$ exists, is defined on the whole space X, and is bounded. The classical results show that $\rho\,(T)$ is open and the natural domain of the analyticity of $\Re$ (a **domain** in the complex plane **C** is an open subset such that every pair of points can be joined by a broken line consisting of finitely many straight line segments such that all points of they belong to it) and $\sigma\,(T)$ is non-avoid closed bounded set. Since T is a bounded linear operator and $\sigma\,(T)$ is bounded, we have the definition of the **spectral radius** $r\,(T)$,

$$r\,(T) = \sup_{\lambda \in \sigma(T)} |\lambda|\,,$$

and it is known that

$$r\,(T) = \lim_{n\to\infty} \|T^n\|^{\frac{1}{n}}\,,$$

which is called Gelfand's formula (see Kreyszig[13]).

An operator $T : X \to X$ is a **compact operator** if bounded sets are mapped in relatively compact sets. If T is a compact linear operator their properties closely resemble those of operators on finite dimensional spaces. For example, if T is a compact linear operator its set of the eigenvalues is countable (perhaps finite or even empty) and $\lambda = 0$ is the only possible point of accumulation of this set, the dimension of any eigenspace of T is finite and every spectral value $\lambda \neq 0$ is an eigenvalue. Furthermore, if $\lambda \neq 0$ is an eigenvalue there exists a natural number $r = r\,(\lambda)$ such that

$$X = N\,(T_\lambda^r) \oplus T_\lambda^r\,(X)\,,$$

where

$$N\left(T_\lambda^0\right) \subset N\,(T_\lambda) \subset N\left(T_\lambda^2\right) \subset \cdots \subset N\,(T_\lambda^r) = N\left(T_\lambda^{r+1}\right) = \cdots\,,$$

and

$$T_\lambda^0\,(X) \supset T_\lambda\,(X) \supset T_\lambda^2\,(X) \supset \cdots \supset T_\lambda^r\,(X) = T_\lambda^{r+1}\,(X) = \cdots\,.$$

The $\dim\,(N\,(T_\lambda^r))$ is the **algebraic multiplicity of** λ and $\dim\,(N\,(T_\lambda))$ is the **geometric multiplicity of** λ (in the case that $X = \mathbf{R}^n$, $\dim\,(N\,(T_\lambda^r))$ and $\dim\,(N\,(T_\lambda))$ are the multiplicities of λ as a zero of the characteristic polynomial and minimal polynomial of T). Particularly, the order of eingevalue $\lambda \neq 0$ as pole of resolvent operator $\Re(\cdot)$ is its algebraic multiplicity[3].

Let us consider the Banach space $C[0,L]$ of all continuous real functions defined on $[0,L]$, the normal solid cone $C[0,L]^+ = \{f \in X; f(s) \geq 0, s \in [0,L]\}$, and the usual norm, that is, $\|f\| = \sup\{|f(s)|; s \in [0,L]\}$.

The steady state solutions of (2), letting zero the derivatives with respect to time, are

$$\begin{cases} X_\infty(a) = X_b e^{-[\mu a + \Lambda(a) + N(a)]} \\ H_\infty(a) = X_b e^{-(\mu+\sigma)a} \int\limits_0^a e^{\sigma\zeta - N(\zeta)} \lambda_\infty(\zeta) e^{-\Lambda(\zeta)} d\zeta \\ Y_\infty(a) \ = X_b e^{-(\mu+\gamma)a} \int\limits_0^a \sigma e^{(\gamma-\sigma)s} ds \int\limits_0^s e^{\sigma\zeta - N(\zeta)} \lambda_\infty(\zeta) e^{-\Lambda(\zeta)} d\zeta, \end{cases}$$

where $\Lambda(\zeta) = \int_0^\zeta \lambda_\infty(s)ds$ and $N(\zeta) = \int_0^\zeta \nu(s)ds$. Substituting the resulting $Y_\infty(a)$ into the equation (1) at equilibrium, after some calculations we obtain

$$\lambda_\infty(a) = \int\limits_0^L B(a,\zeta) \times M(\zeta, \lambda_\infty(\zeta), \nu(\zeta)) \times \lambda_\infty(\zeta) d\zeta, \tag{4}$$

where the function $M(\zeta, \lambda(\zeta), \nu(\zeta))$ is

$$M(\zeta, \lambda(\zeta), \nu(\zeta)) = e^{-\int_0^\zeta \lambda(s)ds} \times e^{-\int_0^\zeta \nu(s)ds}$$

and the kernel $B(a,\zeta)$ is

$$B(a,\zeta) = \sigma X_b e^{-N(\zeta)} \int\limits_\zeta^L e^{-\sigma(s-\zeta)} e^{\gamma s} \left[\int\limits_s^L \beta(a,a') e^{-(\mu+\gamma)a'} da' \right] ds. \tag{5}$$

Equation (4) is a Hammerstein equation[9]. Notice that the force of infection corresponding to the initial conditions, solutions of (3), is

$$\lambda_0(a) = \int\limits_0^L B'(a,\zeta) M(\zeta, \lambda_0(\zeta), 0) \lambda_0(\zeta) d\zeta,$$

from which we characterize the basic reproduction number R_0.

Let us assume that:

(a) $\beta(a,a')$ is continuous and $\beta(a,a') > 0$ for every $a, a' \in [0,L]$, except for $a = a' = 0$, where $\beta(a,a') = 0$.

(b) $\nu(a)$ is continuous or piecewise continuous with only finitely many discontinues and is bounded.

Let us consider the operator T on $C[0,L]$ defined by

$$Tu(a) = \int_0^L B(a,\zeta)\, M(\zeta, u(\zeta), \nu(\zeta))\, u(\zeta)\, d\zeta, \tag{6}$$

where $B(a,\zeta)$ and $M(\zeta,u,\nu)$ are real functions satisfying the conditions:

(c) $B(a,\zeta)$ is defined on $[0,L]\times[0,L]$, which is positive and continuous in a and ζ.

(d) $M(\zeta,u,\nu)$ is defined on $[0,L]\times[0,\infty)\times[0,\infty)$, which is positive, continuous in ζ for each u and v, strictly monotone decreasing for u for each ζ and ν, and there exists $k_1 \geq 0$ such that

$$|M(\zeta, u_1(\zeta), \nu(\zeta)) - M(\zeta, u_2(\zeta), \nu(\zeta))| \leq k_1 \|u_1 - u_2\| + R(u_1, u_2),$$

with $\lim\limits_{\|u_1-u_2\|\to 0} R(u_1,u_2) = 0$.

(e) there exists a real number $m > 0$ such that $|M(\zeta,u,\nu)| \leq m$ for every ζ, u and ν.

Notice that $|M(\zeta,\lambda(\zeta),\nu(\zeta))| \leq 1$ for all $\zeta \in [0,L]$, and

$$|M(\zeta,\lambda_1(\zeta),\nu(\zeta)) - M(\zeta,\lambda_2(\zeta),\nu(\zeta))| \leq \left|1 - e^{-\int_0^L (\lambda_1(s)-\lambda_2(s))ds}\right| \to 0,$$

when $\|\lambda_1 - \lambda_2\| \to 0$.

Definition 3.1. An operator A is **completely continuous** if it is a compact continuous operator.

The following three theorems are used to proof lemmas below.

Theorem 3.1. *(Krasnosel'skii[12]) Let us consider the Banach spaces E_1 and E_2, the operator $\overline{f} : E_1 \to E_2$ which is continuous and bounded and also $\overline{B} : E_2 \to E_1$ which is completely continuous linear operator. Then the operator $A = \overline{B}\,\overline{f} : E_1 \to E_1$ is completely continuous.*

Theorem 3.2. *(**Ascoli's Theorem,** Kreyszig[13]) A bounded equicontinuous sequence $(x_n)_n$ in $C[0,L]$ has a subsequence which converges in the norm on $C[0,L]$ (a sequence $(y_n)_n$ in $C[0,L]$ is said to be **equicontinuous** if for every $\varepsilon > 0$ there is a $\delta > 0$, depending only on ε, such that for all y_n and all $a, a' \in [0,L]$ satisfying $|a - a'| < \delta$ we have $|y_n(a) - y_n(a')| < \varepsilon$).*

Theorem 3.3. *(**Compactness criterion,** Kreyszig[13]) Let $S : Y \to Z$ be a linear operator where Y and Z are normed spaces. Then S is compact operator if and only if it maps every bounded sequence $(y_n)_n$ in Y onto a sequence in Z which has a convergent subsequence.*

Lemma 3.1. *T is completely continuous positive operator.*
Proof: If $u \in C[0,L]^+$ then $Tu \in C[0,L]^+$. Let be $a_1, a_2 \in [0,L]$ then

$$\begin{aligned}|Tu(a_1) - Tu(a_2)| &\leq_0^L |M(\zeta, u(\zeta), \nu(\zeta))| \, |u(\zeta)| \, |B(a_1,\zeta) - B(a_2,\zeta)| \, d\zeta \\ &\leq m \|u\|_0^L |B(a_1,\zeta) - B(a_2,\zeta)| \, d\zeta.\end{aligned}$$

Note that B is continuous on compact set $[0,L] \times [0,L]$, hence, given $\varepsilon > 0$ there is $\delta > 0$ such that if $\|(a_1,\zeta) - (a_1,\zeta)\| = \sqrt{(a_1 - a_2)^2} \leq \delta$, then $|B(a_1,\zeta) - B(a_2,\zeta)| \leq \frac{\varepsilon}{m\|u\|L}$. Therefore, if $|a_1 - a_2| \leq \delta$ then $|Tu(a_1) - Tu(a_2)| \leq \varepsilon$.
We show that T is continuous. Let $u, u_0 \in C[0,L]$ and $a \in [0,L]$, then

$$\begin{aligned}&|Tu(a) - Tu_0(a)| \\ &\leq \int_0^L |M(\zeta, u(\zeta), \nu(\zeta))\, u(\zeta) - M(\zeta, u_0(\zeta), \nu(\zeta))\, u_0(\zeta)| \, |B(a,\zeta)| \, d\zeta.\end{aligned}$$

As B is continuous over a compact set, so there is $m_1 > 0$ such that $|B(a,\zeta)| \leq m_1$ for all $(a,\zeta) \in [0,L] \times [0,L]$. Furthermore,

$$\begin{aligned}&|M(\zeta, u(\zeta), \nu(\zeta))\, u(\zeta) - M(\zeta, u_0(\zeta), \nu(\zeta))\, u_0(\zeta)| \\ &\leq |M(\zeta, u(\zeta), \nu(\zeta))\, u(\zeta) - M(\zeta, u(\zeta), \nu(\zeta))\, u_0(\zeta)| \\ &+ |M(\zeta, u(\zeta), \nu(\zeta))\, u_0(\zeta) - M(\zeta, u_0(\zeta), \nu(\zeta))\, u_0(\zeta)| \\ &\leq |M(\zeta, u(\zeta), \nu(\zeta))| \, |u(\zeta) - u_0(\zeta)| + |M(\zeta, u(\zeta), \nu(\zeta)) \\ &- M(\zeta, u_0(\zeta), \nu(\zeta))| \, |u_0(\zeta)| \leq |M(\zeta, u(\zeta), \nu(\zeta))| \, \|u - u_0\| \\ &+ [k_1 \|u - u_0\| \, \|u_0\| + R(u(\zeta), u_0(\zeta))] \, \|u_0\| \leq m \|u - u_0\| \\ &+ k_1 \|u_0\| \, \|u - u_0\| + \|u_0\| \, R(u(\zeta), u_0(\zeta)),\end{aligned}$$

from which

$$\begin{aligned}&|Tu(a) - Tu_0(a)| \\ &\leq \int_0^L m_1 [m \|u - u_0\| + k_1 \|u_0\| \, \|u - u_0\| + \|u_0\| \, R(u(\zeta), u_0(\zeta))] \, d\zeta \\ &\leq m_1 (m + k_1 \|u_0\|) \|u - u_0\| \, L + m_1 \|u_0\| \int_0^L R(u(\zeta), u_0(\zeta)) \, d\zeta.\end{aligned}$$

Since $\lim_{\|\lambda_1 - \lambda_2\| \to 0} R(\lambda_1, \lambda_2) = 0$, we have that $\|Tu - Tu_0\| \to 0$ when $\|u - u_0\| \to 0$.
Now we show that T is compact. To prove this we will use the Theorem 3.1. Let us consider the operators

$$\begin{cases} \overline{B} : C[0,L] \to C[0,L] \\ \overline{B}u(a) = \int_0^L B(a,\zeta)\, u(\zeta) \, d\zeta \end{cases}$$

and

$$\begin{cases} \overline{f} : C[0,L] \to C[0,L] \\ \overline{f}u(\zeta) = M(\zeta, u(\zeta), \nu(\zeta))\, u(\zeta). \end{cases}$$

It is enough to verify that $\overline{B}$ is completely continuous and $\overline{f}$ is continuous and bounded.
We show that $\overline{B}$ is completely continuous, that is, $\overline{B}$ is compact and continuous. Let be $u \in C[0,L]$ and $a, a_0 \in [0,L]$, then

$$\left|\overline{B}u(a) - \overline{B}u(a_0)\right| = \left|\int_0^L B(a,\zeta)\, u(\zeta)\, d\zeta - \int_0^L B(a_0,\zeta)\, u(\zeta)\, d\zeta\right|$$
$$\leq \|u\| \int_0^L |B(a,\zeta) - B(a_0,\zeta)|\, d\zeta.$$

As B is continuous on compact, given $\varepsilon > 0$ there is $\delta > 0$ such that $|a - a_0| \leq \delta$, then $|B(a,\zeta) - B(a_0,\zeta)| \leq \frac{\varepsilon}{\|u\| L}$. Thus $\overline{B}u \in C[0,L]$.
First we show that $\overline{B}$ is continuous. Let be $u, u_0 \in C[0,L]$ and $a \in [0,L]$, then

$$\left|\overline{B}u(a) - \overline{B}u_0(a)\right|$$
$$= \left|\int_0^L B(a,\zeta)\, u(\zeta)\, d\zeta - \int_0^L B(a,\zeta)\, u_0(\zeta)\, d\zeta\right| \leq m_1 L \|u - u_0\|.$$

So $\left\|\overline{B}u - \overline{B}u_0\right\| \to 0$ when $\|u - u_0\| \to 0$.
Second, $\overline{B}$ is compact. Let $(u_n)_n$ be a bounded sequence in $C[0,L]$, that is, there is $m_2 \in \mathbf{R}$ such that $\|u_n\| = \sup\limits_{a \in [0,L]} |u_n(a)| \leq m_2$ for every n. Hence, given $a_1, a_2 \in [0,L]$, we have

$$\left|\overline{B}u_n(a_1) - \overline{B}u_n(a_2)\right| = \left|\int_0^L B(a_1,\zeta)\, u_n(\zeta)\, d\zeta - \int_0^L B(a_2,\zeta)\, u_n(\zeta)\, d\zeta\right|$$
$$\leq m_2 \int_0^L |B(a_1,\zeta) - B(a_2,\zeta)|\, d\zeta.$$

Being B continuous on compact, given $\varepsilon > 0$ there is $\delta > 0$ such that if $|a_1 - a_2| \leq \delta$, then $|B(a_1,\zeta) - B(a_2,\zeta)| \leq \frac{\varepsilon}{m_2 L}$. So $\left(\overline{B}(u_n)\right)_n$ is equicontinuous. That $\left(\overline{B}(u_n)\right)_n$ is bounded sequence is checked straightforwardly. As $\left(\overline{B}(u_n)\right)_n$ is equicontinuous and bounded sequence on $C[0,L]$, it has a convergent subsequence (see Theorem 3.2). Since $\overline{B}$ maps a bounded sequence into a sequence which has a convergent subsequence it is a compact operator (see Theorem 3.3).

In relation to $\overline{f}$, we show that $\overline{f}u \in C[0,L]$ if $u \in C[0,L]$. Let be $\zeta, \zeta_0 \in [0,L]$, then

$$\left|\overline{f}u(\zeta)-\overline{f}u(\zeta_0)\right|=\left|e^{-\int_0^\zeta u(s)ds}e^{-\int_0^\zeta \nu(s)ds}u(\zeta)-e^{-\int_0^{\zeta_0} u(s)ds}e^{-\int_0^{\zeta_0} \nu(s)ds}u(\zeta_0)\right|.$$

Let us suppose that $\zeta < \zeta_0$, then, when $\zeta \to \zeta_0$, we have

$$\begin{aligned}&\left|\overline{f}u(\zeta)-\overline{f}u(\zeta_0)\right|\\ &\quad = e^{-\int_0^\zeta u(s)ds}e^{-\int_0^\zeta \nu(s)ds}\left|u(\zeta)-e^{-\int_\zeta^{\zeta_0} u(s)ds}e^{-\int_\zeta^{\zeta_0} \nu(s)ds}u(\zeta_0)\right| \to 0.\end{aligned}$$

Now, we will see that $\overline{f}$ is continuous. Let be $u, u_0 \in C[0,L]$ and $\zeta \in [0,L]$, in a way that

$$\begin{aligned}&\left|\overline{f}u(\zeta)-\overline{f}u_0(\zeta)\right|=\left|M(\zeta,u(\zeta),\nu(\zeta))u(\zeta)-M(\zeta,u_0(\zeta),\nu(\zeta))u_0(\zeta)\right|\\ &\leq \{\left|M(\zeta,u(\zeta),\nu(\zeta))u(\zeta)-M(\zeta,u(\zeta),\nu(\zeta))u_0(\zeta)\right|\\ &\quad +\left|M(\zeta,u(\zeta),\nu(\zeta))u_0(\zeta)-M(\zeta,u_0(\zeta),\nu(\zeta))u_0(\zeta)\right|\}\\ &\leq \left|M(\zeta,u(\zeta),\nu(\zeta))\right|\|u-u_0\|+\left|M(\zeta,u(\zeta),\nu(\zeta))-M(\zeta,u_0(\zeta),\nu(\zeta))\right|\|u_0\|\\ &\leq m\|u-u_0\|+\|u_0\|\left[k_1\|u-u_0\|+R(u(\zeta),u_0(\zeta))\right].\end{aligned}$$

Being $\lim\limits_{\|\lambda_1-\lambda_2\|\to 0} R(\lambda_1,\lambda_2)=0$, we have $\left\|\overline{f}u-\overline{f}u_0\right\| \to 0$ when $\|u-u_0\| \to 0$. That $\overline{f}$ is bounded is checked straightforwardly, i.e.,

$$\begin{aligned}\left\|\overline{f}u\right\| &= \sup_{\zeta\in[0,L]}\left|\overline{f}u(\zeta)\right| = \sup_{\zeta\in[0,L]}\left|M(\zeta,u(\zeta),\nu(\zeta))u(\zeta)\right|\\ &= \sup_{\zeta\in[0,L]}\left|M(\zeta,u(\zeta),\nu(\zeta))\right|\left|u(\zeta)\right| \leq m\|u\|.\end{aligned}$$

Lemma 3.2. *T is Fréchet differentiable at the point $0 \in C[0,L]$ in the directions of the cone $C[0,L]^+$ and*

$$T'(0)h(a) = \int_0^L B(a,\zeta)M(\zeta,0,\nu(\zeta))h(\zeta)d\zeta. \tag{7}$$

Furthermore $T'(0)$ is strongly positive completely continuous operator.

Proof: Let be $u, h \in C[0,L]$. Then, from equation (6), we have

$$
\begin{aligned}
&T(u+h)(a) - Tu(a) \\
&\quad = \int_0^L B(a,\zeta)\, M(\zeta,(u+h)(\zeta),\nu(\zeta))\,(u+h)(\zeta)\,d\zeta \\
&\qquad - \int_0^L B(a,\zeta)\, M(\zeta,u(\zeta),\nu(\zeta))\,u(\zeta)\,d\zeta \\
&\quad = \int_0^L B(a,\zeta)\,[M(\zeta,(u+h)(\zeta),\nu(\zeta)) - M(\zeta,u(\zeta),\nu(\zeta))]\,u(\zeta)\,d\zeta \\
&\qquad + \int_0^L B(a,\zeta)\, M(\zeta,(u+h)(\zeta),\nu(\zeta))\,h(\zeta)\,d\zeta.
\end{aligned}
$$

So we have at $u \equiv 0$,

$$
\begin{aligned}
T(h)(a) - T(0)(a) &= \int_0^L B(a,\zeta)\, M(\zeta,h(\zeta),\nu(\zeta))\,h(\zeta)\,d\zeta \\
&= \int_0^L B(a,\zeta)\,[M(\zeta,h(\zeta),\nu(\zeta)) - M(\zeta,0,\nu(\zeta))]\,h(\zeta)\,d \\
&\quad + \int_0^L B(a,\zeta)\, M(\zeta,0,\nu(\zeta))\,h(\zeta)\,d\zeta.
\end{aligned}
$$

Defining $\omega(a,h)$ by equation

$$
\omega(a,h) = \int_0^L B(a,\zeta)\,[M(\zeta,h(\zeta),\nu(\zeta)) - M(\zeta,0,\nu(\zeta))]\,h(\zeta)\,d\zeta,
$$

we observe that

$$
\begin{aligned}
|\omega(a,h)| &= \left| \int_0^L B(a,\zeta)\,[M(\zeta,h(\zeta),\nu(\zeta)) - M(\zeta,0,\nu(\zeta))]\,h(\zeta)\,d\zeta \right| \\
&\le \int_0^L |B(a,\zeta)|\,[k_1|h(\zeta)| + R(h(\zeta),0)]\,|h(\zeta)|\,d\zeta,
\end{aligned}
$$

and $\lim\limits_{\|h\|\to 0} R(h(\zeta),0) = 0$, then

$$
\lim_{\|h\|\to 0} \frac{\|\omega(a,h)\|}{\|h\|} = 0.
$$

Hence, we have (7), the definition of Fréchet derivative,

$$
T'(0)\,h(a) = \int_0^L B(a,\zeta)\, M(\zeta,0,\nu(\zeta))\,h(\zeta)\,d\zeta.
$$

Now we show that $T'(0)$ is strongly positive. Let us consider $0 \neq h \in C[0,L]^{+}$, that is, there exists $\zeta^* \in [0,L]$ such that $h(\zeta^*) \neq 0$. If $T'(0)\,h(a^*) = 0$ for some $a^* \in [0,L]$, then

$$\int_0^L B(a^*,\zeta)\,M(\zeta,0,\nu(\zeta))\,h(\zeta)\,d\zeta = 0.$$

Since $B(a,\zeta)\,M(\zeta,0,\nu(\zeta))\,h(\zeta)$ is positive continuous function in ζ for each a, we have

$$B(a^*,\zeta)\,M(\zeta,0,\nu(\zeta))\,h(\zeta) = 0$$

for all ζ, particularly for ζ^*, we have

$$B(a^*,\zeta^*)\,M(\zeta^*,0,\nu(\zeta^*))\,h(\zeta^*) = 0.$$

Therefore, we have

$$B(a^*,\zeta^*) = 0,$$

which implies that $\beta(a^*,a') = 0$ for all $a' \in [\zeta^*,L]$ and this is not possible (see condition (a) on $\beta(a,a')$).
Since $T'(0)$ is a linear operator, to verify that it is completely continuous, it is sufficient to proceed like the case of operator T in Lemma 3.1.

To demonstrate that $R_\nu = r(T'(0))$, we use three theorems stated below. In their enunciates, X, Y, K and T will be general spaces and operator, respectively. Notice that R_0 is calculated by (7) letting $\nu = 0$.

Theorem 3.4. *(Krasnosel'skii[11]) Let the positive operator T ($T0 = 0$) have a strong Fréchet derivative $T'(0)$ with respect to a cone and a strong asymptotic derivative $T'(\infty)$ with respect to a cone. Let the spectrum of the operator $T'(\infty)$ lie in the circle $|\mu| \leq \rho < 1$. Let the operator $T'(0)$ have in K an eigenvector h_0; then*

$$T'(0)\,h_0 = \mu_0 h_0,$$

where $\mu_0 > 1$, and $T'(0)$ does not have in K eigenvectors to which an eigenvalue equals to 1. Then if T is completely continuous, the operator T has one non-zero fixed point in the cone.

Theorem 3.5. *(Deimling[3]) Let be X a Banach space, $K \subset X$ a solid cone, that is, $int(K) \neq \emptyset$, and $T : X \to X$ strongly positive compact linear operator. Then:*
(i) $r(T) > 0$, $r(T)$ is a simple eigenvalue with eigenvector $v \in int(K)$ and

there is not eigenvalue with positive eigenvector.
(ii) if λ is an eigenvalue and $\lambda \neq r(T)$, then $|\lambda| < r(T)$.
(iii) if $S : X \to X$ is bounded linear operator and $Sx \geq Tx$ on K, then $r(S) \geq r(T)$, while $r(S) > r(T)$ if $Sx > Tx$ for $x \in K, x > 0$.

Definition 3.2. (Deimling[3]) Let X, Y be Banach spaces, $J = (\lambda_0 - \delta, \lambda_0 + \delta)$ a real interval, $\Omega \subset X$ a neighborhood of 0 and $F : J \times \Omega \longrightarrow Y$ such that $F(\lambda, 0) = 0$ for all $\lambda \in J$, then $(\lambda_0, 0)$ will be a **bifurcation point** for $F(\lambda, x)$ if

$$(\lambda_0, 0) \in \overline{\{(\lambda, x) \in J \times \Omega; F(\lambda, x) = 0, x \neq 0\}}.$$

Theorem 3.6. *(Bifurcation Theorem, Griffel[9]) Consider the equation $Au = \eta u$, where A is a compact non-linear operator, Fréchet-differentiable at $u = 0$, such that $A0 = 0$. Then:*
(i) if μ_0 is a bifurcation point of $F(\mu, x) = x - \mu Ax$, then μ_0^{-1} is an eigenvalue of the linear operator $A'(0)$.
(ii) if μ_0^{-1} is an eigenvalue of $A'(0)$ with odd multiplicity, then μ_0 is a bifurcation point of $F(\mu, x)$.

Theorem 3.7. ***(Existence Theorem)*** *Let us consider the operator $T : C[0, L] \to C[0, L]$ described by the equation (6), or*

$$Tu(a) = \int_0^L B(a, \zeta) M(\zeta, u(\zeta), \nu(\zeta)) u(\zeta) d\zeta.$$

If $r(T'(0)) \leq 1$, the only solution of equation (4), that is,

$$\lambda(a) = \int_0^L B(a, \zeta) M(\zeta, \lambda(\zeta), \nu(\zeta)) \lambda(\zeta) d\zeta$$

is the trivial solution. Otherwise, if $r(T'(0)) > 1$ there is at least one non-trivial positive solution for this equation.

Proof: We use the same arguments given in Greenhalgh[7]. Suppose $r(T'(0)) \leq 1$ and the equation (4) has a non-trivial positive solution λ^*, that is,

$$\lambda^*(a) = \int_0^L B(a, \zeta) M(\zeta, \lambda^*(\zeta), \nu(\zeta)) \lambda^*(\zeta) d\zeta.$$

Since $\lambda^* > 0$ and $M(\zeta, \lambda, \nu)$ is strictly monotone decreasing for λ we have that

$$\int_0^L B(a,\zeta) M(\zeta, \lambda^*(\zeta), \nu(\zeta)) \lambda^*(\zeta) d\zeta < T'(0) \lambda^*(a).$$

Since both side of last equation are continuous on compact, then there exists $\varepsilon > 0$ such that

$$\lambda^*(1+\varepsilon) < T'(0) \lambda^*.$$

By finite inducing over n we have that

$$\lambda^*(1+\varepsilon)^n < T'(0)^n \lambda^*.$$

So

$$\|\lambda^*(1+\varepsilon)^n\| < \|T'(0)^n \lambda^*\| \leq \|T'(0)^n\| \|\lambda^*\|,$$

and

$$(1+\varepsilon)^n < \|T'(0)^n\|$$

for every $n = 1, 2, 3, \cdots$. Then $r(T'(0)) > 1$, which is an absurd.
Let us suppose that $r(T'(0)) > 1$. Firstly, we will calculate $T'(\infty)$. For every $u \in K$, since

$$T(tu) = \int_0^L B(a,\zeta) e^{-t\int_0^\zeta u(s)ds} tu(\zeta) d\zeta,$$

where $B(a,\zeta)$ is given by equation (5), we will have

$$\lim_{t\to\infty} \frac{T(tu)}{t} = 0,$$

then $T'(\infty) = 0$. Now, we show that T is strongly asymptotically linear with respect to the cone K,

$$\lim_{R\to\infty} \sup_{\|x\|\geq R, x\in K} \frac{\|Tx - T'(\infty)x\|}{\|x\|} = \lim_{R\to\infty} \sup_{\|x\|\geq R, x\in K} \frac{\|Tx\|}{\|x\|}.$$

We have

$$\begin{aligned}
\|Tx\| &= \sup_{a\in[0,L]} \left| \int_0^L B(a,\zeta) e^{-\int_0^\zeta x(s)ds} x(\zeta) d\zeta \right| \\
&= \sup_{a\in[0,L]} \left| \int_0^L B(a,\zeta) \tfrac{d}{d\zeta} \left(-e^{-\int_0^\zeta x(s)ds}\right) d\zeta \right| \\
&\leq m' \int_0^L \tfrac{d}{d\zeta} \left(-e^{-\int_0^\zeta x(s)ds}\right) d\zeta = m' \left[1 - e^{-\int_0^L x(s)ds}\right],
\end{aligned}$$

where $m' = \sup_{a,\zeta\in[0,L]} |B(a,\zeta)|$. Then

$$\lim_{R\to\infty} \sup_{\|x\|\geq R, x\in K} \frac{\|Tx\|}{\|x\|} \leq \lim_{R\to\infty} \sup_{\|x\|\geq R, x\in K} \frac{m'\left[1-e^{-\int_0^L x(s)ds}\right]}{\|x\|} = 0,$$

that is, T is strongly asymptotically linear with respect to the cone K, with the strong asymptotic derivative with respect to the cone K equals $T'(\infty) = 0$.

Let us consider in Theorem 3.4 $\mu_0 = r(T'(0))$. Following Theorem 3.5, $r(T'(0))$ is a simple eigenvalue of $T'(0)$ with eigenvector in $int(K)$ and there is not other eigenvalue of $T'(0)$ with positive eigenvector. Obviously, being $T'(0)$ a positive operator, 1 can not be a positive eigenvalue of $T'(0)$ by above argument. Since T is completely continuous, all conditions of the Theorem 3.4 are satisfy, and we conclude that the equation (4) has a non-trivial solution.

Moreover, let us consider $0 < \overline{\mu} < \frac{1}{r(T'(0))}$ such that there is a $\overline{x} \in K, \overline{x} \neq 0$, with $F(\overline{\mu}, \overline{x}) = \overline{x} - \overline{\mu}T\overline{x}$. Then $T\overline{x} = \frac{1}{\overline{\mu}}\overline{x}$ and it follows that $\frac{1}{\overline{\mu}} \leq r(T'(0))$, and this is not possible. So such $\overline{\mu}$ does not exist.

Since $r(T'(0))$ is a simple eigenvalue of $T'(0)$, we have that $\frac{1}{r(T'(0))}$ is a bifurcation point of $F(\mu, x) = x - \mu Tx$ (Theorem 3.6).

Let us suppose now that there exists $\mu^* > \frac{1}{r(T'(0))}$ being a bifurcation point of $F(\mu, x) = x - \mu Tx$, that is, there exists $(\mu_n, x_n) \to (\mu^*, 0)$ when $n \to \infty$, where $x_n \in K\setminus\{0\}$ and $F(\mu_n, x_n) = x_n - \mu_n Tx_n = 0$, that is, $Tx_n = \frac{1}{\mu_n}x_n$. Being T Fréchet differentiable at $u = 0$ in the direction of K, we have

$$Tx_n = T(0) + T'(0)x_n + \omega(0, x_n) = \frac{1}{\mu_n}x_n,$$

where $\lim_{n\to\infty} \frac{\|\omega(0,x_n)\|}{\|x_n\|} = 0$. Since $T(0) = 0$ we have

$$T'(0)\frac{x_n}{\|x_n\|} + \frac{\omega(0, x_n)}{\|x_n\|} = \frac{1}{\mu_n}\frac{x_n}{\|x_n\|}.$$

Being $T'(0)$ is a compact operator and $\left(\frac{x_n}{\|x_n\|}\right)$ is a bounded sequence, we can assume that there is $v \in K$ such that

$$\lim_{n\to\infty}\left\{T'(0)\frac{x_n}{\|x_n\|}\right\} = v.$$

On the one hand,

$$\lim_{n\to\infty}\left\{T'(0)\left(\frac{x_n}{\|x_n\|}\right)\right\} = v.$$

From the linearity and continuity of $T'(0)$, we have

$$\begin{aligned}\nu &= \lim_{n\to\infty} T'(0)\frac{x_n}{\|x_n\|} = \lim_{n\to\infty} T'(0)\left(\mu_n \frac{1}{\mu_n}\frac{x_n}{\|x_n\|}\right)\\ &= T'(0)\left(\lim_{n\to\infty} \mu_n \frac{1}{\mu_n}\frac{x_n}{\|x_n\|}\right) = T'(0)(\mu^*\nu),\end{aligned}$$

that is, $T'(0)(\nu) = \frac{1}{\mu^*}\nu$, and $\frac{1}{\mu^*}$ is an eigenvalue of $T'(0)$ with a positive eigenvector, which is not possible by Theorem 3.5. So such μ^* can not exist. Therefore, we have that $\frac{1}{r(T'(0))}$ is a unique bifurcation point of $F(\mu, x) = x - \mu T x$.

4. Discussion and conclusion

A characterization of the basic reproduction number R_0 was done considering fixed point and monotone operators[11,12], and properties regarding to the positive operators and strongly positive operators on cones[3].

We compare our results with those obtained by Greenhalgh[7] and Lopez and Coutinho[14]. Greenhalgh[7] assumed that the contact rate is strictly positive, which is not necessary in our case. Lopez and Coutinho[14] applied Schauder's theorem, which requires that the application acts on a convex set. The definition of convexity given by Griffell[9] has the following geometric meaning: a convex set must contain any line segment joining any two points belonging to it. For the sake of simplicity, let us consider $L = 1$. It is easy to verify that the set $T = C[0, L]^+ \cap \{\varphi; \|\varphi\| = 1\}$ is not convex: From the functions $x(a) = a$, $y(a) = 4a(1-a)$ belonging to T and $z(a) = \frac{1}{2}x(a) + \left(1 - \frac{1}{2}\right)y(a)$, we obtain $\|z\| = \frac{50}{64}$, which shows that z does not belong to T, even that it belongs to the line segment joining points of T, namely, x and y. Remember that the Schauder's theorem establishes the existence of a fixed point with respect to a continuous operator acting on a closed and convex set, which image is contained in a relatively compact subset of the defined domain. When we consider the set $C[0, L]^+ \cap \{\varphi; \|\varphi\| \leq 1\}$, a convex set, we can not disregard the possibility that the null function is the fixed point obtained by applying the Schauder's theorem, due to the fact that for the particular operator considered, the image of the zero function is the zero function itself.

With respect to the generalization of the results obtained using positive core to include non-negative core made by Lopez and Coutinho[14], they defined that a set of positive functions is cone if a function has a finite number of points at which the function is zero plus zero function. However, a cone must be a closed set. Taking again $L = 1$ for the same reason given above, and considering the following sequence

$$f_n(a) = \begin{cases} \frac{1}{n}\left[sen\left(4n\pi a\right) + 1\right], & 0 \le a \le \frac{1}{2} \\ \frac{2(n-1)}{n}\left(a - \frac{1}{2}\right) + \frac{1}{n}, & \frac{1}{2} \le a \le 1, \end{cases}$$

which belongs to positive function, this sequence converges to function

$$f(a) = \begin{cases} 0, & 0 \le a \le \frac{1}{2} \\ 2\left(a - \frac{1}{2}\right), & \frac{1}{2} \le a \le 1, \end{cases}$$

which does not belong to the set.

The characterization of R_ν as the spectral radius of an operator allows us to assess vaccination strategies having as goal the eradication of the disease. It is possible to introduce vaccination rate in the form $\nu(a) = \nu\theta(a - a_1)\,\theta(a_2 - a)$, where ν is a constant vaccination rate and $[a_1, a_2]$ is the age interval of individuals that are vaccinated, and we determine[19,20]: (i) if the vaccination programme is efficient, that is, yields $R_\nu \le 1$, in which case we have $\lambda_\infty \equiv 0$; (ii) the minimum vaccination effort, ν_m such that $R_\nu = 1$; and (iii) the more appropriate vaccinated age interval $[a_1, a_2]$ to control the infection.

In a companion paper[4] we show uniqueness of the non-trivial solution in order to validate R_0 obtained by applying spectral radius as the basic reproduction number. The unique bifurcation value corresponds to the appearance of non-trivial solution corresponding to the endemic level. We also evaluate the basic reproduction number for some functions describing the contact rate. Due to the difficulty and complexity found in the calculation of the spectral radius, we evaluate the upper and lower limits for R_0.

Acknowledgments

We thank Prof. David Greenhalgh for comments and suggestions that contributed to improve this paper.

References

1. R. M. Anderson and R. M. May, *J. Hyg. Camb.* **94**, 365 (1985).
2. R. M. Anderson and R. M. May, *Infectious Diseases of Humans: Dynamics and Control*, Oxford, Oxford University Press (1991).
3. K. Deimling, *Nonlinear Functional Analysis*, Springer-Verlag, Berlin (1985).
4. C. H. Dezotti and H. M. Yang, *Proceedings of Biomat*, submitted (2010).
5. K. Dietz, *Proceedings of a SIMS Conference on Epidemiology, Alta, Utah, July 8-12* **1974**, 104 (1975).
6. C. P. Farrington, *Stat. Med.* **9**, 953 (1990).

7. D. Greenhalgh, *Math. Biosc.* **100**, 201 (1990).
8. B. T. Grenfell and R. M. Anderson, *J. Hyg.* **95**, 419 (1985).
9. D. H. Griffel, *Applied Functional Analysis*, Ellis Horwood Limited, Chichester, England (1981).
10. H. Inaba, *J. Math. Biol.* **28**, 411 (1990).
11. M. A. Krasnosel'skii, *Positive Solutions of Operator Equations*, P. Noorddhoff ltda. Groningen, The Netherlands (1964).
12. M. A. Krasnosel'skii, *Topological Method in the Theory of Nonlinear Integral Equation*, Pergamon Press, Oxford (1964).
13. E. Kreyszig, *Introductory Functional Analysis with Applications*, John Wiley & Sons, New York (1989).
14. L. F. Lopez and F. A. B. Coutinho, *J. Math. Biol.* **40**, 199 (2000).
15. D. Schenzle, *IMA J. Math. App. Med. Biol.* **1**, 169 (1984).
16. E. Trucco, *Math. Bioph.* **27**, 285 (1965).
17. D. W. Tudor, *Math. Biosc.* **73**, 131 (1985).
18. H. M. Yang, *Math. Compt. Model.* **29 (8)**, 39 (1999).
19. H. M. Yang, *Math. Compt. Model.* **29 (7)**, 11 (1999).
20. H. M. Yang, *Appl. Math. Comput.* **122 (1)**, 27 (2001).

CONTROL OF WEST NILE VIRUS BY INSECTICIDE IN THE PRESENCE OF AN AVIAN RESERVOIR

E. BAUMRIN, J. DREXINGER, J. SOTSKY and D. I. WALLACE *

Department of Mathematics,
Dartmouth College,
Hanover, NH, 03755, USA
E-mail: Dorothy.Wallace@Dartmouth.edu

The first cases of West Nile virus (WNV) in North America appeared in New York City in the summer of 1999. By 2002, the disease had spread across the continent, resulting in thousands of infections and deaths. WNV is still easily spread from mosquitoes to humans in densely populated areas and since it can be fatal, WNV continues to attract significant public attention. Although this disease has been acknowledged for some time, there are still no recognized effective treatments and public efforts have focused primarily on preventing transmission of the virus. This paper compares the reliability of several distinct mathematical models in predicting the transmission and population dynamics of the virus in mosquito vectors and avian reservoirs. The most robust model is extended to include humans. Numerical experiments are conducted to establish the most effective quantity and timing of chemical insecticide spray needed to prevent a human epidemic of West Nile virus in large urban areas. This study concludes that early insecticide spraying is essential in preventing an epidemic and that the quantity of insecticide sprayed is less important than the timing. The model suggests that a low concentration of insecticide sprayed at the emergence of human cases is an effective strategy for reducing the level of a human epidemic.

1. Introduction

In the summer of 1999, the first reported cases of West Nile virus (WNV) in North America presented in New York City. The virus, which had previously been confined to Africa, the Middle East, western Asia, and some areas of Europe, immediately attracted public attention in the United States as a result of its ease of transmission from infected mosquitoes to humans and its potentially fatal neurological effects. West Nile virus is an arbovirus and a single stranded RNA virus of the genus Flavivirus and the family Fla-

*Corresponding author

viviridae [1]. Approximately 80% of people infected with West Nile virus are asymptomatic. Approximately 20% report fever, headache, fatigue, and/or rash as their primary symptoms. However, around 1% of people that contract WNV develop a more severe and potentially fatal form of the disease. Symptoms for severe West Nile virus include meningitis, encephalitis, high fever, ataxia, seizures, and death. An uncharacteristically high number of people infected with West Nile in North America have developed severe symptoms, suggesting the presence of a particularly virulent strain and the relative absence of immunity to the disease in the U. S. By the end of 1999, the Center for Disease Control (CDC) reported 59 confirmed WNV infections and 7 deaths in New York State. No standard preventative measures were implemented and by late 2002, a highly virulent strain of the virus had spread across the continent. The yearly CDC West Nile report confirmed that in the U. S. in 2002 alone, over 4,156 people in 40 states had been infected with the virus, and 284 had died. The virus was becoming an epidemic in North America and there were no known effective treatments.

West Nile virus can take several forms of transmission. The disease exploits birds (140 infected species have been found in North America) and equines as reservoirs for surviving throughout and in between outbreaks although the virus has been recorded in several other mammal species throughout the United States. The West Nile virus, transmitted mainly by Culex mosquito vectors from numerous bird reservoirs, has a complex life cycle. Ultimately, humans are used as a secondary carrier, with human infection occurring through vector-human contact. However, humans usually do not produce enough viremia to be able to infect a healthy vector that bites them.

In their paper comparing the effects of biological assumptions on transmission terms and disease predictions, Wonham *et al*, [2], present five models that were constructed following the 1999 New York City outbreak. Each of these epidemiological models takes a different approach to modeling the disease through varying assumptions, and consequently varying transmission terms, although all are based on the standard susceptible-infected (S-I) model. A core model was constructed with three variations on the form taken by the disease transmission terms. In all cases the reproductive number associated to the system is density dependent. The median, high and low values of each parameter were used to construct distributions as a source for a Monte Carlo simulation. The paper reports mean and variance in the reproductive number for low, medium and high populations of reservoirs and vectors. Substantial variation in the reproductive number results from choosing different forms for the transmission term. Their paper, [2], serves as a warning that the choice of reservoir model will have serious consequences for human epidemiology when coupled with equations representing human populations. Changing the form of the model varies more than just the reproductive number, however. Transmission to humans involves not only this measure of virulence, but also the density of the infected mosquito population.

Bowman *et al*, [1], has a full epidemiological model including vector, reservoir, and human host. The vector-reservoir submodel in [1] is one of those mentioned in [2]. However, our analysis will show that it is not the best choice for a model that is consistent across bird species with respect to reproductive number and infected vector density. The authors use their full model to study the effects of outbreak control via adulticide and larvicide of the vector. Their model assumes continuous proportional death of either or both forms of the vector. In this paper we apply this strategy using an improved vector-reservoir model and compare its effectiveness for different starting times and mortality rates.

We will look at the consequences of four of the models studied in [2] across six species of North American birds. In Section 2 we introduce models by Thomas and Urena, [3], Wonham *et al*, [4], Bowman *et al*, [1], and Cruz-Pacheco *et al*, [5]. In Section 3 we describe the range of parameters for the model, including the specific death and recovery rates for the various species of birds. In Section 4 we give mean and standard variation in reproductive number and maximum density of infected vector across these six bird species. We also note qualitative differences in output from these models. On the basis of this discussion we choose the model with least variability across bird species. In Section 5 we couple this vector-reservoir model with the human population, following [1] for the human epidemiology submodel. In Section 6 we show the results of attempting to control West Nile virus with American Crow bird reservoir by continuous applications of insecticide (adulticide) introduced at varying times and with varying effectiveness. In Section 7 we discuss our results. All numerical integrations were performed using BGODEM software (Reid, copyright 2008).

2. Vector-Reservoir Transmission Models

We preserve the notation of [2], referring to the models in this section as WN2-WN5. We refer the reader to Table 1 of [2], which summarizes the basic features of each model, giving here only the equations for each. A more complete description of the final model chosen is in section 5. The differential equations for the various models include the population dynamics of L_V (larval vectors), S_V (susceptible vectors), E_V (exposed vectors, accounting for incubation time of the virus), I_V (infected vectors), S_R (susceptible reservoirs), I_R (infected reservoirs), and R_R (recovered reservoirs). Reproduction numbers are as stated in [2]. The quantities N_V^* and N_R^* are the disease free equilibrium values for vector and reservoir populations respectively. All models are assumed to be valid only for a single season.

2.1. *WN2 (Thomas and Urena, [3])*

This model uses a mass action disease transmission term, which assumes biting rates, the main factor in disease transmission, are limited by both vector and reservoir densities. It does not model the larval vector population, nor does it take into account the species specific disease mortality.

$$\frac{dS_V}{dt} = b_V[S_V + (1 - \rho_V)(E_V + I_V)] - \beta_R I_R S_V - d_V S_V \tag{1}$$

$$\frac{dE_V}{dt} = b_V \rho_V (E_V + I_V) + \beta_R I_R S_V - (d_V + \kappa_V)E_V \tag{2}$$

$$\frac{dI_V}{dt} = \kappa_V E_V - d_V I_V \tag{3}$$

$$\frac{dS_R}{dt} = b_R N_R - \beta_R^* S_R I_V - d_R S_R \tag{4}$$

$$\frac{dI_R}{dt} = \beta_R^* S_R I_V - (d_R + \gamma_R)I_R \tag{5}$$

The reproduction number associated to WN2 is (equation here).

$$R_0 = \frac{\rho_V}{2} + \sqrt{(\frac{\rho_V}{2})^2 + \frac{\phi_V \beta_R^{*2} N_V^* N_R^*}{d_V(d_R + \gamma_R)}} \tag{6}$$

2.2. *WN3 (Wonham et al, [4])*

This model uses a reservoir frequency dependent transmission term, which assumes the vector biting rate is saturated and does not depend on reservoir densities. This model includes all populations, accounting for all constants.

$$\frac{dL_V}{dt} = b_L N_V - (m_L + d_L)L_V \tag{7}$$

$$\frac{dS_V}{dt} = -\alpha_V \beta_R \frac{I_R}{N_R} S_V + m_L L_V - d_V S_V \tag{8}$$

$$\frac{dE_V}{dt} = \alpha_V \beta_R \frac{I_R}{N_R} S_V - (\kappa_V + d_V)E_V \tag{9}$$

$$\frac{dI_V}{dt} = \kappa_V E_V - d_V I_V \tag{10}$$

$$\frac{dS_R}{dt} = -\alpha_R \beta_R \frac{S_R}{N_R} I_V \tag{11}$$

$$\frac{dI_R}{dt} = \alpha_R \beta_R \frac{S_R}{N_R} I_V - (\delta_R + \gamma_R)I_R \tag{12}$$

$$\frac{dR_R}{dt} = \gamma_R I_R \tag{13}$$

The reproduction number associated to WN3 is

$$R_0 = \sqrt{\frac{\phi_V \beta_R^2 \alpha_R \alpha_V N_V^*}{d_V(\delta_R + \gamma_R)N_R^*}} \tag{14}$$

2.3. *WN4 (Bowman et al, [1])*

This model uses a reservoir frequency dependent transmission term, which assumes the vector biting rate is saturated and does not depend on reservoir densities. This model does not include larval or exposed vector populations, does not account for recovery rate of reservoirs, nor incubation time, and uses a "recruitment" rate in place of birth and death rates.

$$\frac{dS_V}{dt} = a_V - \alpha_V \beta_R \frac{I_R}{N_R} S_V - d_V S_V \tag{15}$$

$$\frac{dI_V}{dt} = \alpha_V \beta_R \frac{I_R}{N_R} S_V - a_V I_V \tag{16}$$

$$\frac{dS_R}{dt} = a_R - \alpha_R \beta_R \frac{S_R}{N_R} I_V - d_R S_R \tag{17}$$

$$\frac{dI_R}{dt} = \alpha_R \beta_R \frac{S_R}{N_R} I_V - (\delta_R + d_R) I_R \tag{18}$$

The reproduction number associated to WN4 is

$$R_0 = \sqrt{\frac{\beta_R^2 \alpha_R \alpha_V N_V^*}{d_V (\delta_R + d_R) N_R^*}} \tag{19}$$

2.4. *WN5 (Cruz-Pacheco et al, [5])*

This model uses a reservoir frequency dependent transmission term, which assumes the vector biting rate is saturated and does not depend on reservoir densities. It does not model larval or exposed vector populations, and thus ignores incubation time, although it does account for all constants.

$$\frac{dS_V}{dt} = b_V[S_V + (1 - \rho_V)(E_V + I_V)] - \beta_R I_R S_V - d_V S_V \tag{20}$$

$$\frac{dI_V}{dt} = b_V \rho_V I_V + \alpha_V \beta_R \frac{I_R}{N_R} S_V - d_V I_V \tag{21}$$

$$\frac{dS_R}{dt} = a_R - \alpha_R \beta_R \frac{S_R}{N_R} I_V - d_R S_R \tag{22}$$

$$\frac{dI_R}{dt} = \alpha_R \beta_R \frac{S_R}{N_R} I_V - (\delta_R + d_R + \gamma_R) I_R \tag{23}$$

$$\frac{dI_R}{dt} = \gamma_R I_R - d_R R_R \tag{24}$$

The reproduction number associated to WN5 is

$$R_0 = \frac{\rho_V}{2} + \sqrt{(\frac{\rho_V}{2})^2 + \frac{\alpha_R \alpha_V \beta_R^2 N_V^*}{d_V(d_R + \delta_R + \gamma_R)}} \tag{25}$$

3. Parameter ranges for all models

Wonham *et al* [2] gives values for all parameters in models WN2-WN5 based on quantities reported in the literature or calculated from simple assumptions. We have used these values with the exceptions of parameters a_V, a_R, δ_R and γ_R , which are explained below. We also summarize the notation and values from Wonham *et al* (2006), to which we refer the reader for further information such as ranges and bibliographic sources.

Quantities δ_R and γ_R stand for the disease mortality rate of reservoirs and for the reservoir recovery rate respectively. These two values were the focus of the sensitivity analysis done here and were the only changed values from model to model. Using the values of π_R (survival probability), σ_R (days infectious), and τ_R (days to death) from [2], the values of δ_R and γ_R, (for the six North American species of American Crow, American Robin, Blue Jay, House Sparrow, Northern Mockingbird and Northern Cardinal), were calculated and the resulting rates are given in Table 1.

Table 1. Reservoir Species Specific Parameter Values

Species	survival probability π_R	days infectious σ_R	days to death τ_R	death rate δ_R	recovery rate γ_R
American Crow	0	3.25	5.10	0.20	0
American Robin	1.00	3.00	n.a.	0.00	0.33
Blue Jay	0.25	3.75	4.70	0.29	0.08
House Sparrow	0.47	3.00	4.70	0.16	0.21
Northern Mockingbird	1.00	1.25	n.a.	0.00	.80
Cardinal	0.78	1.50	4.00	0.06	0.89

Parameters a_V and a_R are both titled "recruitment rates" and are specific to the vectors and reservoirs of WN4. The recruitment rate a_R was set to zero, reflecting the assumption that the model is for only one season. These values take the place of a typical birth rate. The value for a_V was taken from [1]. The reported range of 5000-22000 mosquitoes born per day was averaged to a constant value of 13500 mosquitoes born/day.

The value of m_L is 0.07 mosquitoes per day and this value accounts for the maturation rate of the mosquitos. The values of d_V, d_L, and d_R, are 0.03 mosquitoes/day, 0.02 larvae/day, 0.0015 birds/day respectively, accounting for natural death of vectors, larvae, and reservoirs respectively. The birth rates for these particular models (b_V, b_L, and b_R) are equal to their respective death rates (0.03 mosquitoes/day, 0.02 larvae/day, 0.0015 birds/day). The value of ρ_V is 0.001 mosquitoes/day and this value is accounting for the number of mosquitoes born that are already infected with West Nile virus. 0.69 is the probability in a single day that the virus will be transmitted to a vector, and this value is represented by the variable α_k. κ_V is used in several of the models to account for the incubation rate of the virus and is measured as .10 mosquitoes per day. ϕ_V is an account of the proportion of vectors that survive the incubation period, and here it is calculated using the supplied equation as 0.77 mosquitoes/day.

The quantity β_R is the saturated reservoir frequency dependent bite rate measured as a constant 0.44 bites/mosquito/day and can be thought of as "maximum possible number of bites per day made by a single mosquito." This term drives the reservoir frequency dependent disease transmission term. β_R^* is measured in bites per day per unit density bird at disease-free equilibrium values, and this ratio was taken into account as a moving variable in the graphical input programs by using the equation supplied in [2]. This term drives the mass action disease transmission term and varies depending on the density of reservoirs. The constant m_R is the maturation rate of reservoirs and is again set to 0 and not used in the equations due to the single season nature of the models. For models with reservoir frequency dependent transmission terms, 0.74 is the probability in a single day that the virus will be transmitted to a reservoir, and this value is represented by the parameter α_R.

Typical runs for the system WN3 are shown in Figure 1 for American Crow and Figure 2 for American Robin. It is easy to see that the change in death and recovery rates has a big effect on model predictions. Figure 3 shows how model WN3 varies across all six species in terms of the proportion of mosquitos which are infected, which is the key predictor of infection in human populations.

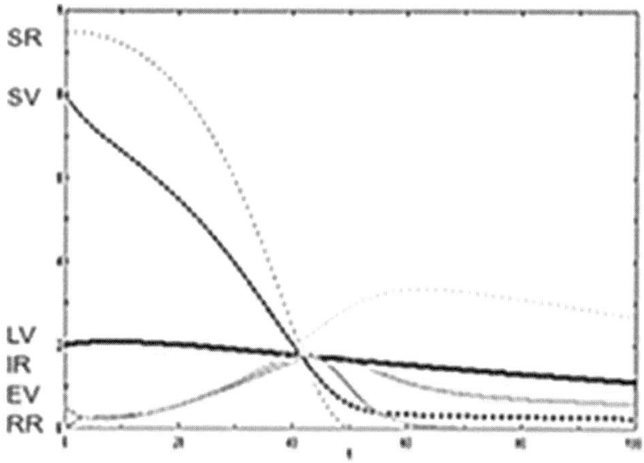

Figure 1. The model WN3 with parameters for American Crow

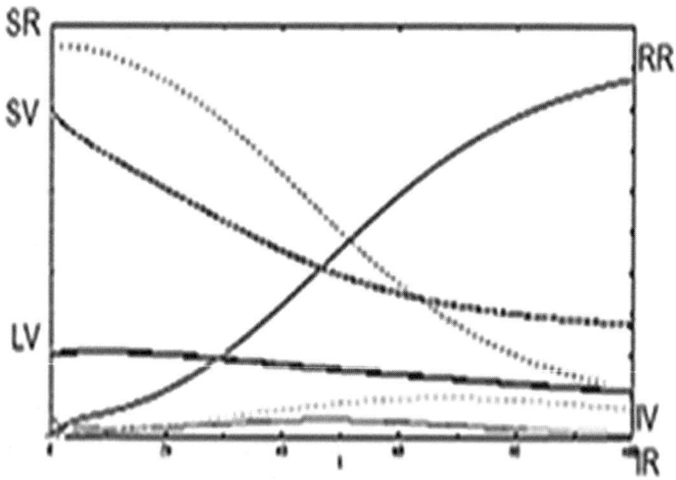

Figure 2. The model WN3 with parameters for American Robin.

Similarly, there is a wide range of predictive outputs across the four models we are considering. Figure 4 shows the infected mosquito output

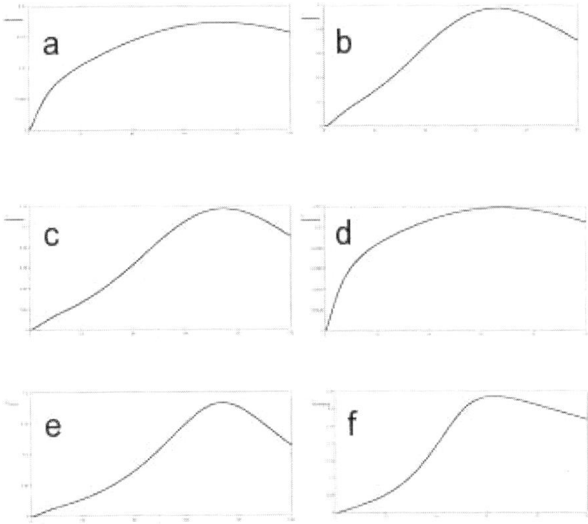

Figure 3. Infected vector populations in model WN3 for all reservoirs considered. a. Northern Mockingbird, b. American Robin, c. House Sparrow, d. Cardinal, e. Blue Jay, f. American Crow.

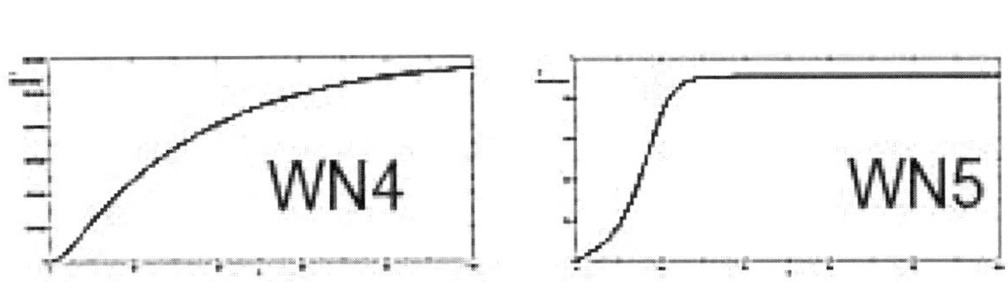

Figure 4. Infected vector populations in models WN2-WN5 with parameters for American Crow.

for all four models, using the death and recovery rates for American Crow. In the next section we analyze the variation among these models as the parameters range across the six species of bird.

4. Variation in reproductive number and density of infected vector across six reservoir species

The reservoir species specific values for disease reproduction rate, R0, which has a direct correlation with human infection rate, were calculated algebraically from the four R_0 equations listed above and the reported parameters. The average and standard deviation were then calculated to test for species specific sensitivity. The values of N_V^* and N_R^* are the population disease-free equilibrium values at low, medium and high densities of vector and reservoir populations. The values for WN2 were low density for numerical reasons and WN3-WN5 were all medium density models. The results are giving in Table 2.

Table 2. Species Specific R_0 and Standard Deviation

Species	death rate δ_R	recovery rate γ_R	R_0 WN2 $N_v =$ 1000 $N_R =$ 100	R_0 WN3 $N_v =$ $10,000$ $N_R =$ 1000	R_0 WN4 $N_v =$ $10,000$ $N_R =$ 1000	R_0 WN5 $N_v =$ $10,000$ $N_R =$ 1000
Crow	0.20	0.00	36.40	11.66	12.79	13.243
Robin	0.00	0.33	2.45	9.08	148.21	10.32
Blue Jay	0.29	0.08	4.94	8.58	10.63	9.75
Sparrow	0.16	0.21	3.07	8.58	14.28	9.75
Mockingbird	0.00	0.80	1.58	5.83	22.78	6.08
Cardinal	0.06	0.89	1.49	5.35	22.78	6.09
Average			8.32	8.18	59.48	9.30
STD			13.81	2.32	68.85	2.62

The reservoir species specific values for maximum density of infected vector, I_V, which has a direct correlation with human infection rate as well, were found using outputs from BGODEM (Reid, 2008) which uses a Runge-Kutta algorithm to numerically integrate systems of ordinary differential equations. The averages and standard deviations were then calculated to test for species specific sensitivity. All values were taken directly from graphical solution proportion values for models WN2, WN3, and WN5. WN4 did not have a population cap and thus did not produce per capita population proportions so the maximum portion of the mosquito population that was infected at any time within the 100 days was calculated through turning the given values of IV and SV, into proportions, ($\frac{I_V}{[I_V+S_V]}$ = the proportion of the mosquito population that is infected), which could then be directly compared to the other graphical outputs. The results are given in Table 3.

Table 3. Species Specific Infected Vector Maximum Values

Species	WN2	WN3	WN4	WN5
American Crow	0.21	0.34	0.093	0.91
American Robin	0.001	0.097	0.093	0.385
Blue Jay	0.17	0.18	0.93	0.63
House Sparrow	0.036	0.117	0.093	0.46
Northern Mockingbird	0.00018	0.017	0.093	0.138
Cardinal	0.00014	0.0119	0.093	0.112
Average	0.0695	0.127	0.093	0.44
STD	0.0948	0.12	0	0.30

In order to determine which model was the most consistent, or in other words, least affected by species specific values while still being biologically viable, both sets of standard deviations as well as general trends will be analyzed. Looking first to model WN2, [3], the standard deviation for the infected vector maximum values was fairly low, at 0.09479, which was larger than the mean value across the six species; however, the graphical outputs were fairly inconsistent in terms of population trends. In addition, the standard deviation for R_0 values at 13.8144 was large, (in fact larger than the mean value), despite the consistency with general trends. This model as previously mentioned also fails to take into account species specific disease mortality, which could certainly cause erroneous outputs as this value ranges from 100 % mortality to 0 % depending on the species.

Examining model WN4, [1], the standard deviation for R_0 values at 68.8504 was extremely high (also larger than the mean), and the trends were inconsistent with those of the other three models. This model, as previously mentioned, fails to take into account any reservoir recovery and uses a "recruitment rate" in place of a birth rate, which has been set to zero in these trials. It also does not account for incubation time. The graphical outputs for infected vectors were very similar and produced no noticeable difference at 100 days across the six bird species. It seems likely, that without a nonzero recruitment rate or recovery rate for the reservoir species, those quantities arrive at equilibrium too quickly to have much effect on the overall dynamics of the system. Computer simulations did confirm the authors' claim of an endemic equilibrium. This model was eliminated due to the conflict between the high variance of the reproduction number, which did not translate into a corresponding effect on the proportion of infected vectors. If a good estimate of the "recruitment rate" for bird species became available, then this model might be more realistic.

Moving to model WN5, [5], the standard deviation for R_0, despite general consistency with trends, was slightly higher than WN3 at 2.6229, and the maximum I_V output deviation was the highest of any model at 0.30353. In addition, this model fails to consider incubation time of the virus, which we felt was an important feature.

Consequently model WN3, [4], was selected as the optimal predictor of real-world disease transmission for multi-species or unknown avian reservoir analysis. Its trends for modeling infected vector population were graphically consistent and it had the second lowest standard deviation at 0.12059, although this is still quite close to the mean value. Not only was this model consistent with the general trends of highest and lowest disease reproduction rates, but it most importantly also had the smallest standard deviation of R_0 at 2.3154. It's nature as a reservoir frequency dependent model also makes it a more believable model than WN2, [3]. Finally, this model accounts for all populations and all constants, which no other model in the study does.

5. The Human/Vector/Reservoir model

The model we construct here combines the vector-reservoir model of [4] with the human submodel in [1]. This hybrid model incorporates all three animal populations and effectively demonstrates transduction of the disease from bird to mosquito to human. The resulting box model is depicted in Figure 5.

The top section of the model is the transmission of West Nile virus between bird (reservoir) and mosquito (vector) while the bottom section is between mosquito and humans. These two schemes are connected by the population of infected vector, which contracts the virus from the reservoirs

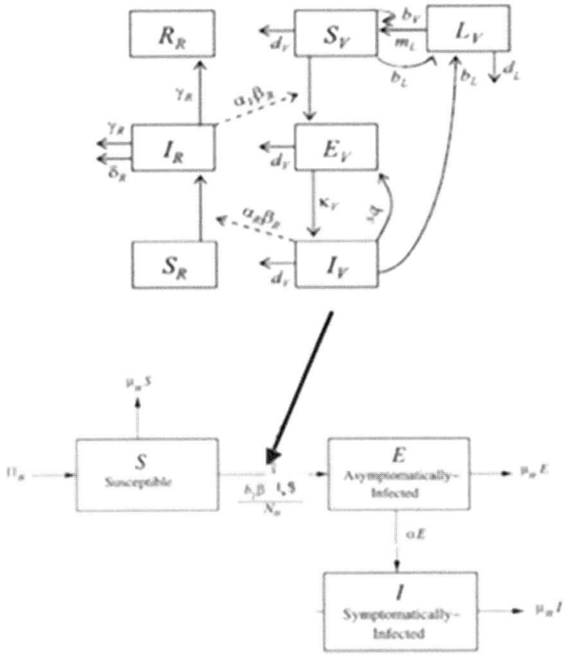

Figure 5. The box model for the full system based on WN3.

and transmits it to humans. Quantities represented in Figure 5 include the populations of larval vector (L_V), susceptible vector (S_V), exposed vector (E_V), infected vector (I_V), susceptible reservoir (S_R), infected reservoir (I_R), recovered reservoir (R_R), susceptible human (S), asymptomatically infected human (E), and symptomatically infected human (I). The resulting system of equations is given here.

Change in larval vector = birth - (death and maturation)

$$\frac{dL_V}{dt} = b_L N_V - (m_L + d_L)L_V \tag{26}$$

Change in susceptible vector = -loss to exposed vectors + maturation - death

$$\frac{dS_V}{dt} = -\alpha_V \beta_R \frac{I_R}{N_R} S_V + m_L L_V - d_V S_V \tag{27}$$

Change in exposed vector =
gain from susceptible - progress to infection - death

$$\frac{dE_V}{dt} = \alpha_V \beta_R \frac{I_R}{N_R} S_V - (\kappa_V + d_V) E_V \tag{28}$$

Change in infected vectors = gain from exposed - death

$$\frac{dI_V}{dt} = \kappa_V E_V - d_V I_V \tag{29}$$

Change in susceptible reservoir = loss due to infection

$$\frac{dS_R}{dt} = -\alpha_R \beta_R \frac{S_R}{N_R} I_V \tag{30}$$

Change in infected reservoir = gain from infection - loss due to recovery or death

$$\frac{dI_R}{dt} = \alpha_R \beta_R \frac{S_R}{N_R} I_V - (\delta_R + \gamma_R) I_R \tag{31}$$

Change in recovered reservoir = gain from recovery of infected individuals

$$\frac{dR_R}{dt} = \gamma_R I_R \tag{32}$$

Change in susceptible humans = -loss due to infection - death

$$\frac{dS}{dt} = -\frac{b_2 \beta_3 I_V S}{N_H} - \mu_H S \tag{33}$$

Change in asymptomatic humans = gain due to infection - death - progression to symptomatic

$$\frac{dE}{dt} = \frac{b_2 \beta_3 I_V S}{N_H} - \mu_H E - \alpha E \tag{34}$$

Change in symptomatic humans = progression to symptomatic - death

$$\frac{dI}{dt} = \alpha E - \mu_H I \tag{35}$$

All of the parameters for vector/reservoir dynamics remain the same as in Section 3. The American Crow was determined to be the most lethal reservoir with the highest transmission rate and constants representing death and recovery for this reservoir were therefore used in our analysis to determine potential human epidemic.

Another set of constants describing the transmission terms between vector and human are reported by Bowman, [1] and are found in Table 4.

Table 4. Constants for vector and human transmission

Parameter	Description	Value
$1/\mu H$ Average human lifespan (days)	70*365	
b_2	Bite rate of humans by mosquitoes per day	0.09
β_3	Probability of mosquito/human transmission per bite	0.88
$1/\alpha$	Incubation period (days)	14
δ	Hospitalization rate (days)	1

A typical output for this system is given in Figure 6.

6. Control of an outbreak through spraying

The analysis was modeled after the New York City outbreak in 1999 in which the disease was carried into the population by bird reservoirs that had migrated to the location. In the model, therefore, the infected reservoir has a small nonzero initial population whereas the infected vectors and infected humans start at zero. The recruitment rate of the susceptible human population was set to zero, reflecting the short time frame of an outbreak. The output allowed a determination of the population curves for all subpopulations and, of interest, the infected reservoir, infected mosquito, and infected human populations.

These trends were then used in the second part of the analysis to determine the effectiveness of prevention techniques. The model focuses on insecticide spraying of adult mosquitoes, which stems from actual treatment used in New York City in 1999 and Sacramento in 2005. These massive sprayings of Pyrethrin, a biodegradable, organic insecticide, were conducted over large masses of urban land and carried out as one-time treatment events [6]. The effects of insecticide persist for a while after spraying, so in this

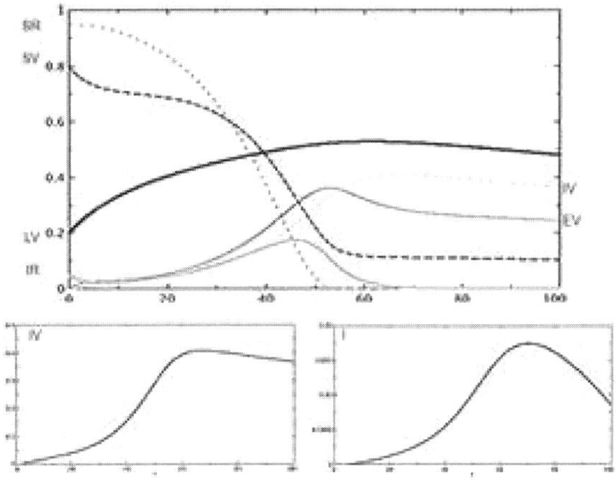

Figure 6. The full vector/reservoir/human model for American Crow parameters. The top graph is the vector/reservoir submodel. The lower left (IV) is the proportion of mosquitoes infected. The lower right (I) is infected humans.

study we incorporate a continuous proportional mortality rate from spraying, varying the mortality rate and the timing of the start of spraying with reference to the disease cycle.

In one treatment, the insecticide concentration killed 30% of the vectors per day while the in the other, it killed 70% of the vectors per day. The two concentrations of insecticide are sprayed at 20 days and at 40 days, when the number of cases is rising as well as at 60 days, the peak of human cases (visible in Figure 6). 20 days represents a scenario in which insectide was immediately sprayed after the first WNV cases were confirmed because the CDC reports that it takes a minimum of 20 days to confirm WNV [7]. Combined, these trials allowed a determination of insecticide concentration and timing in order to maximize effect in decreasing human transmission of the virus while minimizing the use of insecticide.

The spraying trials were conducted by first running a normal transmission cycle. The populations for all relevant species were then determined at the target time for spraying (20, 40, or 60 days). These populations were taken and used as the initial conditions for the subsequent insecticide prevention trials. The spray was incorporated into the model by adding a term to the adult vector populations (S_V, E_V, and I_V), which either expressed 30% killing ($-0.3I_V$) or 70% killing $-0.7I_V$. The trials were restarted with the additional term and the change in infected vectors and infected humans was noted. Table 5 shows the results of these numerical experiments.

Table 5. Effect of Different Prevention Strategies on Infected Vector and Human Populations

Time of spraying	Kill rate	maximum % infected vector	Time of peak IV	maximum % infected human	Time of peak I
n.a.	none	41%	67 days	1.7 %	70 days
20	30%	4.6%	20 days	.14 %	25 days
20	70%	4.5%	20 days	.13 %	22 days
40	30%	14%	40 days	.58 %	44 days
40	70%	14%	40 days	.56 %	42 days
60	30%	40%	60 days	1.5 %	62 days
60	70%	40%	60 days	1.5 %	61 days

7. Discussion

The first trials tested infected vector and symptomatic humans in a population exposed to infected West Nile reservoirs. As seen in Figure 6, infected vector population reaches 41% of total vectors and then levels off to a predicted equilibrium state. This population of infected vectors is very high and leads to a peak of 1.7% of humans showing West Nile virus symptoms. In NYC, with a population of 10 million, this would signify adverse effects in 170,000 people. The peak of symptomatic humans (70 days) also comes after the peak of infected vectors (67 days) confirming the transmission scheme of reservoirs to vectors to humans. With symptomatic populations reaching epidemic levels, the insecticides were implemented to reduce virus transmission.

To determine the effect of spray concentrations, insecticide quantities were used that would either kill 30% of existing vectors or 70% of existing vectors at the time of spraying. When comparing 30% and 70% spray concentrations within the same spray time, the infected vectors did indeed decrease at a faster rate with higher concentration of insecticide. However, the effect that this vector death rate had on the symptomatic human population was of little significance. For example, at the 20 day spray time, the 30% insecticide concentration led to a peak of .14% symptomatic humans while the 70% insecticide concentration led to a peak of .13% symptomatic humans. These small differences in human populations between spray concentrations were also seen in the 40 day and 60 day trials. Therefore, the concentration of insecticide does not greatly effect the variation in human epidemic and can thus be reduced to amounts that will not adversely effect the environment.

To determine the effect of the time at which insecticide was sprayed over the course of human infection trends, spraying was started at either 20, 40, or 60 days after introduction of the virus to the population via infected reservoirs. In contrast to the spray concentration results, the time of insecticide spraying had a significant effect on symptomatic human populations. The 20 day trials represent spraying at the first sign of human infection. When sprayed at this time, the insecticide reduced infected vectors to zero within 20 days and reduced symptomatic humans to a peak of .14%, down from 1.7% without spray. The 40 day spray time was chosen as a midway point between the first symptomatic human showing and the peak of symptomatic humans in the no spray trial. While this trial reduced the peak of symptomatic humans from the no spray trial to .58%, it did not reduce the epidemic as much as the 20 day trial. Lastly, the 60 day trial, which occurred at the peak of symptomatic humans in the no spray trial had the least effect in reducing symptomatic human populations. It left a peak of 1.5%, which was almost equivalent to the original symptomatic human population. These results demonstrate that the timing of the spray is critical to its effectiveness in reducing human disease. Although all spray times reduce the infected vector to zero, the earlier the spray the more it prevents virus transmission to the human population. Our results show that the effect of insecticide depends far more on the early application of spray than on the amount of pesticide used. A concentration of insecticide that reduces 30% of adult mosquitoes at 20 days after the first symptomatic humans are identified is sufficient to reduce the maximum number of human infections by over 90%. It takes approximately 20 days for the CDC to confirm West Nile virus. This model suggests that large urban areas begin preparing to spray as soon as they suspect the first WNV cases, and spray as soon as they are confirmed. During this waiting period, health professionals in the area should be advised to closely monitor the population for West Nile symptoms. Our data shows that this early mosquito reduction is crucial in preventing an epidemic of WNV in humans. We also note that a 30% reduction rather than 70% since 70% reduction does not result in significantly fewer symptomatic humans.

Acknowledgments

The authors wish to acknowledge the generosity of the Neukom Institute, the National Science Foundation Epscor Program, the local chapter of the Association for Women in Mathematics and the Dartmouth Mathematics Department for supporting Jocelyn Drexinger to present this paper at the Society for Mathematical Biology Annual Meeting 2010.

References

1. C. Bowman, A. B. Gumel, P. Van den Driessche, J. Wu, and H. Zhu, *Bulletin of Mathematical Biology* **67 5**, 1107-1133 (2005).
2. M. J. Wonham, M. A. Lewis , J. Rencławowicz and P. van den Driessche. *Ecology Letters* **9 6**, 706-725 (2006).
3. D. M. Thomas and B. Urena *Mathematical and Computer Modelling* **34 7-8**, 771-781 (2001).
4. M.J. Wonham, T. de-Camino-Beck, and M. A. Lewis, *Proc. R. Soc. Lond.* **271 1538**, 501-501 (2004).
5. G. Cruz-Pacheco, L. Esteva, and Montaõ-Hirose, *Bulletin of mathematical biology 67 6*, 1157-1172 (2005).
6. D.E.A. Elnaiem, K. Kelley, S. Wright, R Laffey, G. Yoshimura, M. Reed, G. Goodman, T. Thiemann, L. Reimer, W. K. Reisen *et al Journal of Medical Entomology* **45 4**, 751-757 (2008).
7. Centers for Disease Control and Prevention, Division of Vector-Born Illness, *http://www.cdc.gov/ncidod/dvbid/westnile/surv&control.htm*

A MODIFIED LESLIE-GOWER PREDATOR-PREY MODEL WITH HYPERBOLIC FUNCTIONAL RESPONSE AND ALLEE EFFECT ON PREY

CLAUDIO ARANCIBIA-IBARRA and EDUARDO GONZÁLEZ-OLIVARES

Grupo de Ecología Matemática, Instituto de Matemáticas,
Pontificia Universidad Católica de Valparaíso, Chile
E-mail: clanarib@gmail.com, ejgonzal@ucv.cl

This work deals with a modified Leslie-Gower type predator-prey model considering two important aspects for describe the interaction: the functional response is Hollling type II and the Allee effect acting in the prey growth function.
With both assumptions, we have a modification of the known May-Holling-Tanner model, and the model obtained has a significative difference with those model due the existence of an equilibrium point over y-axis, which is an attractor for all parameter values.
We prove the existence of separatrix curves on the phase plane dividing the behavior of the trajectories, which have different $\omega - límit$. System has solutions highly sensitives to initial conditions To simplify the calculus we consider a topologically equivalent system with a minor quantity of parameters. For this new model, we prove that for certain subset of parameters, the model exhibits biestability phenomenon, since there exists an stable limit cycle surrounding a singularities of vector field or an stable positive equilibrium point.

1. INTRODUCTION

This work deals with a continuous-time predator-prey considering that:

i) The functional response is the hyperbolic type[23],

ii) the growth of the predator population is of logistic form[18,23], and

iii) the growth of the prey population is affected by the Allee effect[7,22].

This last assumption characterize the Leslie-Gower type models[14,19] in which the environmental carrying capacity K_y is a function of prey size x[18,23], that is, it depends on the available resources. Usually it is assumed that $K_y = K(x) = nx$, i.e., the carrying capacity is proportional to the prey abundance, just as it is considered in the May-Holling-Tanner model[1,11,21].

We will assume that the predators have an alternative food when the quantity of prey diminish, this is, $K(x) = nx + c$, obtaining a modified Leslie-Gower model[2,15], being c the measure of their alternative food. For $x = 0$, then $K(x) = c$, concluding that the predator is generalist since it has an alternative food[2].

The predator functional response or consumption function refers to the change in attacked prey density per unit of time per predator when the prey density changes. In many predator-prey models considered in books, it is assumed that the functional response grows monotonic, being the inherent assumption the more prey in the environment, the better for the predator[23].

In this work, the predator functional response is expressed by the hyperbolic function $\varphi(x) = \frac{q\,x}{x+a}$, corresponding to the Holling type II[23]. Here, the parameter a is a measure of abruptness for the functional response[12]. If $a \to 0$, the curve grows quickly, while if $a \to K$, the curve grows slowly, i. e., a bigger amount of prey is needed to obtain $\frac{q}{2}$.

On the other hand, any ecological mechanism that may lead to a positive ratio between measurable component of individual fitness and the population number or density may be called a mechanism of Allee effect[6,22].

This phenomenon has been also called *depensation* in Fisheries Sciences[4,17], or *negative competition effect*, *inverse density dependence* or *positive density dependence* in Population Dynamics[17].

Populations can exhibit Allee effect dynamics due to a wide range of biological phenomena, such as reduced antipredator vigilance, social thermoregulation, genetic drift, mating difficulty, reduced antipredator defense, and deficient feeding at low densities; however, several other causes may lead to this phenomenon (see Table 2.1 in[7]).

Many algebraic forms have been used for describing the Allee effect[13,24], but we use the most common to describe this phenomenon. However it is possible to prove the topological equivalence between them[13].

In turn, the existence and uniqueness of limit cycles are interesting subjects in predator-prey models, but the determination of the amount of limit cycles that may arise by the bifurcation of a center focus singularity[5] is not a easy task; this matter is still without answer for the predator-prey models[16] just as it happens with the Hilbert 16th Problem for polynomial differential equations systems[10,16].

The main goal of this work is to describe the behaviour of model establishing its bifurcation diagram[3] based on the parameter values and to study the number of limit cycles.

2. THE MODEL

We analyze the modified Leslie-Gower type predator-prey model, described by the autonomous bidimensional differential equations system

$$X_\mu : \begin{cases} \frac{dx}{dt} = \left(r\left(1 - \frac{x}{K}\right)(x-m) - \frac{q}{x+a}y\right)x \\ \frac{dy}{dt} = s\left(1 - \frac{y}{nx+c}\right)y \end{cases} \tag{1}$$

which is of Kolmogorov type system[9], where $x = x(t)$ and $y = y(t)$ indicate the prey and predator population sizes, respectively (measure in number of individuals, density or biomass); $\mu = (r, K, q, m, a, s, n, c) \in \mathbb{R}^8_+$ and parameters have the following meanings:

r is the intrinsic prey growth rate,

K is the environmental carrying capacity,

q is the consuming maximum rate per capita of the predators,

a is the amount of prey to reach half of q,

s is intrinsic predator growth rate,

n is the food quality and it indicates how the predators turn eaten prey into new predator births, and

c is the amount of alternative food available for the predators.

The parameter m satisfies that $-K \leq m << K$. When $m > 0$, the population growth rate decreases if the population size is below the level m and the population goes to extinction, i.e., m the minimum of viable prey population or extinction threshold. In this case, the prey population is affected by a *strong* Allee effect[24]. If $m \leq 0$, it is said that the population is affected by a *weak Allee effect*[7]. In fisheries the same phenomena are called *critical* and *pure depensation*, respectively[4,17].

The parameter $c > 0$ indicates that the predator is generalist and that if it does not exist available prey it has a source of alternative food.

As system (1) is of Kolmogorov type, the coordinates axis are invariable sets of the model and it is defined in the first quadrant:

$$\Omega = \{(x, y) \in \mathbb{R}^2 / \ x \geq 0, \ y \geq 0\} = \mathbb{R}^+_0 \times \mathbb{R}^+_0$$

The equilibrium points of the system (1) or singularities of vector field X_μ are $Q_0 = (0, 0)$, $Q_m = (m, 0)$, $Q_K = (K, 0)$, $Q_c = (0, c)$ and $Q_e = (x_e, y_e)$, the positive equilibrium points (in the interior of the first quadrant) satisfying the equations of the isoclines $y = nx + c$ and $y = \frac{r}{qx}\left(1 - \frac{x}{K}\right)(x - m)(x + a)$.

To simplify the calculus we follow the methodology used in[14,19,21], doing the change of variable and the time rescaling, given by the function φ : $\breve{\Omega} \times \mathbb{R} \longrightarrow \Omega \times \mathbb{R}$; so that

$$\varphi(u,v,\tau) = \left(Ku, Knv, \frac{\left(u + \frac{a}{K}\right)\left(u + \frac{c}{Kn}\right)}{rK}\tau \right) = (x,y,t)$$

and $\det D\varphi(u,v,\tau) = \frac{Kn\left(u + \frac{a}{K}\right)\left(u + \frac{c}{Kn}\right)}{r} > 0$; hence, φ is a diffeomorphism[3] and the vector field X_μ, in the new coordinates, is topologically equivalent to the vector field $Y_\nu = \varphi \circ X_\mu$, having the form $Y_\nu = P(u,v)\frac{\partial}{\partial u} + Q(u,v)\frac{\partial}{\partial v}$; the associated differential equation system is given by the polynomial system of fifth degree:

$$Y_\nu : \begin{cases} \frac{du}{d\tau} \ (u + C)\left((1-u)(u-M)(u + A) - Qv \right) u \\ \frac{dv}{d\tau} \ S(u+A)(u + C - v)\ v \end{cases} \tag{2}$$

with $\nu = (A, M, S, C, Q) \in \Delta =]0,1[\times]0,1[\times\mathbb{R}^3_+$, where $A = \frac{a}{K} < 1$, $S = \frac{s}{rK}$, $M = \frac{m}{K} < 1$, $Q = \frac{nq}{rK}$, $C = \frac{c}{Kn}$. System (2) is defined in

$$\breve{\Omega} = \{(u,v) \in \mathbb{R}^2 / \ u \geq 0, \ v \geq 0\}.$$

The equilibrium points of system (2) or singularities of vector field Y_ν are: $P_0 = (0,0)$, $P_M = (M,0)$, $P_1 = (1,0)$, $P_C = (0,C)$ and the points lie in the intersection of the isoclines curves $v = \frac{1}{Q}(1-u)(u-M)(u + A)$ and $v = u + C$, obtaining:

$$u^3 - (M + 1 - A)u^2 - (A(M+1) - Q - M)u + (AM + CQ) = 0 \quad (\alpha)$$

Applying the Descartes signs rule, equation (α) can have two positives roots, one of multiplicity two, or none, for any sign of $A(M+1) - Q - M$, since the factors $M + (1-A) > 0$ $(A < 1)$ and $AM + CQ > 0$. Replacing $-u$ in (α) we have:

$-u^3 - (M + 1 - A)u^2 + (A(M+1) - Q - M)u + (AM + CQ) = 0,$

which has a unique change of sign; then, it exists a unique negative real root that we will denote it by $-H$.

Due to the difficult to determine the exact solutions of equation (α), we divide it by $u + H$. Denoting by $A_1 = 1 - A + H + M$ and $A_0 = Q + M - H(H - A + M + 1) - AM + 1$, the following equation is obtained:

$$u^2 - A_1 u + A_0 = 0 \qquad (\beta)$$

and the rest is $(C - H)\,Q + (H+1)\,(H+M)\,(A-H) = 0$, implying that

$$Q = \frac{1}{C-H}\,(H+1)\,(H+M)\,(H-A) > 0$$

with $C \neq H$. Then, $A > H$ and $H > C$ or else, $A < H$ and $H < C$.

The solutions of equation (β) are:

$$u_1 = \frac{1}{2}\left(A_1 - \sqrt{\Delta}\right) \text{ and } u_2 = \frac{1}{2}\left(A_1 + \sqrt{\Delta}\right)$$

with

$$\Delta = A^2 + 2AH + 2AM + 2A - 3H^2 - 2HM - 2H + M^2 - 2M - 4Q + 1.$$

For equation (β) it has:

a) If $A_1 > 0$ and $A_0 > 0$, we have three possibilities:

i) There are two positive roots u_1 and u_2, if and only if, $\Delta > 0$

ii) There is one positive root of multiplicity 2, if and only if, $\Delta = 0$.

iii) There are no real root, if and only if, $\Delta < 0$.

b) If $A_1 > 0$ and $A_0 < 0$, there is a unique positive root u_2.

c) If $A_1 > 0$ and $A_0 = 0$, there is a unique positive root $u = 1 - A + H + M$.

d) If $A_1 \leq 0$ and $A_0 > 0$ there is no positive root.

e) If $A_1 \leq 0$ and $A_0 < 0$ there is a unique positive root.

The three alternatives a i, ii and iii for equation (β) are shown in the Figure 1 a, b and c.

The Jacobian matrix of system (2) is:

$$DY_\nu(u,v) = \begin{pmatrix} DY_\nu(u,v)_{11} & -Qu\,(C+u) \\ 2Svu + Sv\,(A+C-v) & S\,(A+u)\,(C+u-2v) \end{pmatrix}$$

with

$DY_\nu(u,v)_{11} = -5u^4 + 4\,(M - C - A + 1)\,c_3 u^3 + 3c_2 u^2 + 2c_1 u + c_0$

$c_2 = A\,(M+1) - M + C\,(M - A + 1)$

$c_1 = -C\,(M - A\,(M+1)) - AM - Qv$

$c_0 = -C\,(AM + Qv)$.

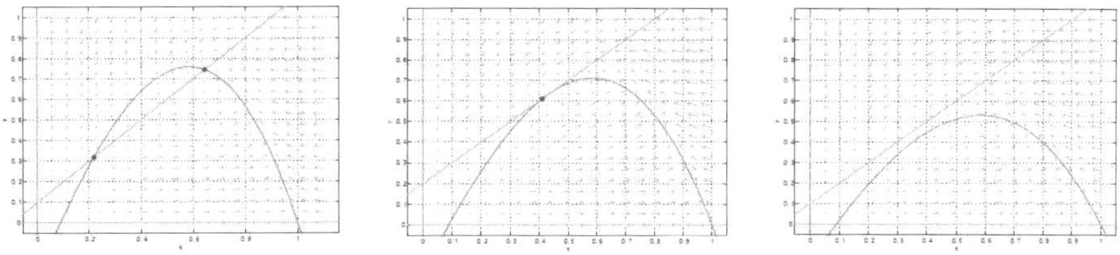

Fig 1a Fig 1b Fig 1c

Figure 1. Figure 1a: Two positive roots for equation (β). Figure 1b: One positive root of multiplicity two. Figure 1c: None real positive root.

3. MAIN RESULTS

For system (2) we have the following results:

Lemma 3.1. *The set* $\tilde{\Gamma} = \left\{ (u,v) \in \tilde{\Omega} / \ 0 \leq u \leq 1, \ v \geq 0 \right\}$ *is an invariant region*[3,20].

Proof. Clearly, the $u - axis$ and the $v - axis$ are invariant sets because the system is a Kolmogorov type[9]. If $u = 1$, we have

$$\frac{du}{d\tau} = -Qv\,(1 + C) < 0$$

and whatever it is the sign of $\frac{dv}{d\tau}$, the trajectories enter and remain in the region $\tilde{\Gamma}$. □

Lemma 3.2. *The solutions are bounded.*

Proof. We use the Poincaré compactification[3,8] to study the behavior of point $(0, \infty)$. Making a change the variables and the time rescaling given by the function $\theta : \tilde{\Omega} \times \mathbb{R} \longrightarrow \breve{\Omega} \times \mathbb{R}$, where

$$\theta\,(X, Y, T) = \left(\tfrac{X}{Y}, \tfrac{1}{Y}, Y^4 T\right) = (u, v, \tau).$$

Doing a tedious algebraic work we get the new vector field

$$\bar{U}_\nu : \begin{cases} \frac{dX}{dT} = -X\left(X^4 + a_3 X^3 + a_2 X^2 + a_1 X + a_0\right) \\ \frac{dY}{dT} = -SY^3\left(X^2 + (AY + CY - 1)\,X + AY\,(CY - 1)\right) \end{cases}$$

with

$a_3 = AY - Y + CY - MY$

$a_2 = MY^2 - CY^2 - AY^2 + SY^2 + ACY^2 - AMY^2 - CMY^2$

$a_1 = QY^2 - SY^2 - ACY^3 + AMY^3 + CMY^3 + ASY^3 + CSY^3 - ACMY^3$

$a_0 = CQY^3 - ASY^3 + ACMY^4 + ACSY^4.$

The Jacobian matrix[1] in the vector field $\bar{U}_\nu$ is

$$D\bar{U}_\nu(X,Y) = \begin{pmatrix} b_3X^3 - 5X^4 + b_2X^2 + b_1X + b_0 & D\bar{U}_{\nu 12} \\ -SY^3\left(2X + AY + CY - 1\right) & D\bar{U}_{\nu 21} \end{pmatrix}$$

with

$$D\bar{U}_{\nu 12} = X\left(c_0 + Xc_1 + X^2c_2 -^3 \left(A + C - M - 1\right)X\right)$$
$$D\bar{U}_{\nu 21} = SY^2\left(3X + 4AY - 3X^2 - 5ACY^2 - 4AXY - 4CXY\right)$$

where

$$b_3 = -4Y\left(A + C - M - 1\right)$$
$$b_2 = -3Y^2\left(-A - C + M + S + AC - AM - CM\right)$$
$$b_1 = 2Y^2\left(S - Q + Y\left(AC - AM - CM - AS - CS + ACM\right)\right)$$
$$b_0 = Y^3\left(AS - CQ - ACMY - ACSY\right)$$
$$c_3 = -2Y\left(-A - C + M + S + AC - AM - CM\right)$$
$$c_2 = Y\left(2S - 2Q + 3Y\left(AC - AM - CM - AS - CS + ACM\right)\right)$$
$$c_1 = Y^2\left(3AS - 3CQ - 4ACMY - 4ACSY\right)$$

Evaluating in the point $(0,0)$ we obtain

$$D\bar{U}_\nu\left(0,0\right) = \begin{pmatrix} 0 & 0 \\ 0 & 0 \end{pmatrix}$$

To desingularize the origin in the vector field $\bar{U}_\nu$, we apply the blowing up method[3,8]. By the change of variables $X = rw$ and $Y = w$ and making the time rescaling given by $\zeta = w^3T$, we obtain the system:

$$\breve{U}_\nu : \begin{cases} \frac{dr}{d\zeta} = r\left(-CQ - Qr - ACMw + ACMrw\right) \\ \frac{dw}{d\zeta} = Sw\left(A + r - rw - ACw - Arw - Crw\right) \end{cases}$$

The Jacobian matrix of the vector field $\breve{U}_\eta$ is:

$$D\breve{U}_\nu(r,w) = \begin{pmatrix} 2ACMrw - 2Qr - ACMw - CQ & ACMr\left(r - 1\right) \\ D\breve{U}_\nu(r,w)_{21} & D\breve{U}_\nu(r,w)_{22} \end{pmatrix}$$

with

$$D\breve{U}_\nu(r,w)_{21} = -Sdw\left(rw - r - A + ACw + Arw + Crw\right)$$
$$D\breve{U}_\nu(r,w)_{22} = -S\left(2rw - r - A + 2ACw + 2Arw + 2Crw\right).$$

Evaluating the matrix $D\breve{U}_\eta(r,w)$ in the point $(0,0)$ we obtain:

$$D\breve{U}_\nu(0,0) = \begin{pmatrix} -CQ & 0 \\ 0 & AS \end{pmatrix}$$

and $\det D\breve{U}_\nu(0,0) = -CAQS < 0$. Therefore, $(0,0)$ is a saddle point of the vector field $\breve{U}_\nu$ and of $\bar{U}_\nu$; then, the point $(0,\infty)$ is a saddle point in the compactified vector field Y_ν. Then, the orbits are bounded. □

Lemma 3.3. *Nature of equilibrium points over the axis*

For all $\eta = (A,B,C,M,Q) \in \mathbb{R}^5_+$, *the singularity*

a) $P_1 = (1,0)$ *is a saddle point.*

b) $P_0 = (0,0)$ *is a saddle point.*
c) $P_M = (M,0)$ *is a repellor point.*
d) $P_C = (0,C)$ *is an attractor point.*

Proof. Evaluating the Jacobian matrix in each point we obtain:

a) In the point $(1,0)$

$$DY_\nu(1,0) = \begin{pmatrix} -(1-M)(C+1)(A+1) & -Q(C+1) \\ 0 & S(A+1)(C+1) \end{pmatrix}.$$

Then, $\det DY_{Y_\nu}(1,0) = -S(1-M)(A+1)^2(C+1)^2 < 0$. Therefore, the singularity $(1,0)$ is saddle point.

b) In the point $(0,0)$

$$DY_\nu(0,0) = \begin{pmatrix} -ACM & 0 \\ 0 & ACS \end{pmatrix}$$

Then, $\det DY_\nu(0,0) = -CAMSAC < 0$. Therefore, the equilibrium $(0,0)$ is saddle point.

c) In the point $(M,0)$

$$DY_\nu(M,0) = \begin{pmatrix} DY_\nu(M,0)_{11} & -QM(C+M) \\ 0 & S(A+M)(C+M) \end{pmatrix}$$

and $DY_\nu(M,0)_{11} = M\left(AC(4-M) + (1-M)\left(M^2 + AM + CM\right)\right) > 0$.

Clearly $\det DY_\nu(M,0) > 0$ and $\mathrm{tr} DY_\nu(M,0) > 0$. Therefore, the singularity $(M,0)$ is a repellor.

d) In the point $(0,C)$

$$DY_\eta(0,C) = \begin{pmatrix} -C(AM+QC) & 0 \\ ACS & -ACS \end{pmatrix}$$

Then, $\det DY_\nu(0,0) > 0$ and $\mathrm{tr} DY_\nu(0,0) < 0$. Therefore, the equilibrium $(0,C)$ is an attractor. □

3.1. *A particular case of weak Allee effect*

Due to the difficulty to determine explicitly the nature of the positive equilibrium point in system (2), now we will consider only the case in which $M = 0$ and $C = 0$, (or $m = 0$ and $c = 0$ in system (1)), obtaining the polynomial system[11]

$$Y_\lambda : \begin{cases} \frac{du}{d\tau} = ((1-u)\,u\,(u+A) - Qv)\,u^2 \\ \frac{dv}{d\tau} = Bv\ (u-v)(u+A) \end{cases} \tag{3}$$

with $\lambda = (A,B,Q) \in \mathbb{R}^3_+$. The equilibrium points of system (3) are $(0,0)$, $(1,0)$ and (u_e, v_e) over the isoclinic curves

$v = \frac{1}{Q}(1-u)u(u+A)$ and $v = u$.

We note that the point over the $u - axis$ (and $y - axis$ in system (1)) is absent; in this case those point coincides with the origin. The abscise of positive equilibrium points satisfy the quadratic equation:

$$u^2 - (1-A)\,u + (Q-A) = 0 \qquad (\gamma)$$

with $1 - A > 0$. The Jacobian matrix of system (3)

$$DY_\lambda(u,v) = \begin{pmatrix} DY_\lambda(u,v)_{11} & -Qu^2 \\ Bv\,(2u+A-v) & B\,(u+A)\,(u-2v) \end{pmatrix}$$

where $DY_\lambda(u,v)_{11} = -u\left(5u^3 - 4\,(1-A)\,u^2 - 3Au + 2Qv\right)$. System (3) is topological equivalent to the modified Leslie-Gower predator system:

$$X_\eta : \begin{cases} \frac{dx}{dt} = \left(r\,\left(1 - \frac{x}{K}\right)x - \frac{q}{x+a}y\right)x \\ \frac{dy}{dt} = s\left(1 - \frac{y}{nx}\right)\,y \end{cases} \qquad (4)$$

with $\eta = (r, K, q, a, s, n) \in \mathbb{R}^6_+$ and parameters have the same meanings to system (1). In system (3), the equilibrium point $(0,0)$ is the collapse of the equilibrium points $(0,0)$ and $(M,0)$ of system (2). Its nature can be determined by the blowing-up method[3,8].

For system (3) we have the following properties:

Lemma 3.4. *The point* $(0,0)$ *is a non-hyperbolic singularity of system* (3), *which has a stable manifold* $W^s\,(0,0)$ *determining a separatrix curve, obtaining an hyperbolic sector and a parabolic sector at the neighborhood of this equilibrium point*[20].

Proof. Based on the properties of the points $(0,C)$, $(M,0)$ and $(0,0)$ of system (2), we can see that the point $(0,0)$ has a sector attractor and other sector repellor. □

This result implies that the point $(0,0)$ is an attractor for all the trajectories with initial conditions that are above the separatrix curve and for the trajectories with initial conditions that are below this curve, then the origin is a non-hyperbolic saddle point.

Lemma 3.5. *i) It exists two positive equilibrium points* (u_1, u_1) *and* (u_2, u_2) *with*

$$u_1 = \frac{1}{2}\left(1 - A - \sqrt{(1+A)^2 - 4Q}\right) \text{ and } u_2 = \frac{1}{2}\left(1 - A + \sqrt{(1+A)^2 - 4Q}\right),$$

if and only if, $Q < \frac{(A+1)^2}{4}$ *and* $Q > A$.

ii) It exists a unique positive equilibrium point $(u_c, u_c) = (1 - A, 1 - A)$, *if and only if,* $Q = A$.

iii) It exists a unique positive equilibrium point (u_2, u_2) *with*
$u_2 = \frac{1}{2}\left(1 - A + \sqrt{(1 + A)^2 - 4Q}\right)$, *if and only if,* $Q < A$.

iv) It exists a unique positive equilibrium point (E_c, E_c) *with* $E_c = \frac{1}{2}(1 - A)$, *if and only if,* $Q = \frac{(A+1)^2}{4}$.

v) It doesn't exist a positive equilibrium point if and only if, $Q > \frac{(A+1)^2}{4}$.

Proof. The number of equilibrium points of system (3) depends of equation $(\gamma)^{11}$. By the Descartes sign rule, this equation can have two positive real roots, one of multiplicity two or none, according the sign of $Q - A$.

a) If $Q - A > 0$, equation (γ) has

i) Two positive roots, if and only if, $\Delta = (1 - A)^2 - 4Q > 0$.

ii) One positive root of multiplicity two, if and only if, $\Delta = 0$.

iii) None real root if and only if, $\Delta < 0$.

b) If $Q - A = 0$, there exists a unique positive root $u = 1 - A$ and $u = 0$.

c) If $Q - A < 0$, there are one positive and one negative root. □

When it doesn't exist positive equilibrium point we have the following:

Lemma 3.6. *If* $Q > \frac{(A+1)^2}{4}$, *then the point* $(0, 0)$ *is globally asymptotically stable.*

Proof. As the point $(1, 0)$ is always a saddle point and it does not exist positives equilibrium points, we have that $(0, 0)$ is an attractor for all the trajectories. □

3.1.1. *Nature of positive equilibrium points*

As the positives singularities lie at the curve $v = u$; then,

$$Q = (1 - u)(u + A)$$

and we have the Jacobian matrix is:

$$DY_\lambda(u, u) = \begin{pmatrix} -u^2\left(-A - 2u + 2Au + 3u^2\right) - (1 - u)(u + A)u^2 \\ Bu(u + A) & -Bu(u + A) \end{pmatrix},$$

and $\det DY_\lambda(u, u) = Bu^4 (A + u)(A + 2u - 1)$.

a) If $u > \frac{1-A}{2}$, then $\det DY_\lambda(u, u) > 0$ and the nature of singularity depends on the sign of the $\operatorname{tr} DY_\lambda(u, u)$.

b) If $u < \frac{1-A}{2}$, then $\det DY_\lambda(u,u) < 0$, the point (u,u) is saddle.

When it exists an equilibrium point by collapse of two of them we have that $Q = \frac{(1+A)^2}{4}$

Theorem 3.1. *The equilibrium point* (E_c, E_c) *with* $E_c = \frac{1-A}{2}$, *is:*

i) a saddle-node attractor if $B > \frac{1-A^2}{4}$,

ii) a saddle-node repellor if $B < \frac{1-A^2}{4}$,

iii) a cusp point if $B = \frac{1-A^2}{4}$?.

Proof. The Jacobian matrix evaluated in (E_c, E_c) is

$$DY_\lambda(E_c, E_c) = \begin{pmatrix} \frac{1}{16}(A+1)^2(1-A)^2 & -\frac{1}{16}(A+1)^2(1-A)^2 \\ \frac{B}{4}(A+1)(1-A) & -\frac{B}{4}(A+1)(1-A) \end{pmatrix};$$

then, $\det DY_\lambda(E_c, E_c) = 0$ and the sign of $\mathrm{tr} DY_\lambda(E_c, E_c) = \frac{1}{16}(1-A)(A+1)\left(1-A^2-4B\right)$ depends on the factor $f(A,B) = 1-A^2-4B$. When $B = \frac{1-A^2}{4}$ the Jacobian matrix is

$$DY_\lambda(E_c, E_c) = \frac{1}{16}(A+1)^2(1-A)^2 \begin{pmatrix} 1 & -1 \\ 1 & -1 \end{pmatrix}$$

whose Jordan form matrix is[1]

$$DY_\lambda(E_c, E_c) = \frac{1}{16}(A+1)^2(1-A)^2 \begin{pmatrix} 0 & 1 \\ 0 & 0 \end{pmatrix}$$

and the Bogdanov-Takens bifurcation[25] (a codimension 2 bifurcation) is obtained and the point (E_c, E_c) is a cusp point. □

If $Q = A$ and $u_c = 1-A$, also a unique positive equilibrium point (u_c, u_c) exists.

Theorem 3.2. *The equilibrium point* $(u_c, u_c) = (1-A, 1-A)$ *is:*

i) an attractor, if and only if, $B > A(2-A)^2 - 1$,

ii) a repellor, if and only if, $B < A(2-A)^2 - 1$; *then, a limit cycle is generated by Hopf bifurcation.*

iii) a weak focus, if and only if, $B = A(2-A)^2 - 1$.

Proof. As the Jacobian matrix is

$$DY_\lambda(u_c, u_c) = \begin{pmatrix} -\left(A^2-3A+1\right)(1-A)^2 & -A(1-A)^2 \\ (1-A)B & -(1-A)B \end{pmatrix}.$$

Then $\det DY_\lambda(u_c, u_c) = B(1-A)^5 > 0$ and the behavior depends on

$$\mathrm{tr} DY_\lambda(1-A, 1-A) = (1-A)\left(A(2-A)^2 - 1 - B\right).$$ □

We note that in the last case, a one-parameter system is obtained described by:

$$Y_\eta : \begin{cases} \frac{du}{d\tau} = ((1-u)u(u+A) \; - \; Av)\, u^2 \\ \frac{dv}{d\tau} = \left(A\,(2-A)^2 - 1\right) v\; (u-v)(u+A) \end{cases} \tag{5}$$

Moreover, a bistable phenomenon exists in system (5), since the point $(1-A, 1-A)$ can be a local attractor or repellor surrounded of a limit cycle, meanwhile the point $(0,0)$ is local attractor for a wide subset of trajectories. In system (2) or in system (1) it must be a similar phenomenon when it exists a unique positive equilibrium point, since $(0, C)$ is an attractor.

Now, supposing that

$u_1 = E = \frac{1}{2}\left(1 - A - \sqrt{(1+A)^2 - 4Q}\right) < 0,$

the unique equilibrium point is given by

$u_2 = F = \frac{1}{2}\left(1 - A + \sqrt{(1+A)^2 - 4Q}\right)$

Assuming that $Q < A$, we obtain $(1-F)(F+A) < A$, i.e., $F(1-A-F) < 0$. Clearly $F > \frac{1-A}{2}$, and system (3) becomes

$$Y_\rho : \begin{cases} \frac{du}{d\tau} = ((1-u)(u+A)\,u \; - (1-F)(F+A)\,v)\,u^2 \\ \frac{dv}{d\tau} = Bv\,(u-v)(u+A) \end{cases} \tag{6}$$

Theorem 3.3. *In system (6) the unique equilibrium point (F,F) is:*

i) an attractor, if and only if, $B < \frac{F(A(1-2F)+F(2-3F))}{(A+F)}$,

ii) a repellor, if and only if, $B > \frac{F(A(1-2F)+F(2-3F))}{(A+F)}$*; then, a limit cycle is generated by Hopf bifurcation.*

iii) a weak focus, if and only if, $B = \frac{F(A(1-2F)+F(2-3F))}{(A+F)}$.

Proof. The Jacobian matrix is

$DY_\rho\,(F,F) =$

$$\begin{pmatrix} -F^2\left(-A-2F+2AF+3F^2\right) & -(1-F)(A+F)\,F^2 \\ (A+F)\,FB & -(A+F)\,FB \end{pmatrix},$$

with $\det DY_\rho(F,F) = BF^4\,(A+F)\,(A+2F-1) > 0$, and

$\operatorname{tr} DY_\eta(F,F) = -(A+F)\,B - F\,((2F-1)\,A + F\,(3F-2))$

i) When $\operatorname{tr} DY_\eta(F,F) < 0$, the point (F,F) is local attractor

a) if $F \geq \frac{2}{3}$, for all value of B, and

b) If $F <$, then it must fulfill that $B > \frac{F(A(1-2F)+F(2-3F))}{(A+F)}$.

ii) When $\operatorname{tr} DY_\eta(F,F) > 0$ and $F\,(A\,(1-2F) + F\,(2-3F)) > 0$, then $B < \frac{F(A(1-2F)+F(2-3F))}{(A+F)}$; newly, $F\,(A\,(1-2F) + F\,(2-3F)) > 0$, then

a) $F < \frac{1}{2}$ and $0 < A < 1$, or

b) If $\frac{1}{2} < F < \frac{2}{3}$, then $A < \frac{F^2(2-3F)}{2F-1}$ and (F, F) is an unstable singularity

iii) When $\text{tr}DY_\eta(F, F) = 0$, then $B = \frac{F(A(1-2F)+F(2-3F))}{(A+F)}$ and $F(A(1-2F) + F(2-3F)) > 0$,

a) this is fulfilled for all $F < \frac{1}{2}$ and $0 < A < 1$.

b) If $\frac{1}{2} < F < \frac{2}{3}$, then $A < \frac{F(2-3F)}{2F-1}$ and (F, F) is a weak focus. □

We note the existence of at least one limit cycle, via Hopf bifurcation, remaining pendent the existence of others. Now, we consider the case in which it exists a second singularity at interior of the first quadrant as it was establish in lemma 3.

Theorem 3.4. *The equilibrium point* $(u_1, u_1) = (E, E)$ *is a hyperbolic saddle.*

Proof. The determinant of Jacobian matrix evaluated at (E, E) is

$$\det DY_\rho(E, E) = BE^4(A + E)(A + 2E - 1)$$

As $E = \frac{1}{2}\left(1 - A - \sqrt{(1+A)^2 - 4Q}\right) < \frac{1-A}{2}$, $\det DY_\rho(E, E) < 0$ and the singularity (E, E) is saddle point. □

Theorem 3.5. *a) It exists conditions on the parameter values for which an homoclinic curve is determined by the stable and unstable manifold of point* (u_1, u_1).

b) It exists an limit cycle that bifurcates of the homoclinic, surrounding the point (u_2, u_2).

Proof. Let $W_-^s(u_1, u_1)$ and $W_+^s(u_1, u_1)$ the upper and under stable manifolds of (u_1, u_1), respectively; let $W_+^u(u_1, u_1)$ and $W_-^u(u_1, u_1)$ the right and left unstable manifolds of (u_1, u_1), respectively.

a) Since $\bar{\Gamma}$ is an invariant region, the trajectories can not cross the line $u = 1$ towards the right. As consequence of the existence and uniqueness theorem, the trajectory determined by the right unstable manifolds $W_+^u(u_1, u_1)$ can not intersect the trajectory determined by the upper stable manifold W_+^u of this point.

Moreover, the $\alpha - limit$ of the upper stable manifold $W_+^s(u_1, u_1)$ must be the point (u_2, u_2) or lies at infinity; hence, the $\omega - limit$ of the trajectory determined by $W_+^u(u_1, u_1)$ must be:

i) the point (u_2, u_2), when this point is an attractor or else, a stable limit cycle, if it exists, when (u_2, u_2) is a repellor.

ii) the point $(0, 0)$, when the point (u_2, u_2) is a node repellor and this point is the $\alpha - limit$ of the upper stable manifold $W_+^s(u_1, u_1)$.

Then, it exists a set of parameter values for which the trajectories determined by $W_+^u(u_1, u_1)$ and $W_+^s(u_1, u_1)$ intersect and the homoclinic curve is obtained.

b) When the homoclinic is broken, a limit cycle is generated, whose stability must be determined (bifurcation of cycle graph[10]). □

We note that, the limit cycle generated by the broken of the homoclinic curve, whose existence was showed in the above theorem, could to coincide with the limit cycle obtained via Hopf bifurcation.

4. CONCLUSIONS

In this work, we partially analyzed a modified Leslie-Gower predator-prey model[2,15] due the complexities algebraic arising by consider that the prey population is affected by a Allee effect or depensation[7,4] and predators have a alternative food.

Using a diffeomorphism[3], we analyzed a topologically equivalent system, depending of five parameters establishing the local stability of the equilibrium points that lies over the axis. We proved that the point $(1, 0)$ in system (2) (and its associated $(K, 0)$ in system (1)) is always a saddle point for all parameter values.

Due the difficulty of calculus for studying the behavior of positive equilibrium points, we consider a particular case in which we assume $c = 0$ and $m = 0$. In this particular case, we have that the point $(0, 0)$ is non-hyperbolic singularity due the collapse of the attractor point $(0, c)$ and the saddle point $(0, 0)$ in the original system (1); then, the origin determines a separatrix curve dividing the behavior of trajectories in the phase plane.

The existence of the separatrix implies that the trajectories are highly sensitive to initial conditions, some of which having $(0, 0)$ as its $\omega - limit$ and other a stable positive equilibrium point or a stable limit cycle, depending on the parameter values.

If the ratio prey-predator is high (many prey and little predator) then the populations can coexist for a wide set of parameter values, but this ratio is low then it exists a great possibility that two populations can go to extinction.

We also show that the dynamics of the model with Allee effect differs of the May-Holling-Tanner model[1,21], in which the Allee effect is absent, since our model can have two equilibrium points at interior of the first quadrant, one of which (u_1, u_1), is always a saddle point.

We conjecture of this model that has two limit cycles for a certain parameter values and that the model considering strong Allee effect has similar dynamics, as it is shown in the following simulations.

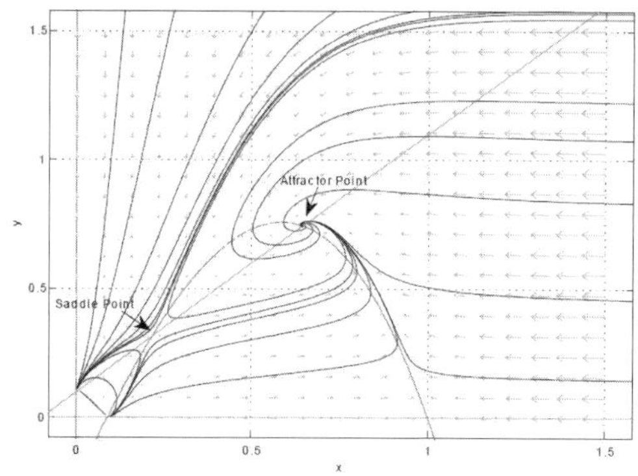

Figure 2. The point (u_2, u_2) is a local attractor and (u_1, u_1) is saddle point. Moreover $(0, 0)$ is also local attractor and $(M, 0)$ is repellor.

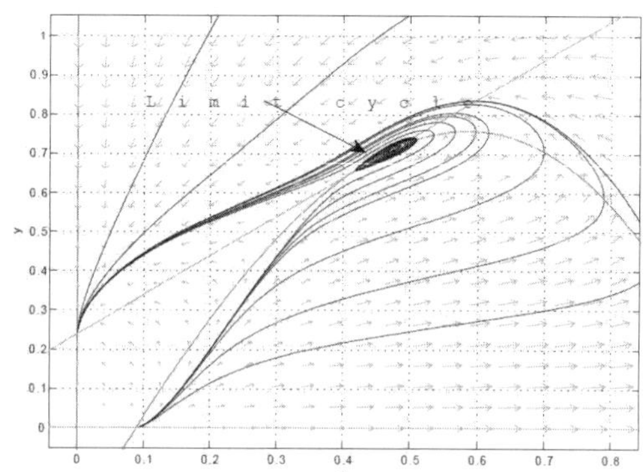

Figure 3. The point (u_2, u_2) is a repellor surrounded by a limit cycle, (u_1, u_1) is saddle point, $(0, 0)$ is attractor and $(M, 0)$ is repellor.

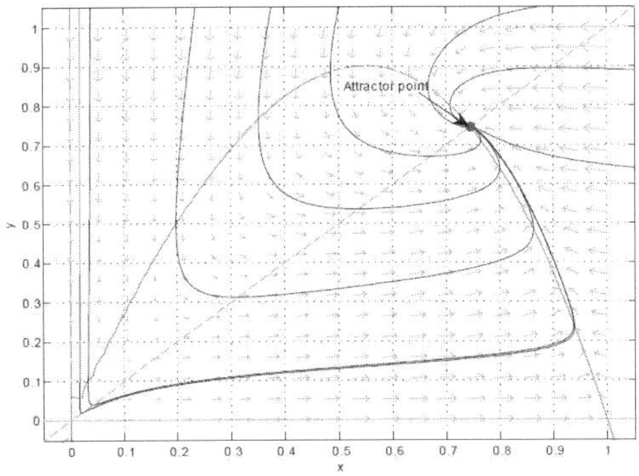

Figure 4. The unique positive equilibrium point (u_2, u_2) is a local attractor, $(0,0)$ is attractor and $(M,0)$ is repellor.

Acknowledgments

Authors thank to the members of the Mathematics Ecology Group from the Mathematics Institute at Pontificia Universidad Católica de Valparaíso for their valuable comments and suggestions.

References

1. D. K. Arrowsmith and C. M. Place, *Dynamical Systems. Differential equations, maps and chaotic behaviour*, Chapman and Hall (1992).
2. M. A. Aziz-Alaoui and M. Daher Okiye,*Applied Mathematics Letters* **16**, (2003) 1069-1075.
3. C. Chicone, *Texts in Applied Mathematics* **34**, Springer (2006).
4. C. W. Clark, *Mathematical Bioeconomic: The optimal management of renewable resources* (2nd edition), John Wiley and Sons (1990).
5. C. S. Coleman, *Differential Equations Model*, Springer Verlag, (1983) 279-297.
6. F. Courchamp, T. Clutton-Brock and B. Grenfell, *Trends in Ecology and Evolution* **14** (1999) 405-410.
7. F. Courchamp, L. Berec and J. Gascoigne, *Allee effects in Ecology and Conservation*, Oxford University Press 2007.
8. F. Dumortier, J. Llibre and J. C. Artés, *Qualitative theory of planar differential systems*, Springer, 2006.
9. H. I. Freedman, *Deterministic Mathematical Model in Population Ecology*, Marcel Dekker (1980).
10. V. Gaiko, *Global Bifurcation Theory and Hilbert s Sexteenth Problem*, Kluwer Academic Press (2003).
11. L. M. Gallego-Berrío, *Consecuencias del efecto Allee en el modelo de depredación de May-Holling-Tanner*, Tesis de Maestría, Maestría en Biomatématicas Universidad del Quindío, Armenia, Colombia (2004).

12. W. M. Getz, *Ecology* **77**(7) (1996) 2014-2026.
13. E. González-Olivares, B. González-Yañez, J. Mena-Lorca and R. Ramos-Jiliberto. *Proceedings of the 2006 International Symposium on Mathematical and Computational Biology BIOMAT 2006*, Rio de Janeiro (2007) 53-71.
14. B. González-Yañez, E. González-Olivares, and J. Mena-Lorca, *BIOMAT 2006 - International Symposium on Mathematical and Computational Biology*, World Scientific Co. Pte. Ltd., (2007) 359-384.
15. A. Korobeinikov, *Applied Mathematical Letters* **14** (2001) 697-699.
16. Y. Li and D. Xiao, *Solitons and Fractals* **34** (2007) 8606–8620.
17. M. Liermann and R. Hilborn, *Depensation: evidence, models and implications, Fish and Fisheries* **2** (2001) 33-58.
18. R. M. May, *Stability and complexity in model ecosystems* (2nd edition), Princeton University Press (2001).
19. J. Mena-Lorca, E. González-Olivares and B. González-Yañez, *Proceedings of the 2006 International Symposium on Mathematical and Computational Biology BIOMAT 2006*, E-papers Serviços Editoriais Ltda. (2007) 105-132.
20. L. Perko, *Differential equations and dynamical systems*, Springer (1991).
21. E. Sáez and E. González-Olivares, *SIAM Journal of Applied Mathematics* **59** (1999) 1867-1878.
22. P. A. Stephens, W. J. Sutherland and R. P. Freckleton, *Oikos* **87** (1999) 185-190.
23. P. Turchin, *Monographs in Population Biology* **35** Princeton University Press (2003).
24. G. A. K. van Voorn, L. Hemerik, M. P. Boer and B. W. Kooi, *Mathematical Biosciences* **209** (2007) 451–469.
25. D. Xiao and S. Ruan, Bogdanov-Takens *Field Institute Communications* **21** (1999) 493-506.

CONTROL AND SYNCHRONIZATION OF CHEMOTAXIS PATTERNING AND SIGNALING

HECTOR PUEBLA, SERGIO A. MARTINEZ-DELGADILLO

Departamento de Energía, Universidad Autonoma Metropolitana, Azcapotzalco
Av. San Pablo No. 180, Reynosa-Tamaulipas
Azcapotzalco, 02200, D.F. México
E-mail hpuebla@correo.azc.uam.mx

ELISEO HERNANDEZ-MARTINEZ

Departamento de IPH, Universidad Autonoma Metropolitana, Iztapalapa
Apartado Postal 55-534
Iztapalapa, 09340, D.F. México
E-mail cbi2063808074@xanum.uam.mx

AMERICA MORALES-DIAZ

Robótica y Manufactura Avanzada, Cinvestav Saltillo,
km 13 Carr. Saltillo-Monterrey,
Ramos Arizpe, Coahuila 25900, México
E-mail america.morales@cinvestav.edu.mx

Chemotaxis patterns and chemotaxis signaling are crucial for proper functioning of cellular systems. In this work we explore the external forcing for control and synchronization of chemotaxis pattern formation and chemotaxis signaling via a feedback control approach. Two nonlinear, one-dimensional, parabolic equations are used to describe chemotaxis considering volume filling and quorum sensing effects. The chemotaxis signaling in *Dictyostelium discoideum* is described with a set of coupled nonlinear ODE. In the first case, applying suitable external forcing to the cell population variable, we can successfully control chemotaxis patterns to a desired behavior. In the second case, via the manipulation of the substrate influx to the cell signaling system we can control the ATP concentration dynamics.

1. Introduction

An essential characteristic of living organisms is the ability to sense signals in the environment and adapt their movement accordingly [3]. The signals are processed by the intracellular signaling network [4]. A signal that is produced in the wrong time or place will lead to inappropriate responses, which can be dangerous and cells must be protected against this [5,24]. Obviously, signaling events must be precisely regulated. Many biological systems have the ability to sense the direction of external chemical sources and respond by polarizing and migrating toward chemoattractants or away from chemorepellants [18]. This phenomenon, referred to as chemotaxis, is crucial for proper functioning of single-cell organisms, such as bacteria and amoebae, and multi-cellular systems as complex as the immune and nervous systems [3,4]. Chemotaxis also appears to be important in wound healing and tumor metastasis [16,17]. This central role of chemotactic movement in the cell makes it an especially attractive cellular process for feedback control studies. The potential benefits of controlling the chemotaxis patterning are known to be significant [5,13,19,25,11]. For instance, it is recognized that many biological systems use quorum sensing systems to regulate behavior, and failure of such mechanisms can result in abnormal functioning [20].

The chemotactic response of cells to a single attractant or repellent has been characterized experimentally for many cell types and has been extensively studied from a theoretical standpoint [3,4]. For instance, the cellular slime mold *Dictyostelium discoideum* (Dd hereafter) is a widely used system for the study of many developmental processes, including extracellular communication, signal transduction, chemotaxis, pattern formation, and differentiation [7]. In the presence of an adequate food supply the amoebas exist as free-ranging individuals, but when the food supply is exhausted an elaborate developmental program is initiated. After a period of starvation, cells attain relay competence, by which it is meant that they can detect an external CAMP pulse and respond to it by synthesizing and releasing CAMP [7,14]. Chemotaxis has attracted a great deal of computation and modeling attention [8,15,17]. Models for chemotaxis have been successfully applied to bacteria, slime molds, skin pigmentation patterns, leukocytes and many other examples [12,16,17,18].

In this work, using both a PDE model of chemotactic movement [9,20], which displays a rich array of patterns, and a model of coupled Dd cells, we introduce a simple model-based feedback control approach to control the spatiotemporal behavior in chemotaxis. An interesting feature of the PDE model, which is a coupled set of non-linear partial differential equations, is that the model includes two mechanisms to prevent blow up: a volume filling effect and quorum sensing considerations. Numerical simulations on both a 1D chemotaxis patterns and 10 coupled Dd cells shows the effectiveness of the proposed control approach.

This work is organized as follows: In Section 2, we present the PDE model of chemotaxis patterning and the basic model of Dd cells and its coupling. In Section 3 we introduce our control approach for control and synchronization of chemotaxis patterns. Numerical simulations in Sections 4 shows the control and synchronization capabilities of our control approach. Finally, some concluding remarks are given in Section 5.

2. Mathematical models of chemotaxis patterns and signaling

2.1. *PDE model of chemotaxis patterns*

The most known and used model for chemotaxis is the Keller-Segel model, which has been introduced by Keller and Segel in 1971. The possibility of blow up has been shown in the Keller-Segel model [9]. Blow up denotes a solution with a maximum that grows to infinity in finite-time. After blow up has occurred the model is no longer appropriate, for instance to describe bacteria aggregation [20]. Painter and Hillen (2001) proposed a modification of the Keller-Segel equations such that density effects can be included, *e.g.*, some bacteria release a quorum sensing molecule, which allows for detection of the local cell density and prevent blow up.

We consider the following model for chemotactic movement proposed by Painter and Hillen (2002),

$$\frac{\partial u}{\partial t} = D_u \nabla^2 u - \nabla \cdot \{u\chi(u,v)\nabla v\} + f(u,v)$$

$$\frac{\partial v}{\partial t} = D_v \nabla^2 v + g(u,v) \qquad (1)$$

where $u(x,t)$ represents the density of the cell population, and $v(x,t)$ represents the chemoattractant (repellent) concentration. The chemotactic component is represented by the negative-cross diffusion term in the cell density equation, where $\chi(u,v)$ is commonly referred to as the chemotactic sensitivity. Zero-flux boundary condition on a bounded domain are considered. D_u and D_v are the diffusion coefficients for cell population and the chemical respectively. The functions $f(u,v)$ and $g(u,v)$ express local kinetics of the variables u and v. For the functions $\chi(u,v)$, $f(u,v)$ and $g(u,v)$ we consider the following forms [20],

$$\chi(u,v) = \chi_0(1-u)$$

$$f(u,v) = ru(1 - u/u_m)$$

$$g(u,v) = \gamma u - \delta v \tag{2}$$

This model has been studied in Hillen and Painter, (2001) and Painter and Hillen, (2002) where global existence of the solution has been proven and interesting phenomena of pattern formation and formation of stable aggregates was shown. Besides, in Painter and Hillen, (2002) it was demonstrated how model (1) can naturally arise by incorporating appropriate biological details, including the space available for cells to migrate and the response of cells to quorum-sensing molecules.

We consider the one dimensional case of the model (1) with both zero and non-zero cell kinetics. These simulations demonstrate interesting spatial and temporal behavior. Initially we set $u(x,0) = u_0$ constant and a spatial perturbation about the homogeneous steady state for the chemical concentration, *i.e.*, $v(x,0) = v_0 + 0.005\sin(2\pi x)$. We have focus on the one-dimensional case as a first step to gain insights for design effective and simple feedback control schemes for two and three dimensional cases of model (1).

The following comments are in order:

- The chemotaxis model excludes blow-up and permits global existence independently of thresholds or of space dimensions. Blow-up is undesirable from a biological standpoint, since it implies the formation of cell aggregates of infinite density [9]. Furthermore, the highly stiff nature of the problem can create many difficulties from a computational standpoint [20]. In the context of pattern formation, a process leading to "blow-up" indicates permissibility of aggregative behavior: *i.e.*, self-organization is possible [16].
- The derivation of model (1) relies on the hypothesis that populations (*e.g.*, of bacteria) possess some form of regulatory mechanism which allows them to control the size of the aggregate [20]. Indeed, such behavior is crucial for populations such as *Dictyostelium*, which accumulate into fruiting bodies of up to 10^5 cells [4].
- Pattern formation due to chemosensitive movement is seen in bacteria such as *E. coli* and *S. typhimurium*. *E. coli* have been shown to form a range of patterns, including rings and spots as they spread out in a nutrient environment [4,16,18]. In other applications chemotaxis equations have been applied to several processes of pattern

formation in development, for example in the formation of pigmentation patterns [17].

2.2. *Coupled ODE model of chemotaxis signaling*

The mechanism that underlies the periodic synthesis of cAMP in Dd involves a positive feedback loop: extracellular cAMP binds to a cell surface receptor and thereby, via the action of G proteins, activates adenylate cyclase, the enzyme that transforms ATP into cAMP. Intracellular cAMP thus synthesized is transported into the extracellular medium where it becomes available for binding to its receptor [7,14]. As indicated by models for cAMP signaling in Dd, the self-amplification resulting from this positive feedback loop is at the core of the instability mechanism leading to oscillations.

A model for cAMP signaling has been proposed, on the basis of receptor desensitization and self-amplification in cAMP synthesis [7]. This three-variable model for cAMP signaling in Dd accounts for periodic oscillations of cAMP and for relay of supra threshold cAMP pulses. The model also predicts the possibility of bursting and of aperiodic synthesis of cAMP signals [14]. While simple periodic oscillations can occur when the level of ATP remains constant, complex oscillations were found only when the level of ATP was allowed to vary, even though the range of variation of ATP remained reduced. The mechanism leading from simple periodic behavior to bursting and chaos relies here on the interplay between two endogenous oscillatory mechanisms [6].

The time evolution of the model is governed by the following system of three nonlinear differential equations [7]:

$$\frac{d\alpha}{dt} = v - \sigma\phi(\alpha, \rho_T, \gamma)$$

$$\frac{d\rho_T}{dt} = f_2(\gamma) - \rho_T[f_1(\gamma) + f_2(\gamma)]$$

$$\frac{d\gamma}{dt} = q\sigma\phi(\alpha, \rho_T, \gamma) - k_e\gamma \qquad (3)$$

with

$$f_1(\gamma) = \frac{k_1 + k_2\gamma^2}{1+\gamma^2}, \qquad f_2(\gamma) = \frac{k_1 L_1 + k_2 L_2 c^2 \gamma^2}{1 + c^2\gamma^2}$$

$$\acute{q} = \frac{qk_t}{h(k_i + k_t)}, \qquad \phi(\rho_T, \gamma, \alpha) = \frac{\alpha(\lambda\theta + \frac{\epsilon\rho_T\gamma^2}{1+\gamma^2})}{(1 + \alpha\theta) + (\frac{\epsilon\rho_T\gamma^2}{1+\gamma^2})(1 + \alpha)}$$

where, α represents the dimensionless intracellular concentration of the substrate ATP, ρ_T denotes the fraction of active cAMP receptor, and γ the dimensionless concentration of extracellular cAMP. The parameters υ and σ are the substrate input and maximum rate of cAMP production, k_i and k_e relate to cAMP hydrolysis by intracellular and extracellular phosphodiesterase, respectively, while k_t measures the rate of cAMP transport into the extracellular medium, q, λ, θ, c, and ϵ are ratios of kinetic or dissociation constants, while the parameters L_1 and L_2 are ratios of rate constants relating to receptor desensitization and resensitization [7].

We consider that there are N subsystems in a lattice x_{ij}, $i = 1,..,3$, and $j = 1, 2, .., N$, with diffusive coupling as follows

$$\dot{x}_{1j} = \varepsilon(x_{1j+1} - 2x_{1j} + x_{1j-1}) + f_{1j}(x_i) + u_j, \; j = 1, ..., N$$

$$\dot{x}_{kj} = f_{kj}(x_i), \qquad k = 2, 3 \tag{4}$$

The following comments are in order:

- The analysis of the model based on receptor desensitization showed that there exists an optimal pattern of pulsatile signaling that maximizes the capability of cells to synthesize cAMP in response to an extracellular cAMP pulse [7,14].
- With regard to the physiological situation, the domain of interest is that of chaos and of simple periodic oscillations. Indeed, only there is the period of the order of minutes, as observed in experiments [6].

3. Robust Feedback Control Design

In this section, it is presented a robust feedback control approach for both control and synchronization of the dynamic behavior of the PDE model and ODE coupled oscillators as described above. The control methodology is based on a modelling compensation approach, (MEC) [21,1]. The underlying idea behind MEC control designs is to lump the input-output uncertainties into a term, which is estimated and compensated via a suitable algorithm. The key feature of MEC control is that simple practical control design with good robustness and performance capabilities is obtained.

3.1. *Control and synchronization problems*

In both cases, the PDE model and the ODE coupled oscillators, the control problem and synchronization problems are related to explore the forcing of chemotaxis patterns and signaling. Control and synchronization problems addressed in this paper can be stated as the regulation or synchronization of chemotaxis patterns to a desired dynamic behavior via manipulation of an external input.

The control and synchronization problems description are completed by the following assumptions:

- **A1** The measurement of the variable to be controlled or synchronized y, is available for control or synchronization design purposes.

- **A2** Nonlinear functions are uncertain.

- **A3** Diffusive and the coupling function are unknown.

3.2. *Control of chemotaxis patterns*

In order to control the spatiotemporal behavior of the cell density, we introduced an external input, $c(t)$ which can be considered as a variable influx of $u(t, x)$ across a semipermeable membrane. The influx of $u(t, x)$ can be controlled by adjusting the concentration of an external reservoir of u. Then, to introduce an external input in the cell population variable, Eq. (1) is changed by

$$\frac{\partial u}{\partial t} = D_u \nabla^2 u - \nabla \cdot \{u\chi(u, v)\nabla v\} + f(u, v) + c(t)$$

$$\frac{\partial v}{\partial t} = D_v \nabla^2 v + g(u, v) \tag{5}$$

An external manipulation of such processes is interesting from a methodical perspective related to control of complex spatiotemporal systems and may eventually even turn out to be of practical importance in biological applications like tissue engineering or drug development [13].

The parameters of model (1) are the diffusion coefficients D_u and D_v, parameters of local kinetics and the chemotactic sensitivity which in some situations are to abstract and hard to estimate from experiments [8,20]. In order to obtain a simple input-output model for control design purposes we define a modeling error as follows

$$\eta(t) = D_u \nabla^2 u - \nabla \cdot \{u\chi(u,v)\nabla v\} + f(u,v) \tag{6}$$

such that

$$\frac{\partial u}{\partial t} = \eta(t) + c(t) \tag{7}$$

It can be seen that the modeling error function contains the uncertain terms (*i.e.*, diffusion coefficient, local kinetics, and the chemotactic term).

Let u_r be a desired reference, for the control law we proposed the following simple inverse feedback action

$$c(t) = -(\eta(t) + \tau_c^{-1}(u - u_r) - \dot{u}_r) \tag{8}$$

such that

$$\frac{\partial u}{\partial t} - \frac{\partial u_r}{\partial t} = -\tau_c^{-1}(u - u_r) \tag{9}$$

Such that the closed-loop dynamics (9) is stable and $u \to u_r$ asymptotically with τ_c as the mean convergence time.

Since $c(t)$ can not be implemented just as it is because the modeling error signal $\eta(t)$ contains uncertain terms, we propose the following gradient-like estimator in order to use an estimate signal $\widetilde{\eta}$ instead the real one η,

$$\frac{\partial \widetilde{\eta}}{\partial t} = \tau_e^{-1}(\eta - \widetilde{\eta}) \tag{10}$$

where τ_e is the estimation time constant. From (7), we know that $\eta = \partial u/\partial t - c$. Hence

$$\frac{\partial \widetilde{\eta}}{\partial t} = \tau_e^{-1}(\frac{\partial u}{\partial t} - c(t) - \widetilde{\eta}) \tag{11}$$

introduce the variable $w \overset{def}{=} \tau_e \widetilde{\eta} - u$. Then, the estimator (11) can be realized as follows:

$$\frac{\partial w}{\partial t} = -c(t) - \widetilde{\eta}, \qquad \widetilde{\eta} = \tau_e^{-1}(w + u) \tag{12}$$

which is initialized as follows. Since η is unknown, we have that $\widetilde{\eta}_0 = 0$. Therefore, $w_0 = -u_0$. The computed control law is given by

$$c(t) = -(\widetilde{\eta}(t) + \tau_c^{-1}(u - u_r) - \dot{u}_r) \tag{13}$$

for consider possible physical restrictions in the magnitude of external stimulus we include a saturation function given by

$$c(t)^r = Sat(c(t))$$

$$Sat(c(t)) = \begin{cases} c_{\min} \; if & c(t) \leq c_{\min} \\ c(t) \; if & c_{\min} < c(t) < c_{\max} \\ c_{\max} \; if & c(t) \geq c_{\max} \end{cases}$$

thus, the control input is limited by $c_{\min}$ for the minimum external signal and $c_{\max}$ for the maximum external signal for the application of an external stimuli on the population variable.

3.3. *Control of chemotaxis signaling*

The control objective is the regulation or synchronization of the coupled dynamics of the dimensionless intracellular concentration of the substrate ATP α_j, y^j to a desired reference y^j_{ref} via the manipulation of substrate input $v_j = c_j$. Let $e_j = y^j - y^j_{ref}$ be the regulation or synchronization error, and define the modeling error function η_j as,

$$\eta_j = \varepsilon(x_{1j+1} - 2x_{1j} + x_{1j-1}) + f_{1j}(x_i) \tag{14}$$

notice that the complete functions $f_{1j}(x)$ are considered unknown, as the worst-case design. System (4) and (14) can be written as,

$$\frac{dy^j}{dt} = \eta_j + c_j \tag{15}$$

where η_j are the modeling error functions of the N lattice subsystems. In this way, similar to the control design based for the PDE model presented above, the modeling error functions are estimated with uncertainty estimators, which after some algebraic manipulations can be written as,

$$\dot{w}_j = c_j + \dot{y}^j_{ref} - \widetilde{\eta}_j$$

$$\widetilde{\eta}_j = \tau_{ej}^{-1}(w_j + e_j) \tag{16}$$

where $e_j(t) = y^j(t) - y^j_{ref}(t)$ is the synchronization error. The inverse dynamics feedback control function is given by,

$$u_i = -\left[\eta_j + \tau_{cj}e_j + \dot{y}^j_{ref}\right] \tag{17}$$

Notice that the resulting feedback control design for synchronization (16) and (17) purposes depends only on measured signals $\{y^j, \dot{y}^j_{ref}\}$ and estimated values of parameter C_{mj}, and do not relies on a good mathematical model of system.

The following comments are in order:

- The following tuning guidelines can be borrowed from Alvarez-Ramirez, (1999): In a first stage, set the value of the closed-loop time constant $\tau_c > 0$. τ_c can be chosen as the inverse of the dominant frequency of the chemotaxis patterns or oscillations. In a second stage, set the estimation time constant $\tau_e > 0$. The closed-loop time constant τ_e determines the smoothness of the modeling error and the velocity of the time-derivative estimation respectively.
- The stability analysis of the closed-loop systems is beyond of the scope of this work. However, this can be borrowed with stability arguments from singular perturbation theory and energy methods for distributed parameter systems [10,23].

4. Numerical Simulations

In this section, simulation results are presented for the control of the PDE chemotaxis model a the synchronization of 10 coupled Dd cells. Our simulation results indicate good control and synchronization performance of the closed-loop system. Although a rigorous robustness analysis is beyond the scope of this study, numerical simulations will show that our feedback control approach is able to control and synchronize chemotaxis dynamics despite significant parameter uncertainties and external disturbances.

4.1. *Control of chemotaxis patterns*

Control tasks include the regulation to a constant reference, *i.e.* suppression of oscillatory behavior, and tracking of a sinusoidal signal, *i.e.* enforcing to a new oscillatory behavior. In both cases the control action is activated at $t =$ 50 time units. We have set the control design parameters ω_c and ω_e as 2.0 and 5.0 respectively. We have taken the case of model (1) with cell kinetics and the following cases in order to illustrate the control performance: (i) suppression of the "merging and emerging" pattern to a constant reference, *i.e.*, $u_r = 0.5$, and (ii) enforcing of the pattern dynamics to a desired controlled periodic pattern, *i.e.*, $u_r(x,t) = 0.5 + 0.25\sin(0.2t)$. The control law is turn on at $t = 500$ units and $\tau_c = 15.0$ and 5.0 and $\tau_e = 5.0$ and 0.25 respectively.

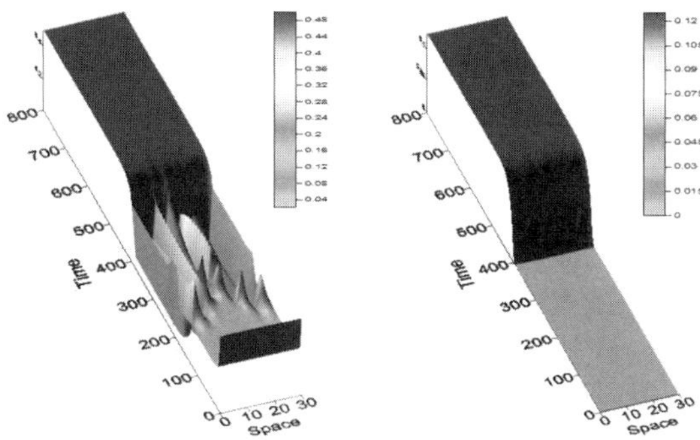

Figure 1. (a) Surface map of the controlled evolution of u to a constant reference. (b) Corresponding surface map of the control input c.

Figures 1 and 2 shows surface maps of the cell density $u(x,t)$ variable and the control input $c(x,t)$ for cases (i) and (ii) respectively. Color range form clear at minimum values of u and c to dark at maximum values of u and c. It can be seen from Fig. 1 that we can successfully perform suppression of the pattern dynamics with a simple step input of the influx of u. Figure 2 shows that in order to obtain a periodic pattern of the cell population it is required the periodic influx of u. Besides, Figures 1 and 2 shows that control inputs no require much effort to drive the system to the new patterns given by references. The mechanism of wave suppression can be explained as follows: The influx of u leads to a chemotactic response that gives rise to a waves that collide and annihilate. This effect is combinated with a gradually decrease due to diffusion processes to finally attain cell population uniformity. For the wave forcing to a desired periodic pattern the periodic influx of u leads to a periodic behavior of the cell density and prevents a constant chemotactic response that could suppress the wave propagation phenomena and the influx of u is module in order to maintain the desired periodic pattern. It should be stressed that chemical wave activity also depends on the dimensions of the space available for cells, competing waves and local and global forcing or other perturbation.

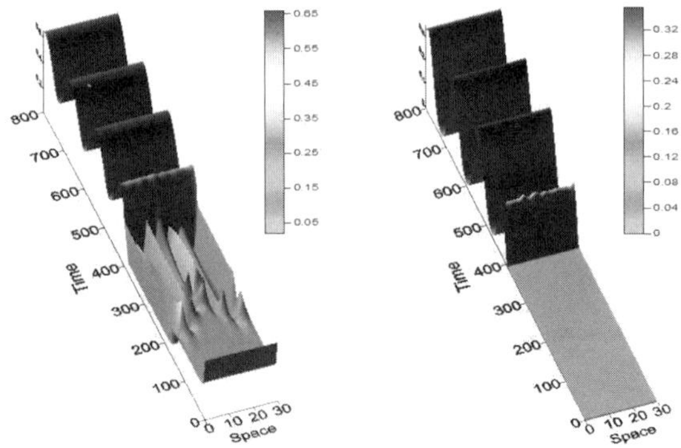

Figure 2. (a) Surface map of the controlled evolution of u to a periodic reference. (b) Corresponding surface map of the control input c.

4.2. *Synchronization of chemotaxis signaling*

We consider the synchronization of individual Dd cells in an array of coupled Dd cells to a desired periodic oscillations. The simulation results is shown in Fig. 3. The control is connected at $t = 50$ and $y^j_{ref}(t) = 1.13 + 0.1\sin(0.5t)$, $i = 1, \ldots, N$. We have set the control design parameters τ_c and τ_e as 0.35 and 0.1 respectively. It can be seen from Fig. 4 that the control input is a simple pattern of the influx of v. The mechanism of the synchronization of the array of Dd cells via the periodic influx of the substrate input is a consequence of the periodic production of the cAMP attractant. Moreover, numerical results can be related to the observation that tactic locomotion in Dd cells requires an oriented pulse of attractant, which suggests that the underlying intracellular reactions may be coupled to an oscillator [13].

5. Conclusions

In this work, we have presented a simple feedback control design for the control of chemotaxis patterns in 1D and the synchronization of nonlinear dynamics in Dd cells. Since some parameters of the chemotaxis model are strongly uncertain, we have proposed a control law structure with modeling error compensation. We have shown via numerical simulations how the chemotaxis patterning and the chemotaxis signaling can be controlled or

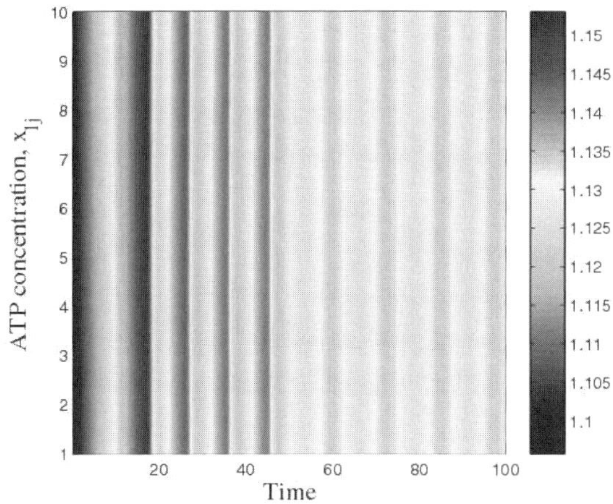

Figure 3. Synchronization of 10 coupled Dd cells.

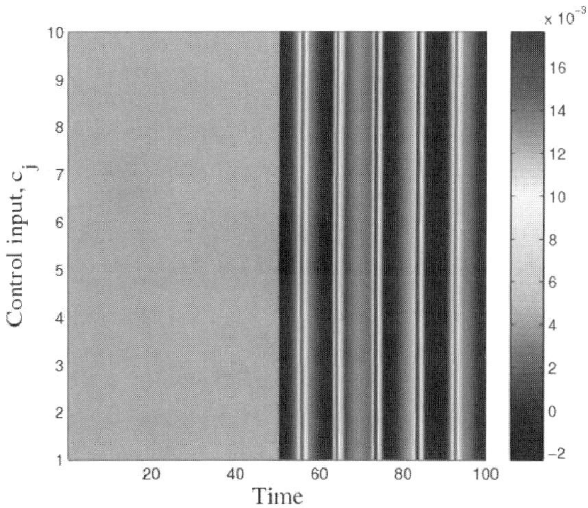

Figure 4. Corresponding control input for Figure 3.

synchronized to a desired dynamics. Numerical simulations shown that external forcing in chemotaxis can be efficiently exploited to induce a great variety of dynamical behavior. In spite our results have been obtained for the 1D case and a simple model of chemotaxis signaling, we expect that our conclusions will be valid for 2D and 3D cases, and more complex models of chemotaxis signaling. It should be stressed that we have focus on the control of chemotaxis patterns by using an external input applied on the cell population variable. However, feedback control can be also achieved by variations that are caused either by processes within the system or by appropriately designed parametric perturbations with the control approach developed in this work.

References

1. Alvarez-Ramirez, J. *Int. J. Robust Nonlinear Cont.*,**9**, 361 (1999).
2. Alvarez-Ramirez, J., Puebla, H., Ochoa-Tapia, J.A. *Syst. Cont. Lett.*, **44** 395 (2001) 395.
3. Bonner, J.T. *Princeton Univ. Press, Princeton* (1967).
4. Chung, C.Y., Firtel, R.A. *Humana Press Inc., Totowa, NJ.*
5. Freeman, M. *Nature* **408,** 313 (2000).
6. Goldbeter, A. *Nature* **420**, 238 (2002).
7. Goldbeter, A. *Cambridge University Press, Cambridge, UK* (1996).
8. Hillen, T. *Math. Models and Methods in Appl. Sci.* **12**, 1007 (2002).
9. Hillen, T., Painter, K.J. *Adv. Appl. Math.* **26,** 280 (2001).
10. Hoppensteadt, F. *J. Diff. Eq.* **15**, 510 (1974).
11. Iglesias, P. *European J. Control.* **9**, 227, (2003).
12. Levine, H., Ben-Jacob, E. *Phys. Biol.* **1**, 14 (2004).
13. Lebiedz, D., Brandt-Pollmann, U. *Chaos* **15**, 23902 (2005).
14. Li, Y.X., Halloy, J., Martiel, J.L., and Goldbeter, A. *Chaos* **2**, 501 (1992).
15. Keller, E.F., Segel, L.A. *J. Theor. Biol.* **30**, 225 (1971).
16. Maini, P.K., Painter, K.J., Phong-Chau, H.N. *J. Chem. Soc. Faraday Trans.* **93**, 3601 (1997).
17. Murray, J.D. *Springer-Verlag, Berlin* (1988).
18. Meinhardt, H. *Academic Press, London* (1982).
19. Mohanty, S., Firtel, R. *Cell & Developmental Biology*, **10**, 597 (1999).
20. Painter, K.J., Hillen, T. *Canadian Appl. Math. Quart.* **10**, 502, (2002).
21. Puebla, H. *J. Biol. Sys.* **13**, 173 (2005).
22. Segel, L.A. *Cambridge University Press, Cambridge, U.K.* (1980).
23. Straughan, B. *Springer-Verlag, New York* (1992).
24. Walleczek, J. *Cambridge University Press, Cambridge, UK* (2002).
25. Yi, T.-M., Huang, Y., Simon, M. I., Doyle, J. *Proc. Natl. Acad. Sci.*, 4649 (2000).

THE OPTIMAL THINNING STRATEGY FOR A FORESTRY MANAGEMENT PROBLEM

ALEJANDRO ROJAS-PALMA and EDUARDO GONZÁLEZ-OLIVARES
Grupo de Ecología Matemática, Instituto de Matemáticas
Pontificia Universidad Católica de Valparaíso, Chile
E-mails: alejandro.rojas.p@mail.ucv.cl, ejLgonzal@ucv.cl

The forests represent an important type of renewable resource and its optimal management has become a important present problem.
The thinning is a process that simply reduces the volume of standing timber. Besides the wood obtaining by means of the operation, the thinning process improves the index of growth and the quality of the forest.
In this paper, we study a bioeconomic model, where the control is the thinning variation rate, rather than the thinning in the classical model. This consideration leads to a nonlinear problem of optimal control. We also obtain the optimal strategy of the model.

1. Introduction

In recent years, bioeconomics has been developed quickly in different areas from the modeling of natural resources, so much renewable as non-renewable. The fast increase of the world population causes to the corresponding increment in the exploitation and the consumption of the natural resources[9].

For a long time, the natural resources such as fisheries and forests, were considered inexhaustible resources, reason why they have been over-exploited, causing a considerable reduction of the stock resources. As a result of this fact, the use and the optimal management of the resources acquire major importance[7].

In some regions of the world still the forests without control or in open access regime are exploited, although almost all the countries have forestal policies whose explicit objective is to conserve forests. The politicians in charge agree to stop the uncontrolled exploitation, but not all understand the same thing about conservation and management[6]. The interpretation of the term and the implantation of forestal policies to conservation have given rise to numerous controversies. The forests as fisheries have always had a complex function in the national economies and local economies, because they provide very variable goods and services, and, inevitably, some uses enter in conflict with others. Some historians have even seen in these conflicts over the use of the forests the origin of the conservation concept[6].

The forests provide an ample range of benefits at local, national and world-wide level. Some of these benefits depend on which the forest remains unperturbed or undergoes a minimum alteration. Others can only be obtained extracting of their woods and other products. The absolute protectionism that rejects the access to renewable resources and leaves them untouchable is so pernicious for the sustainable development of countries like the uncontrolled explotation that does not allow the regeneration of these resources

In this work, a model for the optimal management of forests is formulated. An optimal policy of management looking to reconcile the always present conflict between the forest like an economic value, this is, a capital value to be exploited, and like a biological resource that must be protected and be preserved, as much for ecological reasons as for social reasons.

2. The Model

In a study of optimal thinning and rotation of forests of Scottish pine in Finland, Kilkki and Vaisanen[12] developed a nonlinear discrete-time model which was analyzed by means of standard techniques of dynamic programming. Clark and De Pree[4] studied a linearized continuous-time version which was analyzed by means of the optimal control theory. Let $x = x(t)$ denoting the volume of timber in a given forest stand and we suppose that the growth of this forest is modeled by the differential equation

$$\frac{dx}{dt} = g(t)\, f(x) \tag{1}$$

where $g(t)$ represents the growth coefficient, which is considered decreasing in the time, $f(x)$ represents the growth function, which is positive and has a unique maximum. In[12] the growth coefficient of forest used is

$$g(t) = at^{-b}$$

and the growth function has the form

$$f(x) = xe^{-cx}$$

where a, b, c are positive constants.

Thinning is a process that simply reduces the volume $x(t)$ of standing timber by means of its exploitation[4]. If $u(t)$ represents the rate of thinning effort, then the differential equation that models this dynamics is

$$\frac{dx}{dt} = g(t) f(x) - u(t) \tag{2}$$

It is assumed that the forest is thined to a rate $u(t) \geq 0$ for $t_0 \leq t \leq T$ and the remaining stand is then clearcut at age T, moment at which begins a new rotation period (successive replantation).

The objective functional of the control problem was defined by

$$J_1 = \int_{t_0}^{T} R(t) u(t) dt + e^{-\delta T} q(T) x(T) \tag{3}$$

where $R(t)$ represents the utility obtained by means of the reduction process and $q(T)$ represents the unitary value of timber in the time T. It is assumed in addition that the cutting costs differ from the reduction costs, reason why $R(T) \neq q(T)$. The form of the functional objective (3) of the optimal control problem defined in[4] is known as Bolza form[11].

If it is assumed that all the subsequent rotations are characterized by the same growth functions and the same economic relations of the first rotation, the maximization of the present value of all the future rotations implies that the optimal thinning and the rotation policy will be the same for each rotation. Then, the total present value is determined by the costs functional

$$J = \sum_{k=0}^{\infty} J_1 e^{-k\delta t} = \frac{J_1}{1-e^{-\delta t}} \tag{4}$$

The singular solution $x^*(t)$ of the control problem given by (3) subject to (2) is

$$f'(x^*) = \frac{1}{g(t)} \left[\delta - \frac{R'(t)}{R(t)} \right] \tag{5}$$

If $f'(x)$ is monotone, the equation (5) determines a unique singular path $x^*(t)$. The optimal thinning policy then consists of a combination of bang-bang and singular controls.

In[14] the same problem is analyzed, but without considering rotation. Defining the cost functional

$$J = \int_{t_0}^{\infty} e^{-\delta t} R(t) u(t) dt \tag{6}$$

where

$$R(t) = p - c \tag{7}$$

represents the difference between the constant price p obtained by unit of exploited timber and the constant cost c proportional to the thinning rate.

The functional (7) is most frequently used in economic optimization problems and is described in Lagrange form[11] and together to the system (2), they represent the following optimal control problem of infinite horizon.

Maximize:

$$J = \int_{t_0}^{\infty} e^{-\delta t} (p - c) u(t) \, dt \tag{8}$$

subject to:

$$\begin{aligned} \tfrac{dx}{dt} &= g(t) f(x) - u(t) \\ x(t) &\geq 0 \\ u(t) &\geq 0 \end{aligned} \tag{9}$$

By means of the direct application of the Pontryaguin's Maximum Principle[13] a optimal thinning policy can be obtained.

The Hamiltonian of the control problem is

$$H = (p - c)u(t) + \lambda(t)(g(t) f(x) - u(t)) \tag{10}$$

and the adjoint equation is given by

$$\tfrac{d\lambda(t)}{dt} = \lambda(t)\left[\delta - g(t) f(x)\right].$$

A unique optimal level of the resource exists, which is determined by the equation (5) considering the function defined in (7). This level is

$$f'(x^*(t)) = \tfrac{\delta}{g(t)} \tag{11}$$

then

$$u^*(t) = g(t) f(x^*(t)) - \tfrac{dx^*(t)}{dt} \tag{12}$$

gives the singular control.

The following optimal thinning strategy

$$u^*(t) = \begin{cases} 0 & \text{si } p - c > \lambda(t) \\ g(t) f(x^*) - \tfrac{dx^*(t)}{dt} & \text{si } p - c = \lambda(t) \\ \infty & \text{si } p - c < \lambda(t) \end{cases} \tag{13}$$

leads the solution, from a nonoptimal population level x_0 to the optimal trajectory.

But this thinning policy, although it is a solution of the control problem, is hardly applicable. Because, if the forest management decision maker operates with a given capital (for example: number of machinery, the flow of labor) at one time, he cannot reduce it significantly some other time. With the purpose of to evade these discontinuous jumps, in the model to analyze, we will consider like control variable of the optimization problem, the variation of the thinning rate, this implies that the thinning effort becomes a second variable of state.

It will be assumed that the dynamics of the model will be governed by a growth function of autocatalytic or logistic type

$$\frac{dx}{dt} = ax^{\gamma} - bx^{\alpha}$$

where the parameters γ and α respectively represent the constants of anabolism and catabolism[15]. For the logistic case $\alpha = 2$ and $\gamma = 1$.

Contrary to the model studied in[14], the thinning rate will be considered proportional at the level of stock of the forest resource, which seems to be a realistic hypothesis and is analogous to the capture by unit of effort in fishery models[3,4].

Let the following model:

$$X : \begin{cases} \frac{dx}{dt} = ax - bx^2 - vx \\ \\ \frac{dv}{dt} = u \end{cases} \tag{14}$$

where $x = x(t)$ denotes the volume of timber in the time, $v = v(t)$ denotes the continuous thinning effort rate, $u = u(t)$ denotes the instantaneous variation rate of the thinning effort, the parameter a represents the intrinsic growth rate of resource and $\frac{a}{b}$ represents the environment carrying capacity.

With the purpose to avoid variations too quickly, the variation rate of the thinning effort must be bounded, by a small value, this is

$$0 \leq u(t) \leq u_{\max} \quad \forall t \in \mathbb{R}_0^+ \tag{15}$$

If it is assumed that the exploited volume of timber can be sold at any time in a market, this assumption implies that there is no terminal date at which the optimization must end, the forest management decision maker maximizes the profit discounted over time by δ, which is the maximal real rate of interest in any segment of the capital market

$$J = \int_{t_0}^{\infty} e^{-\delta t} \left[(px - c)\, v(t) \right] dt \tag{16}$$

where p represents the constant price obtained by unit of exploited timber, c represents the constant cost proportional to the rate of reduction and δ represents the annual rate of discount.

The optimization problem to solve is to maximize (16) subject to the restrictions given by (14) and (15). We will demonstrate the existence and stability of the equilibrium point of the control problem, identified by the necessary conditions of optimality.

3. Main Results

Next the existence of a equilibrium point and its asymptotic stability through a permissible fedback control will be demonstrated and an optimal strategy will be described. The proof of the optimality of the strategy will be considered in a future work.

3.1. *Equilibrium existence*

In order to obtain some information about optimality candidates and to prove the existence of an equilibrium point, we apply the Pontryagin's Maximum Principle[13]. The Hamiltonian associated to the control problem defined by (14), (15) and (16) it is given by

$$H(x, v, \lambda_1, \lambda_2, u, t) = e^{-\delta t} \left[(px - c)\, v \right] + \lambda_1 \left[ax - bx^2 - vx \right] + \lambda_2 u \tag{17}$$

where $\lambda_1 = \lambda_1(t)$ and $\lambda_2 = \lambda_2(t)$ are called coestate or adjoints variables.

The adjoints equations are givens by

$$\begin{cases} \frac{d\lambda_1}{dt} = \lambda_1 \left[2bx - a + v \right] - e^{-\delta t} pv \\ \\ \frac{d\lambda_2}{dt} = \lambda_1 x - e^{-\delta t} (px - c) \end{cases} \tag{18}$$

By the maximization principle[14], if the triple (u^*, x^*, v^*) represents an optimal solution of the problem, then

$$H(x^*, v^*, \lambda_1, \lambda_2, u^*, t) = \max_{u \in [0, u_{\max}]} H(x^*, v^*, \lambda_1, \lambda_2, u, t) \tag{19}$$

which implies

$$H\left(x^*, \upsilon^*, \lambda_1, \lambda_2, u^*, t\right) \geq H\left(x^*, \upsilon^*, \lambda_1, \lambda_2, u, t\right) \qquad \forall t \in \mathbb{R}_0^+. \tag{20}$$

From the previous inequality and the linearity of the Hamiltonian with respect to the control variable, is obtained a bang-bang control given by

$$u^*\left(t\right) = \begin{cases} u_{\max} & \text{if } \lambda_2 > 0 \\ 0 & \text{if } \lambda_2 < 0 \end{cases} \tag{21}$$

The condition of the maximum does not give any information on optimal control when $\lambda_2 = 0$. Now, Derivating this equality with respect to t it is obtained

$$\tfrac{d\lambda_2}{dt} = 0 \tag{22}$$

substituting (22) in the second equation of (18) it is

$$0 = \lambda_1 x - e^{-\delta t}\left(px - c\right) \tag{23}$$

we obtain

$$\lambda_1 = e^{-\delta t}\left(p - \tfrac{c}{x}\right). \tag{24}$$

From the equation (23), derivating with respect to t it is obtained

$$0 = \tfrac{d\lambda_1}{dt}x + \lambda_1 \tfrac{dx}{dt} + e^{-\delta t}\left[\delta\left(px - c\right) - p\tfrac{dx}{dt}\right] \tag{25}$$

and substituting (24), in the systems (14) and (18) is obtained the quadratic equation

$$2bpx^2 - \left(bc + p\left(a - \delta\right)\right)x - c\delta \tag{26}$$

which has a positive solution given by

$$x^* = \frac{1}{4bp}\left(\left(bc + p\left(a - \delta\right)\right) + \sqrt{\left(bc + p\left(a - \delta\right)\right)^2 + 8bpc\delta}\right) \tag{27}$$

from the equation (26) we can define the function

$$\Phi\left(x\right) = 2bpx^2 - \left(bc + p\left(a - \delta\right)\right)x - c\delta \tag{28}$$

Furthermore, by means of its evaluation, it is possible to see that if $x < x^*$ then $\Phi\left(x\right) < 0$ and if $x > x^*$ then $\Phi\left(x\right) > 0$.

Lemma 3.1. *If $\frac{a}{b} > \frac{c}{p}$ then $x^* < \frac{a}{b}$ and the equilibrium point x^* is attainable*[9].

Proof. In order to demonstrate that x^* is attainable, it must to show that $x^* < \frac{a}{b}$. It is clear that x^* is a root of the function defined in (28), $\Phi(x^*) = 0$, furthermore, if $x > x^*$ then $\Phi(x) > 0$. It is had that if $\Phi\left(\frac{a}{b}\right) > 0$, necessarily $x^* < \frac{a}{b}$.

To determine the sign of $\Phi\left(\frac{a}{b}\right)$, we evaluate the function defined in (28) for $x = \frac{a}{b}$

$$\Phi\left(\tfrac{a}{b}\right) = (a+\delta)\left(\tfrac{ap}{b} - c\right)$$

where $\Phi\left(\frac{a}{b}\right) > 0$ if and only if $\frac{ap}{b} - c > 0$ so, if $\frac{a}{b} > \frac{c}{p}$ then $x^* < \frac{a}{b}$. □

When the equilibrium point (x^*, v^*) is reached, should stay in this level by means of the election

$$\begin{aligned} v^* &= a - bx^* \Rightarrow \frac{dx^*}{dt} = 0 \\ u^* &= 0 \qquad\quad\; \Rightarrow \frac{dv^*}{dt} = 0 \end{aligned} \tag{29}$$

3.2. *Stability of Equilibrium*

First the local stability of the equilibrium point will be demonstrated

Lemma 3.2. *The equilibrium point (x^*, v^*) is locally stable.*

Proof. Supposing that $\frac{dv}{dt} = 0$ it is had $v = k$ with $k \in \mathbb{R}$, then $k - v = 0$, reason why $\frac{dv}{dt} = k - v$ and the system (14) can be written

$$\begin{cases} \frac{dx}{dt} = ax - bx^2 - vx \\ \frac{dv}{dt} = k - v. \end{cases} \tag{30}$$

The Jacobian matrix associated to the previous system is

$$DX(x, v) = \begin{pmatrix} a - 2bx - v & -x \\ 0 & -1 \end{pmatrix} \tag{31}$$

Evaluating the Jacobian matrix in the equilibrium point $(x^*, a - bx^*)$ is obtained

$$DX(x^*, v^*) = \begin{pmatrix} -bx^* & -x^* \\ 0 & -1 \end{pmatrix}$$

Since $\det D_X(x^*, v^*) = bx^* > 0$ and $tr\ D_X(x^*, v^*) = -bx^* - 1 < 0$ the equilibrium point (x^*, v^*) is locally stable. □

Now, we will demonstrate that the solutions of the system (14) are bounded in the plane

Lemma 3.3. *The region*

$$\Omega = \left\{(x, v) \in \mathbb{R}^2 /\ 0 < x \leq \tfrac{a}{b}\right\}$$

is an invariant set[2] *and the solutions of the system (14) are bounded*

Proof. In first place, the v-axis is an invariant set. The nullclines of the system are given by

$$\begin{aligned} \tfrac{dx}{dt} &= ax - bx^2 - vx = 0 \\ \tfrac{dv}{dt} &= u = 0 \end{aligned}$$

from the first equation of the system (32) $v = a - bx$ and the second equation $v = v^*$ is the coordinate of intersection with the nullcline, $v^* = a - bx^*$ which implies that

$$\frac{dv}{dt} = v^* - v. \tag{32}$$

Now, if $v \geq v^*$ then $\frac{dv}{dt} \leq 0$ and if $v < v^*$ then $\frac{dv}{dt} > 0$.

On the other hand, if $v \geq a - bx$ then $\frac{dx}{dt} \leq 0$ and if $v < a - bx$ then $\frac{dx}{dt} > 0$.

Therefore, all trajectory with initial conditions in the region $\Omega = \left\{(x, v) \in \mathbb{R}^2 /\ 0 < x \leq \frac{a}{b}\right\}$ remains there, in fact, the region Ω is a invariant set and the solutions remain bounded. □

Graphically, the previous lemma can be represented by the following figure

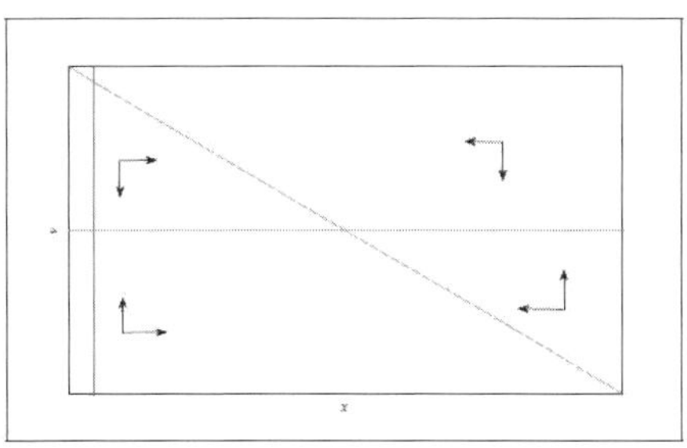

Figure 1. Trajectories behavior in the phase portrait. The red and orange lines are the nullclines of the system.

In order to determine the global stability of the equilibrium point, we will use the direct method of Lyapunov[2,5,8,10].

Theorem 3.1. *The function*

$$V(x,v) = (x - x^*)^2 + \frac{a^2}{b^2 u_{\max}} (v - v^*)^2$$

is a Lyapunov function and the equilibrium point (x^*, v^*) is globally asymptotically stable.

Proof. Is easy to verify that the function $V(x,v)$ defined in the theorem is always positive and it vanishes only at the equilibrium point (x^*, v^*). to demonstrate that it is a Lyapunov function it must proof

$$\langle grad(V), X(x,v) \rangle \leq 0 \tag{33}$$

this condition can be written, in our case, as

$$\left(2(x - x^*), \frac{2a^2}{b^2 u_{\max}} (v - v^*)\right) \begin{pmatrix} ax - bx^2 - vx \\ u \end{pmatrix} \leq 0 \tag{34}$$

or of equivalent way

$$2x(x - x^*)(a - bx - v) + \frac{2a^2}{b^2 u_{\max}} (v - v^*) u \leq 0 \tag{35}$$

for this reason, the sign of the last expression is determined by the values of (x, v) in the phase portrait.

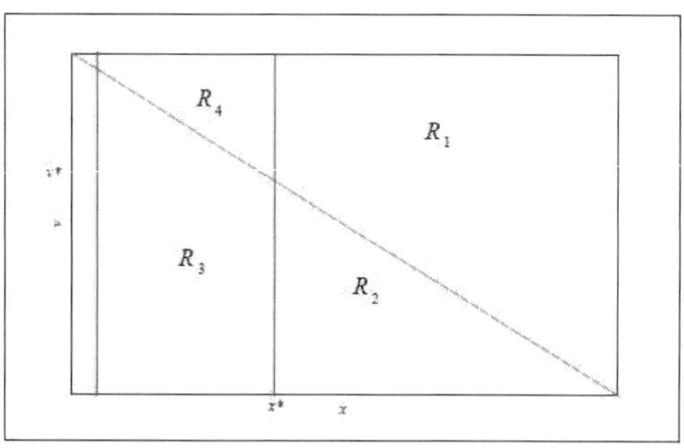

Figure 2. Illustration of the different regions of the phase portrait, determined by the lines $x = x^*$ and $v = a - bx$.

From this, we can define a admissible feedback control[1] $u^o = u^o(x,v)$ in order to prove the inequality.

If $(x, v) \in R_1$, we have $x \geq x^*$ and $v \geq a - bx$, in this case, is sufficient to define the following control

$$u^o(x, v) = \begin{cases} u_{\max} & \text{si } v \leq v^* \\ 0 & \text{si } v > v^* \end{cases} \tag{36}$$

If $(x, v) \in R_2$, we have $x \geq x^*$ and $v \leq a - bx$, in this case, is sufficient to define the following control

$$u^o(x, v) = \tfrac{b^2}{a^2} u_{\max} x (x - x^*) \tag{37}$$

If $(x, v) \in R_3$, we have $x \leq x^*$ and $v \leq a - bx$, in this case, is sufficient to define the following control

$$u^o(x, v) = \begin{cases} u_{\max} & \text{si } v \leq v^* \\ 0 & \text{si } v > v^* \end{cases} \tag{38}$$

If $(x, v) \in R_4$, we have $x \leq x^*$ and $v \geq a - bx$, in this case, is sufficient to define the following control

$$u^o(x, v) = \tfrac{b^2 u_{\max}}{a^2} x (x - x^*) \tag{39}$$

By the Cauchy-Lipschitz theorem[2] applied to the system (14), where u^o is defined previously from the regions R_i, $i = 1, 2, 3, 4$. (see Fig. 2) we can prove a local existence for the solution. Furthermore, where $(x(t), v(t))$ is bounded (see Lemma 3), this solution is global. □

In the previous theorem, we demonstrate that, for any initial condition, i.e., for any level of volume timber and thinning effort, is possible to construct a admissible feedback control, discontinuous, such that the corresponding trajectory $(x(t), v(t))$ reaches the equilibrium point at finite or infinite time. This is a desired situation, because, in first place, is possible to maintained the thinning levels, achieving the objective to sustain the explotation in the time; on the other hand, it keeps the forest volume at a good threshold, this objective has as goal the preservation.

From the last theorem, we can finding control strategies such that the corresponding trajectory reaches the equilibrium point (x^*, v^*) at finite time, for any initial condition in the phase plane. For example, we can define the following control

$$u = \begin{cases} u_{\max} & \text{si } v \leq v^* \\ 0 & \text{si } v > v^* \end{cases} \tag{40}$$

Case 1: *If $v > v^*$ is due to consider $u = 0$. The timber volume and the thinning effort rate will decrease until meeting the nullcline $v = a - bx$.*

Under this nullcline, the volume timber grows and the thinning decrease until meeting the nullcline $v = v^*$. From this moment, maintaining the control, the equilibrium (x^*, v^*) will be reached.

Case 2: *if $v \leq v^*$ is due to consider $u = u_{\max}$. The thinning effort rate grows quickly causing that the timber volume, growing at first, begins to decrease quickly until to reach the nullcline $v = v^*$. From this point, it would be necessary to change the strategy, by taking $u = 0$ in order to remain on the horizontal line and will decrease the timber volume and the thinning effort rate until to reach the equilibrium (x^*, v^*).*

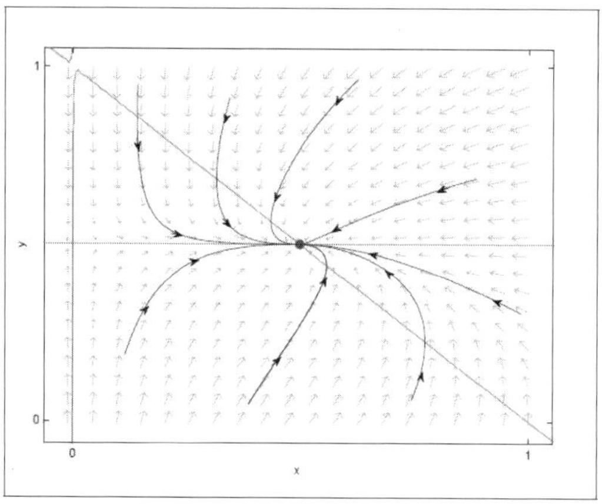

Figure 3. The equilibrium point of the system (14) is globally asymptotically stable for $u = 0$, with parameters conditions $a = 1$ and $b = 1$.

With the help of the classic theorems of the optimal control theory, we proved the existence of a unique equilibrium point for the problem, which is asymptotically globally stable. This result allows us to think of the existence of an optimal strategy for the problem. The object of the next section is to sketch a control strategy that we conjecture is optimal. The optimality of this strategy will be analyzed in a future work.

3.3. *Optimal Strategy Description*

By means of numerical simulations it is possible to identify two curves C_1 and C_2 which reach to the equilibrium point (x^*, v^*). Where C_1 represents the locus of points (x_0, v_0) in the phase plane with $x_0 < x^*$ and $v_0 < v^*$ such that the trajectories $(x(t), v(t))$ with initial conditions (x_0, v_0) and with $u = u_{\max}$ passes through of the equilibrium point (x^*, v^*) and C_2 represents the locus of points (x_0, v_0) in the phase portrait with $x_0 \geq x^*$ and $v_0 \geq v^*$ such that the trajectories $(x(t), v(t))$ with initial conditions (x_0, v_0) and with $u = 0$ passes through of the equilibrium point (x^*, v^*).

These two curves permit us to divide the phase plane into two regions, indicated for I and II, the region I is the area above $C_1 \cup C_2$ and the region II is the area under the same curve.

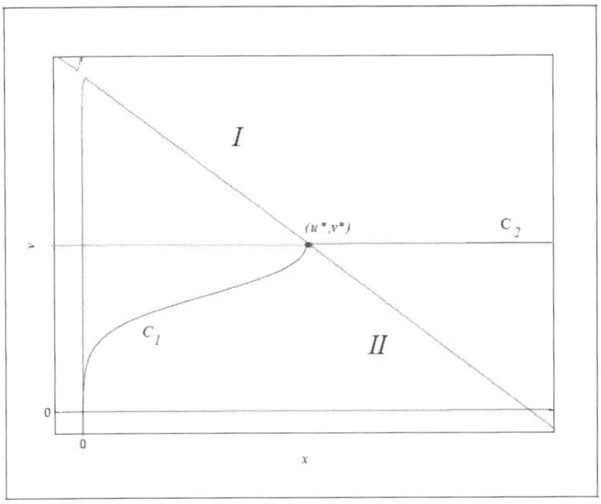

Figure 4. Illustration of the curves C_1 and C_2 in the plane.

Case 1: *If* (x_0, v_0) *is in area* I*, the thinning effort rate is very large. The best choice of control consists of taking* $u = 0$ *in order to reduce the value of the thinning effort as fast as possible. By maintaining the same policy, the trajectory will be below until reaching the equilibrium point* (x^*, v^*).

Case 2: *If* (x_0, v_0) *is in area* II *and under the nullcline* $v = a - bx$*, the thinning effort rate is weak. The best choice of control consists of taking* $u = u_{\max}$ *in order to increase the thinningg effort as fast as possible until the trajectory meets the nullcline* $v = a - bx$*. By maintaining the same policy, the trajectory reaches the isocline* $v = a - bx$*. Over this nullcline, we have* $\frac{dx}{dt} < 0$*, so the timber volume decrease, but at the same time the thinning effort will continue to increase until reaching the curve* C_2*. From this point, it is necessary to change the strategy by taking* $u = 0$ *to stay in* C_1 *until reaching the equilibrium point* (x^*, v^*).

Case 3: *If* (x_0, v_0) *is in area* II *and over the nullcline* $v = a - bx$*, we use the same steps described in the second case until the trajectory will be* *be below the curve C_2. By maintaining the same policy, the trajectory will stay on the curve C_2, will decrease the timber volume, but at the same time the thinning effort will stay constant until reaching the equilibrium point (x^*, v^*).*

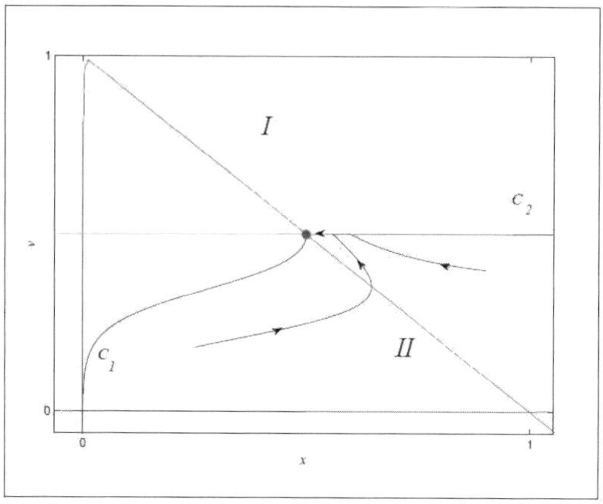

Figure 5. Control strategy of the equilibrium for the case 2 and 3, where parameters have been chosen as $a = 1$, $b = 1$, $u_{\max} = 0.04$.

4. Discussion

In this work, we have elaborated a control strategy which reaches the equilibrium point of the model with a simple feedback control guaranteeing the stability. But the mathematical analysis of the problem is not easy, because, in this model, the reduction effort becomes a second variable of state, and the maximum principle does not identify a candidate explicitly to the optimality. However, it gives information on the strategy of optimal exploitation.

The obtained strategies constitute a base of interesting study in order to determine the impact and the efficacy of some tools of control and deduce a tendency for forest management. But the analized model involves simplifications on biological and economics parameters. One of the most interesting prospect consists on the optimality analysis of the control policy of this model. This will be one of the goals in our next work.

References

1. A. E. Bryson, Yu-Chi Ho, *Applied Optimal Control. Optimization, Estimation and Control.* Taylor and Francis. (1975).
2. C. Chicone, *Texts in Applied Mathematics* **34**, Springer. (1999).
3. C. W. Clark, *Mathematical Bioeconomics: The Optimal Management of Renowable Resources.* 2nd ed. Wiley Interscience, New York, (1990).
4. C. W. Clark and J. D. De Pree, *Applied Mathematics and Optimization* **5**, 181-196, (1979).
5. F. Dumortier, J. Llibre, J. C. Artés, *Qualitative Theory of Planar Differential Systems.* Springer-Verlag Berlin Heidelberg, (2006).
6. C. Elliott, *Unasylva, Revista internacional de silvicultura e industrias forestales.* FAO - Organización de las Naciones Unidas para la agricultura y la alimentación, (1996).
7. W. M. Getz and R. G. Haight, *Population Harvesting: Demografic Models of Fish, Forest and Animal Resources.* Princeton University Press, (1989).
8. B. S. Goh, *Bulletin of Mathematical Biology,* **40**, 525-533, (1978).
9. M. Jerry and N. Raissi, *Mathematical and Compùter Modelling* **36**, 1293-1306, (2002).
10. D. W. Jordan and P. Smith, *Nonlinear Ordinary Differential Equations An introduction for Scientists and Engineers.* Fourth Edition, Oxford University Press Inc., New York, (2007).
11. M. Kamien and N. Schwartz, *Dinamic Optimization, The Calculus of Variations and Optimal Control in Economics and Management.* Elsevier Science B.V. (1991).
12. P. Kilkki and U. Vaisanen, *Acta Forestalia Fennica,* **102**, S-23, (1969).
13. L. S. Pontryagin, V. G. Boltyanskii, R. V. Gamkrelidze and E. F. Mishchenko, *The Mathematical Theory of Optimal Processes.* London: Pergamon Press. (1964).
14. S. P. Sethi, *Optimal control theory: applications to management science and economics.* Kluwer Academic Publisher, (2000).
15. J. K. Vanclay, *Modelling Forest Growth and Yield Applications to Mixed Tropical Forests.* CAB International, Wallingford UK, (1994).

MATING STRATEGIES AND THE ALLEE EFFECT: A COMPARISON OF MATHEMATICAL MODELS

D. I. WALLACE*, R. AGARWAL and M. KOBAYASHI
Department of Mathematics,
Dartmouth College, HB 6188
Hanover, NH 03755, USA
E-mail: Dorothy.Wallace@Dartmouth.edu

Population models are developed that incorporate three breeding strategies. The presence of Allee effects and the strength of such effects are shown to be dependent on breeding strategy. For each model it is shown what constraints on the parameters lead to unconditional extinction and what constraints allow survival. Two of the three models exhibit Allee effects for all parameter choices. For intermediate choices of parameter, one of these two displays unconditional extinction while the other exhibits conditional survival. For suitable choice of parameters, the third strategy shows no Allee effect at all. Comparison of the Allee basins demonstrates the advantages of behavioral adaptations that modify breeding strategies when the population falls in a critical region.

1. Introduction

In 1954, Allee hypothesized that a species may become extinct when its numbers fall below a particular threshold, [1]. This phenomenon, now known as the Allee effect, has been invoked to explain the decline populations of plants ([2], [3], [4]) birds ([5]) and marine organisms ([6],[7]) as well as mammals ([8]).

Researchers have postulated many mechanisms for the Allee effect, ranging from the difficulty of finding mates at low populations ([9],[10]). In this paper we investigate the hypothesis that breeding strategies that make it easier to locate a mate reduce the severity of the Allee effect.

Usually discussions about evolutionary "advantage" center on the notion of competition for resources. In this paper we are looking at a different kind of advantage, namely resilience of a population to shocks. A population

*Corresponding author

that is at a stable equilibrium may suffer a reduction in numbers due to some unusual event. If the population is subject to an Allee effect, the shock may be enough to push the population into the basin of the attractor at the origin, the "Allee basin" as we shall call it, resulting in eventual extinction of the population. The type of mating strategy used by the population affects how large this basin is and therefore affects the ability of the population to withstand shocks. The size of the Allee basin essentially quantifies the advantage one strategy has over another in protecting the population from extinction due to sudden reduction in numbers.

Section 2 of this paper provides some useful background and references for the Allee effect and describes the software used for simulations. Section 3 introduces three different models incorporating mating strategies reflecting three different sets of assumptions. Sections 4, 5 and 6 analyze each of the three models separately. Section 7 compares the three mating strategies expressed by these models against one another. Section 8 summarizes the conclusions of the study.

2. Background

Both the Allee effect and the mechanisms described above are amenable to mathematical modeling. There is a large literature of single species models that use various growth rate terms reflecting a variety of assumptions but that do not distinguish populations of males and females. In many of these models an extinction threshold is built directly into the growth equation ([11],[12],[13],[14],[15],[16],[17],[18],[19]). Not surprisingly, these models exhibit an Allee effect which drives populations below this threshold to zero. Building the threshold directly into the model, however, does not elucidate the mechanism offered to explain the effect. Other single population models exhibit an Allee effect without explicitly building in a threshold below which growth rate is always negative ([20],[21],[22],[23]). All of these models use a growth rate that is a sum of at least two terms, one of which is negative. Mathematically, this change effectively separates the processes of birth and death. For population values where the birth rate drops below the death rate, populations decline. Thus one can calculate which parameters give an Allee effect in these models and also the value of the threshold for a given set of parameters. Again, because these are single population models, they cannot test explanations given in terms of mating, size of predator population, or other interactive scenarios.

One class of models attempts to address mating issues with single population models incorporating various forms of mating probability ([24],[25],[26],[27],[28],[29],[30],[31], [32]). All of these models assume that the population of males is a fixed proportion of the female population, in order to reduce the problem to a single population.

In the study that follows, BGODEM software (Reid, 2008) was used to produce the time series graphs, Maple software produced the phase plane attractor basins, and Adobe Photoshop was used to superimpose Maple plots and find the basin boundaries.

3. A comparative study of mating

In this paper we will compare three models for population growth in the presence of a particular mating strategy. Each model is two dimensional, incorporating male and female populations separately. All models assume that males and females are in competition for resources.

3.1. *Model 1: probabilistic mating*

The first model uses a probabilistic form for mating that reflects the assumption that the act of mating as a strictly probabilistic event.

X=females, Y=males

X'= (growth constant)*(logistic regulator)*(female density)*(male density) - (death constant)*(female density)

Y'= (growth constant)*(logistic regulator)*(female density)*(male density) - (death constant)*(male density)

$$X' = a(1 - X - Y)XY - bX \tag{1}$$

$$Y' = a(1 - X - Y)XY - bY \tag{2}$$

Model 1 assumes little agency on the part of the organism, as for example is the case with wind pollinated plants.

3.2. *Model 2: completely efficient monogamous mating*

In the second model we assume that males and females do not have to work to find one another. All possible mating pairs are formed in what is assumed to be a monogamous fashion. The probabilistic term X*Y is therefore replaced by the number of potential pairs, min(X,Y).

X=females, Y=males

X'= (growth constant)*(logistic regulator)*(number of potential pairs) - (death constant)*(female density)

Y'= (growth constant)*(logistic regulator)*(number of potential pairs) - (death constant)*(male density)

$$X' = a(1 - X - Y)(min(X, Y)) - bX \tag{3}$$

$$Y' = a(1 - X - Y)(min(X, Y)) - bY \tag{4}$$

Model 2 is completely efficient but relies equally on the presence of males and females. Monogamous species living in high density groups, where no searching is required to find a mate, would be an example of the behavior reflected in model 2.

3.3. *Model 3: partially efficient probabilistic mating*

The third model is for situations where the reproductive value of the female has more effect on the growth rate, with the utility of the males approaching 1 at a low percentage of the carrying capacity. This model still has a probabilistic aspect, but assumes a type of mating behavior, such as polygamy, which allows a small number of males to encounter many females. We use a Holling term for the male contribution:

Male utility $= U(Y) = Y/(c + Y)$

Notice that the ratio approaches 1 as Y goes to infinity, and is zero when Y is zero. Choosing $c = .01$, for example, gives $U(.1) = 10/11$, very close to 1 at only ten percent capacity for Y. The constant c is called the "half saturation" constant because when $Y = c$ the utility U equals one half (of the maximum possible utility, which is 1). Note also that, as Y approaches zero, the term $Y/(c + Y)$ representing male utility is $O(Y/c)$, giving a growth rate near zero of a/c. Thus a low proportion of males in the population can effectively serve as many females as are available, and the marginal utility of further males is low. As c approaches zero the model approaches one in which the birth rate only depends on the number of females and the available resource. Thus c describes a range of probabilistic mating behaviors in which the encounter rate is disproportionately large at low male population sizes. We will call this model "partially efficient probabilistic mating".

X=females, Y=males

X'= (growth constant)*(logistic regulator)*(female density)*(male utlity) - (death constant)*(female density)

Y'= (growth constant)*(logistic regulator)*(female density)*(male utility) - (death constant)*(male density)

$$X' = a(1 - X - Y)XY/(c + Y) - bX \tag{5}$$

$$Y' = a(1 - X - Y)XY/(c + Y) - bY \tag{6}$$

In the next three sections we give the basic mathematical results for each of these models.

4. Analysis of model 1

$$X' = a(1 - X - Y)XY - bX \tag{7}$$

$$Y' = a(1 - X - Y)XY - bY \tag{8}$$

The Jacobian for this system is given by

$$\begin{pmatrix} aY - 2aXY - aY^2 - b & aX - aX^2 - 2aXY \\ aY - aY^2 - 2aXY & aX - 2aXY - aX^2 - b \end{pmatrix}.$$

At extinction, the value $(0, 0)$ gives a diagonal Jacobian, with eigenvalues $-b, -b$. By the Hartman-Grobman theorem (Hartman, 1960), extinction is always an attracting state for this system. Therefore if survival of the species is possible, it is always conditional on the size of the populations of males and females. That is, the Allee effect is always present. Solving for critical points other than $(0, 0)$, we get the solutions

$$X = Y = (1/2)(1 + \sqrt{(1 - 8b/a)}) \tag{9}$$

and

$$X = Y = (1/2)(1 - \sqrt{(1 - 8b/a)}) \tag{10}$$

From these we see that survival, which requires critical points with positive real values, is only possible if $a > 8b$. That is, the inherent growth rate must be eight times the natural death rate for the species to be viable. Furthermore, when $a > 8b$ at one of the critical points X and Y are both greater than .5, and at the other critical point they are both less than .5. In the following discussion we assume $a > 8b$. Note that with strict inequality there will be no repeated roots.

Computing the Jacobian at $X = Y$ and using the relation $X^2 = (aX - b)/(2a)$ gives a Jacobian $J(T)$ of the form:

$$\begin{pmatrix} aT & T+b \\ T+b & T \end{pmatrix},$$

where

$$T = -(a/2)X + b/2 \tag{11}$$

Notice that T is a real number.

Computing the eigenvalues of J is straightforward and yields $(2T + b, -b)$. For $X > 1/2$ and $a > 8b$, we have

$$2T + b = -aX + 2b < -a/2 + 2b = (-a + 4b)/2 < 0 \tag{12}$$

Thus by the Hartman-Grobman theorem this is an attracting fixed point. Thus, sufficiently large populations will reach a nontrivial equilibrium. Figure 1 shows a typical simulation of this system with starting populations large enough to avoid the Allee effect.

If we make the common assumption (which Figure 1 demonstrates is not justified) that the male population is a constant proportion of the female population, $Y = mX$, we get the single population system studied in Aviles (1999)

$$X' = a(1 - (1 + m)X)X^2 - bX \tag{13}$$

The basin of attraction for the trivial fixed point, hereafter called the Allee basin, is shown for several choices of constant in Figure 2.

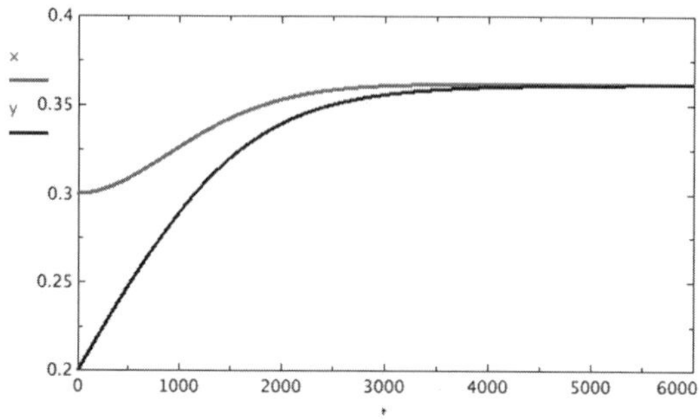

Figure 1. Model 1, with $b = .001$ and $a = .01$

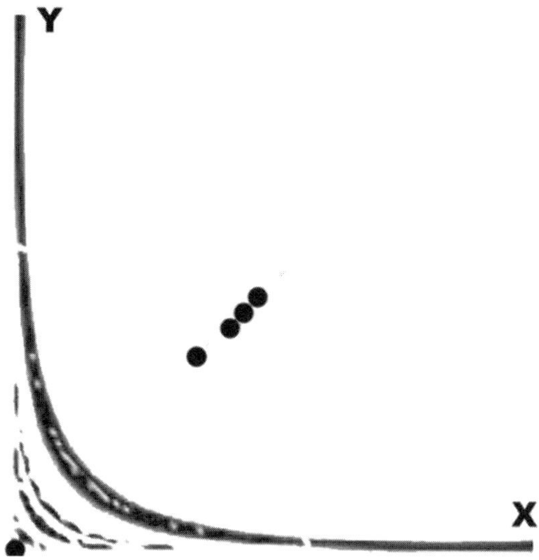

Figure 2. Model 1, with $b = .001$ and $a = .01, .015, .02, .04$

As the inherent growth rate, a, decreases, the basin grows. The fixed points corresponding to conditional survival are shown along with the boundaries for the basins of attraction. Note that the largest basin corresponds to the lowest fixed point.

5. Analysis of model 2

$$X' = a(1 - X - Y)(min(X, Y)) - bX \tag{14}$$

$$Y' = a(1 - X - Y)(min(X, Y)) - bY \tag{15}$$

In this situation the function $min(X, Y)$ is not differentiable. However, it is always equal to the simple function X or Y, depending on which is greater. So we will investigate the dynamics of these two simple systems, each of which represents the behavior of the system on half of the phase space. For ease of discussions we will refer to the following two systems:

System 1, valid for $X > Y$:

$$X' = a(1 - X - Y)Y - bX \tag{16}$$

$$Y' = a(1 - X - Y)Y - bY \tag{17}$$

System 2, valid for $X < Y$:

$$X' = a(1 - X - Y)X - bX \tag{18}$$

$$Y' = a(1 - X - Y)X - bY \tag{19}$$

We can also consider

System 3, valid for $X = Y$:

$$X' = a(1 - 2X)X - bX \tag{20}$$

For both systems 1 and 2 we obtain fixed points at $(0, 0)$ and $X = Y = .5 - b/(2a)$. The second fixed point will be in the positive quadrant if $a > b$. Computing the Jacobians for systems 1 and 2 and the relevant derivative for system 3 we have:

$$J(system1) = \begin{pmatrix} -aY - b & a - aX - 2aY \\ -aY & a - aX - 2aY - b \end{pmatrix}.$$

$$J(system2) = \begin{pmatrix} a - aY - 2aX - b & -aX \\ a - aY - 2aX & -aX - b \end{pmatrix}.$$

$$J(system3) = a - 4aX^2 - b$$

At the critical point $(0,0)$, the eigenvalues for systems 1 and 2 are $-b$ and $a-b$, whereas the value of the Jacobian for system 1 is $a-b$. Thus if $a < b$, the equilibrium at $(0,0)$ is attracting, as all of these values are negative. However if $a > b$, the eigenvalues have mixed signs. Because this model is really three different systems in the regions $X > Y$, $X < Y$ and $X = Y$ we cannot say automatically whether extinction is an attracting state or not. However if we look at the value of Y' near the origin for system 1, we have

$$Y' = a(1-X-Y)Y - bY = Y(a(1-X-Y)-b) = Y(a-b-a(X+Y)) \quad (21)$$

Thus if $X+Y$ is sufficiently small and $a > b$, we see that $Y' > 0$.

We have shown that near the origin, as long as Y does not equal 0, the flow always has a component in the Y direction away from $(0,0)$ in system 1. For the situations where $Y > X$ in system 2 and $Y = X$ in system 3, a similar calculation shows that $X' > 0$ for X sufficiently small. So this system does not display an Allee effect for $a > b$. The assumption that males are a constant proportion of females yields a version of the logistic equation, displaying unconditional survival or extinction depending on choices of a, b, and the constant of proportionality chosen.

For model 2 there is a second fixed point when $a > b$, given by $X = Y = 1/2 - b/(2a)$. The Jacobian for system 1 representing this model may be evaluated at this value to give eigenvalues of $-b$ and $-(a+3b)$. The Jacobian for system 2 is just the inverse transpose of that of system 1 and thus has the same eigenvalues. The Jacobian of system 3 is the second of these eigenvalues. All are negative. Invoking Hartman-Grobman for each of these systems, and noting that our model is always equivalent to one of the three on every region of the phase plane, we can conclude that the nonzero equilibrium is indeed an attracting state. To summarize, model 2 displays unconditional extinction if $a < b$ and unconditional survival when $a > b$.

6. Analysis of model 3

$$X' = a(1-X-Y)XY/(c+Y) - bX \quad (22)$$

$$Y' = a(1 - X - Y)XY/(c + Y) - bY \tag{23}$$

In these equations all parameters are positive and $c < 1$.

Besides the equilibrium at (0,0), these equations also yield fixed points when $Y = X$ at

$$X = Y = ((a - b) + \sqrt{((a - b)^2 - 8abc)})/4a \tag{24}$$

and

$$X = Y = ((a - b) - \sqrt{((a - b)^2 - 8abc)})/4a \tag{25}$$

Several things are evident from this expression. If $a < b$, the only fixed point of the system is $(0, 0)$. There will be real roots only when $(a - b)^2 > 8abc$. In this case the roots will both be positive when $a > b$. Henceforth for this model we assume that a, b and c fall into this range. The Jacobian of the system is given by

$$\begin{pmatrix} \frac{aY}{(c+Y))}(1 - 2X - Y) - b & (\frac{-aXY}{(c+Y)}) + \frac{(a(1-X-Y)Xc)}{((c+Y)^2)} \\ \frac{(aY}{(c+Y))}(1 - 2X - Y) & (\frac{-aXY}{(c+Y)}) + \frac{(a(1-X-Y)Xc)}{((c+Y)^2)} - b \end{pmatrix}.$$

At $(0, 0)$ the Jacobian is diagonal with both eigenvalues equal to $-b$. Thus the origin is an attracting fixed point and the system displays the Allee effect for all parameters.

We now turn to the computation of stability of the larger nonzero fixed point. It suffices to show that the determinant of the Jacobian at this point is positive and its trace is negative.

Computing the trace of the Jacobian matrix under the constraints for the nonzero fixed points given by

$$0 = a(1 - 2X)X^2/(c + X) - bX \tag{26}$$

and

$$X = Y \tag{27}$$

yields an expression for the trace T at nonzero equilibrium X of

$$T(X) = 1/(c + X)^2(-b/2a)(X(a - b + 2ac) - bc) \tag{28}$$

Let

$$L(X) = (-b/2a)(X(a - b + 2ac) - bc) \tag{29}$$

Then $T(X) < 0$ if and only if $L(X) > 0$. The slope of the line given by the expression $L(X)$ is negative and it is easy to check that $T((a-b)/4a)$ is also negative.

To compute the determinant, we use the fact that at equilibrium $X = Y$ and X solves $2aX = (a - b)X - bc$ to reduce the Jacobian to a positive multiple of

$$\begin{pmatrix} -aX & a - 3X - b \\ a - 3X & -aX - b \end{pmatrix}.$$

Computing the determinant and plugging in the larger of the two nonzero equilibrium values gives the determinant as a positive multiple of the expression

$$2(a + b)\sqrt{((a - b)^2 - 8abc))} - 2(a - b)^2 + 16abc \tag{30}$$

It is an easy exercise to show that this is positive under the assumptions on a, b, c given above.

The fixed point given by equation (24) is therefore stable. Thus if model 3 has positive fixed points other than the origin, at least one of them is stable. Survival of the population is conditional due to the Allee effect that is always present in this model. Figure 3 shows a typical simulation of the two populations of this model.

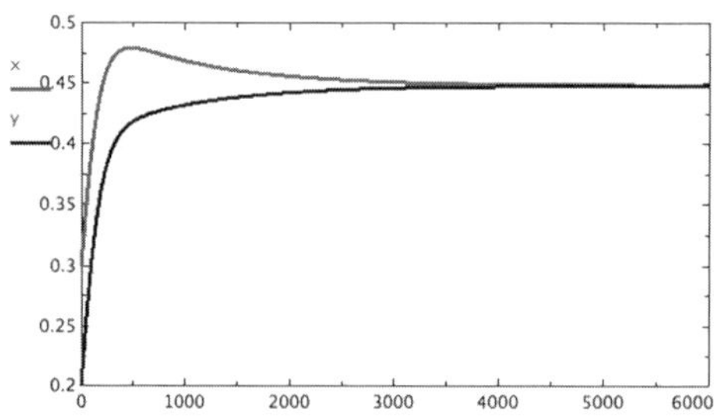

Figure 3. Model 1, with $b = .001$, $a = .01$ and $c = .0125$

Figure 4 shows the change in the Allee basin as the inherent growth rate, a is increased, while other parameters remain fixed.

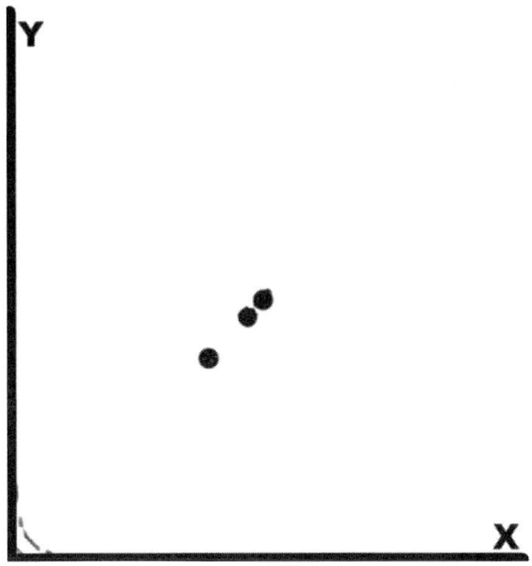

Figure 4. Model 3, with $b = .001$, $c = .0125$ and $a = .005, .01, .015$

If we assume that males are a constant proportion of females (which Figure 3 shows is not justified), we obtain the equation:

$$X' = ma(1 - X - mX)X^2/(c + mX) - bX \tag{31}$$

Note that in this case the lack of symmetry gives a different reduction if we start with the equation for Y'. In any case, this equation has been studied by Takeuchi (1996).

7. Comparison of three strategies

In all three models we have used a and b as birth and death parameters respectively. For model 1 (distributed monogamy) and model 3 (polygamy) we see that the Allee effect is always present. For model 1 a must be greater than $8b$ for conditional survival of the population, whereas for model 3, $a > b$ suffices although there is also a condition on the third constant. For model 2 (enhanced monogamy), $a > b$ is required for survival of the population, but if $a > b$ there is no Allee effect present. Notice that in all models a viable population tends to equal numbers of males and females, even in the "partially efficient mating" model. Figure 5 provides a visual description of the relative advantage of model 3 over model 1.

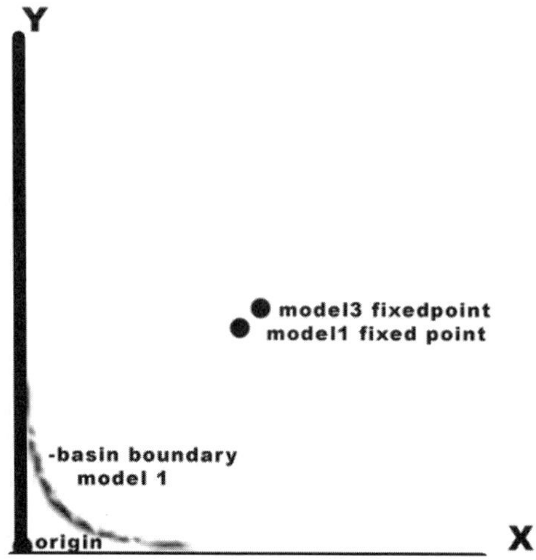

Figure 5. Models 1 and 3, with $b = .001$, $a = .01$ and $c = .0125$

In Figure 5 we see a visible Allee basin for model 1, whereas the Allee basin for model 3 is too small to be visible in the diagram.

The question addressed by this study is how a population with given mating dynamics responds to a shock to the system. A shock, for this purpose, is a sudden reduction of population. Suppose the population has reached its nonzero equilibrium and then experiences an instantaneous symmetric (in X and y) reduction in numbers. We would like to compare the maximal possible shock that the population could sustain and still be viable.

In both models 1 & 3 the diagonal distance from the nonzero stable equilibrium to the Allee basin boundary is just the distance between the two nonzero fixed points. The lower, unstable, nonzero fixed point sits on the Allee basin boundary. So the distance between the two nonzero fixed points is exactly the maximum possible shock the system can handle if the populations start at the stable equilibrium and are reduced equally in X and Y. We will call these distances MSS_1 and MSS_3 respectively, for "maximum symmetric shock"

$$MSS_1 = \sqrt{2}2\sqrt{(1 - 8b/a)} \tag{32}$$

$$MSS_3 = \sqrt{2}2\sqrt{((a - b)^2 - 8abc))}/4a \tag{33}$$

Figure 5 shows a typical run in which MSS_3 is greater than MSS_1. In order for this to happen, it suffices to show that $MSS_3^2 - MSS_1^2 > 0$. A calculation shows that this is equivalent to

$$3 + 4b^2/a^2 - 32bc/a > 0 \tag{34}$$

Recall that, in order for model 1 to have a stable nonzero equilibrium, it must be that $a > 8b$. Making this assumption it suffices to show that

$$3 + 4b^2/a^2 - 4c > 0 \tag{35}$$

The left expression will be positive for all a and b if $c < 3/4$. Remember that reducing c improves the efficiency of mating at low population levels. When $c = 1$ model 3 has no better reproduction rates than model 1 at low levels and does not represent any sort of improvement of efficiency. Our computer runs used $c = .0125$. We see then that model 3 is more resistant to shock than model 1 over a large range of biologically reasonable values for c, the parameter that controls the efficiency of the mating strategy.

8. Conclusions

The mating strategy an organism uses has a quantifiable effect on its ability to withstand reductions in number. We summarize the impact of mating strategies modeled in this paper below.

(1) *The existence and strength of the Allee effect depend on mating strategies of the organism.* Comparison of models 1-3 shows that the Allee effect can disappear entirely or the size of the Allee basin can be reduced by the choice of mating strategy.
(2) *Perfectly efficient mating eliminates the Allee effect (unless due to other causes).* Model 2 describes a situation where all mating pairs are utilized and exhibits no Allee effect when the birth rate exceeds the death rate.

(3) *Partially efficient probabilistic mating strategies enhance the ability of a population to resist shocks.* Even when not terribly efficient, model 3 is more resistant to instantaneous symmetric reduction of its population, as measured by the diagonal distance from its stable nonzero fixed point to the Allee basin boundary.

These models raise interesting questions regarding interspecies competition in the presence of probabilistic mating strategies. In particular, running numerical experiments where a small number of organisms are introduced into an existing population of competitors (as is often the case in nature) show inevitable extinctions of the introduced organism due to the Allee effect. That is, the introduced organism would die out whether the competitor was there or not. The outcome of multiple species models will look very different when mating strategies are included, suggesting many areas for future research.

References

1. Odum, H. T. & Allee, W. C., *Ecology* **35**, 95-97 (1954).
2. Groom, M. J., *Am. Nat.* **151**, 487-496 (1998).
3. Hackney, E. E. & McGraw, J. B., *Conservation Biology* **15** 129 (2001).
4. Forsyth, S., *Oecologia* **136** 551-557 (2003).
5. Veit, R. R. & Lewis, M. A., *Am. Nat.* **148** 255-274 (1996).
6. Dulvy, N., Sadovy, Y. & Reynolds, J. D., *Fish and Fisheries* **4** 25 (2003).
7. Levitan, D. R., Sewell, M. A. & Chia, F. H., *Ecology* **73** 248-254 (1992).
8. Morris, D. W., *Ecology 83* **1** 14-20 (2001).
9. Kuussaari, M., Saccheri, I., Camara, M. and Hanski, I., *Oikos* **82** 384-392 (1998).
10. Berec, L., Boukal, D. S and Berec, M., *Am. Nat.* **157** 217-230 (2001).
11. Edelstein-Keshet, L., *Mathematical Models in Biology* New York: Random House (1988).
12. Lewis, M. A. & Kareiva, P., *Theor. Popul. Biol.* **43** 141-158 (1993).
13. Amarasekare, P., *Am. Nat.* **152** 298-302 (1998).
14. Keitt T. H., Lewis, M. A. & Holt, R. D., *Am. Nat.* **157** 203-216 (2001).
15. Gruntfest, Y., Arditi, R. & Dombrovsky, Y., *J. Theor. Biol.* **185** 539-547 (1997).
16. Courchamp F., Clutton-Brock, T. & Grenfell, B., *Trends Ecol. Evol.* **14** 405-410 (1999).
17. Courchamp, F., Grenfell, B. & Clutton-Brock, T., *Proc. R. Soc. London* **266** 557-563 (1999).
18. Brassil, C. E., *Ecol. Model.* **143** 9-13 (2001).
19. Takeuchi, Y., *Global Dynamical Properties of Lotka-Volterra Systems* Singapore: World Scientific Publishing Company (1996).
20. Jacobs, J., *Oecologia* **64** 389-395 (1984).

21. Asmussen, M. A., *Am. Nat.* **114** 796-809 (1979).
22. Aviles, L., *Evol. Ecol.* **1** 459-477 (1999).
23. Hoppensteadt, F. C., *Mathematical Methods of Population Biology* Cambridge, MA: Cambridge University Press (1982).
24. Dennis, B., *Nat. Resource Modeling* **3** 481-538 (1989).
25. Hopper, K. R. & Roush, R. T., *Ecol. Entomol.* **18** 321-331 (1993).
26. Klomp, H., van Montfort, M. A. & Tammes, P. M. L., *Arch. Neerl. Zoologie* **16** 105-110 (1964).
27. Kuno, E., *Res. Popul. Ecol.* **20** 50-60 (1978).
28. McCarthy, M. A., *The Allee effect finding mates and theoretical models* **Ecol. Model.** 103, 99-102 (1997).
29. Philip, J. R., *Ecology* **38** 107-111 (1957).
30. Stephan, T. & Wissel, C., *Ecol. Model.* **75/76** 183-192 (1994).
31. Grevstad, F. S., *Ecol. Appl.* **9** 1439-1447 (1999).
32. Wells, H., Strauss, E. G., Rutter, M. A. & Wells, P. H., *Biol. Conserv.* **86** 317-324 (1998).

CULTURAL EVOLUTION IN A LATTICE: THE MAJORITY-VOTE MODEL AS A FREQUENCY-DEPENDENT BIAS MODEL OF CULTURAL TRANSMISSION

J. F. FONTANARI and L. R. PERES

Instituto de Física de São Carlos
Universidade de São Paulo
Caixa Postal 369, 13560-970 São Carlos SP, Brazil
E-mail: fontanari@ifsc.usp.br

The pertinence of culture for understanding the hominid evolutionary process has been persuasively advocated by many authors using the biological concept of niche construction. Hence understanding how opinions and, more generally, cultural traits disseminate through a population is crucial to fully comprehend human evolution. Of particular interest here is the understanding of the mechanisms that lead to the appearance of stable domains characterized by distinct cultural traits, given that people's beliefs have a tendency to become more similar to each other's as people interact repeatedly. Here we study an extreme version of the frequency-dependent bias model in which an individual adopts the trait/opinion shared by the majority of its neighbors - the majority-vote rule model. We assume that the individuals are fixed in the sites of a square lattice of linear size L and that they can interact with their four nearest neighbors only. Within a mean-field framework, we derive the equations of motion for the density of individuals adopting a particular opinion in the single-site and pair approximations. Although the single-site approximation predicts a single cultural domain that takes over the entire lattice, the pair approximation yields a qualitatively correct picture with the coexistence of different opinions and a strong dependence on the initial conditions. In addition, extensive Monte Carlo simulations indicate the existence of a rich distribution of cultural domains or clusters, the number of which grows with L^2 whereas the size of the largest cluster grows with $\ln L$. The analysis of the sizes of the cultural domains shows that they obey a power-law distribution for not too large sizes but they are exponentially distributed in the asymptotic limit.

1. Introduction

The existence of juxtaposed regions of distinct cultures in spite of the fact that people's beliefs have a tendency to become more similar to each other's as the individuals interact repeatedly[1,2] is a puzzling phenomenon in the social sciences[3,4]. This issue has been addressed by somewhat idealized agent-based simulations of human behavior producing highly nontrivial insights on the nature of the collective behavior resulting from homogenizing local interactions[5].

In this line, a particularly successful model is Axelrod's model for the dissemination of culture or social influence[6], which is considered the paradigm for idealized models of collective behavior that seek to reduce a collective phenomenon to its functional essence[7]. In Axelrod's model, the agents are placed at the sites of a square lattice of size $L \times L$ and can interact with their neighbors only. The culture of an agent is represented by a string of F cultural features, where each feature can adopt a certain number q of distinct traits. The interaction between any two neighboring agents takes place with probability proportional to the number of traits they have in common. Although the result of such interaction is the increase of the similarity between the two agents, as one of them modifies a previously distinct trait to match that of its partner, the model exhibits global polarization, i.e., a stable multicultural regime[6].

The key ingredient for the existence of stable globally polarized states in Axelrod's model is the rule that prohibits the interaction between completely different agents (i.e., agents which do not have a single cultural trait in common). Relaxation of this rule so as to permit interactions regardless of the similarity between agents leads to one of the q^F distinct absorbing homogeneous configurations[8]. In addition, introduction of external noise so that traits can change at random with some small probability[9] as well as the increase of the connectivity of the agents[10,11,12] also destabilizes the polarized state.

Although Axelrod's model enjoyed great popularity among the statistical physics community due mainly to the existence of a non-equilibrium phase transition[13] that separates the globally homogeneous from the globally polarized regimes[14,15], the vulnerability of the polarized absorbing configurations was considered a major drawback to explaining robust collective social behavior. In this vein, Parisi *et al.*[16] have proposed a lattice version of the classic frequency bias mechanism for cultural or opinion change[3,4], which assumes that the number of people holding an opinion is the key factor for an agent to adopt that opinion, i.e., people have a tendency to espouse cultural traits that are more common in their social environment. Actually, almost any model of cultural transmission admits that the probability that an individual acquires a cultural variant depends on the frequency of that variant in the population. The frequency-dependent bias mechanism requires that the individual be *disproportionately* likely to acquire the more common variant[3].

More to the point, in the model of Parisi *et al.* the culture of an agent is specified by a binary string of length F (so $q = 2$) and each bit of that string takes the value which is more common among its neighbors[16]. This is essentially the well-known majority-vote model of statistical physics[17] and so we found the claim by those authors that such model exhibits a polarized regime most intriguing, as we expected the dynamics to freeze into one of the two low-temperature homogeneous steady states of the Ising model[18,19].

In order to check whether the multicultural absorbing configurations reported by Parisi *et al.* were not artifacts of the small lattice size used in their study ($L = 20$), here we carry out a detailed analysis of the effects of the finite size of the lattice by considering square lattices of linear size up to $L = 4000$. In addition, we focus on the simplest case $F = 1$ which exhibits the same features of the arbitrary F case. Our findings indicate that the polarized regime is indeed stable in the thermodynamic limit $L \to \infty$ and that the element accountable for this stability is the procedure used to calculate the majority, which includes the site to be updated (target site) in addition to its nearest neighbors. For sites in the bulk of the lattice, this procedure essentially specifies the criterion of update in case of a tie, i.e., in case there is no majority among the neighbors. In this case, the majority-vote variant used by Parisi *et al.* leaves the state of the agent untouched, whereas the most common variant used in the literature sets that state at random with probability 1/2 [18]. It is surprising that such (apparently) minor implementation detail produces nontrivial consequences in the thermodynamic limit.

The rest of this paper is organized as follows. In Sect. 2 we describe the variant of the majority-vote model proposed by Parisi *et al.*, which henceforth we refer to as the extended majority-vote model since its key ingredient is the stretching of the neighborhood to include the target site[16]. In Sect. 3 we offer an analytical approach to this model based on the single-site and pair approximations of statistical mechanics[13]. In Sect. 4 we use Monte Carlo simulations to study several properties of the absorbing configurations such as the average number of cultural domains or clusters, the size of the largest cluster and the distribution of cluster sizes, giving emphasis to their dependences on the lattice size. Finally, in Sect. 5 we present our concluding remarks.

2. Model

As pointed out before, in the extended majority-vote model we consider here each agent is characterized by a single cultural feature which can take on the binary values 0 or 1[16]. The agents are fixed at the sites of a square lattice of size $L \times L$ with open boundary conditions (i.e., agents at the corners interact with two neighbors, agents at the sides with three, and agents in the bulk with four nearest neighbors). The initial configuration is completely random wherein the cultural feature of each agent is specified by a random digit 0 or 1 with equal probability. At each time we pick an agent at random and then verify among its extended neighborhood, which includes the target agent itself, what is the more frequent cultural feature (1 or 0). The cultural feature of the target agent is then changed to match the corresponding majority value. We note that there are no ties for agents in the bulk or at the corners of the square lattice since in these cases the extended neighborhood comprises 5 and 3 sites, respectively. However, agents at the sides of the lattice have an extended neighborhood of 4 sites (i.e., 3 neighbors plus the target agent) and so in case of a tie, the feature of the target agent remains unchanged. Of course, in the limit of large lattices the contribution of these boundary sites will be negligible. This procedure is repeated until the system is frozen in an absorbing configuration.

Although the majority-vote rule or, more generally, the frequency bias mechanism for cultural change[3] is a homogenizing assumption by which the agents become more similar to each other, the above-described model does seem to exhibit global polarization, i.e., a non-trivial stable multicultural regime[16]. Since the study of Parisi *et al.* was based on a small lattice of linear size $L = 20$, a more careful analysis is necessary to confirm whether this conclusion holds in the thermodynamic limit as well. It is interesting to note that for the more popular variant of the majority-vote model, in which the state of the target site is not included in the majority reckoning, and ties are decided by choosing the cultural feature of the target agent at random with probability 1/2, the only absorbing states in the thermodynamic limit are the two homogeneous configurations. For finite lattices, however, we find multicultural absorbing states characterized by stripes that sweep the entire lattice.

3. Mean-field analysis

In this section we offer an analytical approximation to the extended majority-vote model introduced by Parisi *et al.*[16]. The state of the agent at site i of the square lattice is represented by the binary variable $\eta_i = 0, 1$ and so the configuration of the entire lattice comprising $N = L^2$ sites is denoted by $\eta \equiv (\eta_1, \eta_2, \ldots, \eta_N)$. The master equation that governs the time evolution of the probability distribution $P(\eta, t)$ is given by

$$\frac{d}{dt}P(\eta, t) = \sum_i \left[W_i\left(\tilde{\eta}^i\right) P\left(\tilde{\eta}^i, t\right) - W_i(\eta) P(\eta, t)\right] \tag{1}$$

where $\tilde{\eta}^i = (\eta_1, \ldots, 1 - \eta_i, \ldots, \eta_N)$ and $W_i(\eta)$ is the transition rate between configurations η and $\tilde{\eta}^i$. For the extended majority-vote model we have

$$W_i(\eta) = \left|\Theta\left[\sum_\delta \eta_{i+\delta} + \eta_i - 3\right] - \eta_i\right| \tag{2}$$

where the notation $\sum_\delta(\ldots)$ stands for the sum over the 4 nearest neighbors of site i and $\Theta(x) = 1$ if $x \geq 0$ and 0 otherwise.

We are interested in determining the fraction of sites in state 1, which we denote by ρ. Since all sites are equivalent we have $\rho \equiv \sum_i \eta_i / N = \langle \eta_i \rangle$ and this mean value is given by the equation

$$\frac{d}{dt}\langle \eta_i \rangle = \langle (1 - 2\eta_i) W_i(\eta) \rangle \tag{3}$$

with $\langle (\ldots) \rangle \equiv \sum_\eta (\ldots) P(\eta, t)$ as usual. To carry out this average we need to make approximations. In what follows we study in detail two such approximation schemes, namely, the single-site approximation and the pair approximation.

3.1. *The single-site approximation*

This the simplest mean-field scheme which assumes that the sites are independent random variables so that Eq. (3) becomes

$$\frac{d\rho}{dt} = -\rho \sum_{n=0}^{4} B_n^4(\rho) \left|\Theta[n-2] - 1\right| + (1 - \rho) \sum_{n=0}^{4} B_n^4(\rho)\, \Theta[n-3] \tag{4}$$

where B_n is the Binomial distribution

$$B_n^4(\rho) = \binom{4}{n} \rho^n (1 - \rho)^{4-n}. \tag{5}$$

Carrying out the sums explicitly yields

$$\frac{d\rho}{dt} = -\rho(1 - \rho)(2\rho - 1)\left(3\rho^2 - 3\rho - 1\right). \tag{6}$$

This equation, which is invariant to the change $\rho \leftrightarrow 1-\rho$, has three fixed points, namely, $\rho^* = 0$, $\rho^* = 1$ and $\rho^* = 1/2$. The first two fixed points are stable and the third one is unstable. This means that in the single-site approximation the only stable configurations are the homogeneous ones. Finally, we note that ρ contains the same information as the single-site probability distribution $p_1(\eta_i)$. In fact, $p_1(\eta_i = 1) = \rho$ and $p_1(\eta_i = 0) = 1-\rho$.

3.2. *The pair approximation*

In this scheme we assume that nearest neighbors sites are statistically dependent random variables so that, in addition to the single-site probability distribution $p_1(\eta_i)$, we need to compute the pair probability distribution $p_2(\eta_i, \eta_{i+\delta})$ as well. Of particular importance for the subsequent calculations is the conditional probability distribution $p_{1|1}(\eta_{i+\delta} \mid \eta_i)$ which is given simply by the ratio between the pair and the single-site probability distributions.

For the sake of concreteness, let us denote the states of the 4 neighbors of site i by η_1, η_2, η_3 and η_4. These are independent random variables since they are not nearest neighbors in the square lattice, so the sum $n = \eta_1 + \eta_2 + \eta_3 + \eta_4$ that appears in the argument of the Theta functions is a sum of independent variables. With these notations we can rewrite Eq. (3) as

$$\frac{d\rho}{dt} = -\rho \sum_{\eta 1,\ldots,\eta_4} p_{4|1}(\eta_1, \eta_2, \eta_3, \eta_4 \mid 1) \left|\Theta(n-2) - 1\right| + (1-\rho) \sum_{\eta 1,\ldots,\eta_4} p_{4|1}(\eta_1, \eta_2, \eta_3, \eta_4 \mid 0)\, \Theta(n-3) \tag{7}$$

where we have explicitly carried out the sum over $\eta_i = 0, 1$. Using

$$p_{4|1}(\eta_1, \eta_2, \eta_3, \eta_4 \mid \eta_i) = p_{1|1}(\eta_1 \mid \eta_i) \times \ldots \times p_{1|1}(\eta_4 \mid \eta_i) \tag{8}$$

we finally obtain

$$\frac{d\rho}{dt} = -\rho \sum_{n=0}^{4} B_n^4 \left[p_{1|1}(1 \mid 1)\right] \left|\Theta[n-2] - 1\right| + (1-\rho) \sum_{n=0}^{4} B_n^4 \left[p_{1|1}(1 \mid 0)\right] \Theta[n-3] \tag{9}$$

which is identical to Eq. (4) except for the arguments of the binomial distributions. Using the notation $\phi \equiv p_2(1,1)$ we write

$$p_{1|1}(1 \mid 1) = \frac{\phi}{\rho} \tag{10}$$

$$p_{1|1}(1 \mid 0) = \frac{\rho - \phi}{1 - \rho} \tag{11}$$

so that Eq. (9) involves two unknowns, ρ and ϕ. Carrying out the summations explicitly so as to get ridden of the Theta functions yields

$$\frac{d\rho}{dt} = -\rho\left[B_0^4\left(\frac{\phi}{\rho}\right) + B_1^4\left(\frac{\phi}{\rho}\right)\right] + (1-\rho)\left[B_3^4\left(\frac{\rho-\phi}{1-\rho}\right) + B_4^4\left(\frac{\rho-\phi}{1-\rho}\right)\right]. \tag{12}$$

We note that this equation reduces to Eq. (6) in the case the neighboring sites are assumed independent, i.e., $\phi = \rho^2$. It is also useful to check whether Eq. (12) is invariant to the change $0 \leftrightarrow 1$, which amounts to the following interchanges

$$p_2(1,1) \leftrightarrow p_2(0,0) \tag{13}$$

$$p_2(1,0) \leftrightarrow p_2(0,1) \tag{14}$$

$$p_1(1) \leftrightarrow p_1(0) \tag{15}$$

with $p_2(1,1) = \phi$, $p_2(1,0) = p_2(0,1) = \rho - \phi$, $p_2(0,0) = 1 - \rho - (\rho - \phi)$, $p_1(1) = \rho$ and $p_1(0) = 1 - \rho$. In fact, this can be easily verified using the results

$$B_0^4\left(\frac{\phi}{\rho}\right) \rightarrow B_0^4\left(1 - \frac{\rho-\phi}{1-\rho}\right) = B_4^4\left(\frac{\rho-\phi}{1-\rho}\right), \tag{16}$$

and

$$B_4^4\left(\frac{\rho-\phi}{1-\rho}\right) \rightarrow B_4^4\left(1 - \frac{\phi}{\rho}\right) = B_0^4\left(\frac{\phi}{\rho}\right). \tag{17}$$

Similar expressions allows us to change B_1^4 into B_3^4 and vice-versa under the symmetry operation $0 \leftrightarrow 1$.

Next, our task is to determine an equation for $\phi = \langle \eta_i \eta_j \rangle$ where j labels one of the four nearest neighboring sites of i, we can write

$$\frac{d}{dt}\langle \eta_i \eta_j \rangle = 2\left\langle \eta_j (1 - 2\eta_i) W_i(\eta) \right\rangle. \tag{18}$$

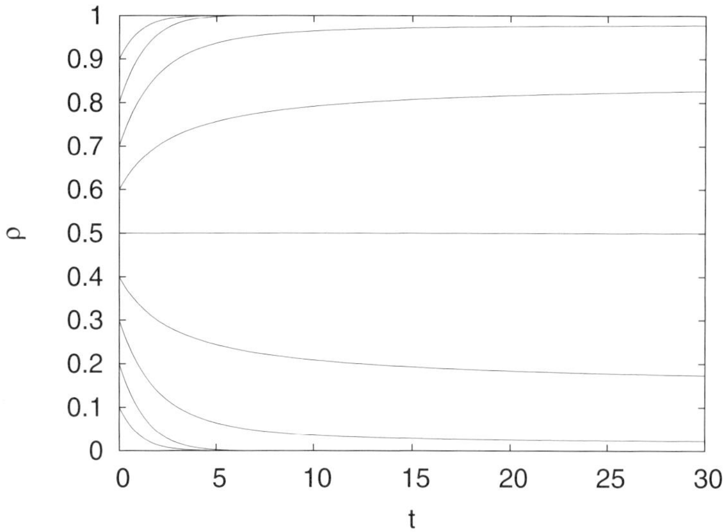

Figure 1. Time evolution of the fraction of sites in state 1, ρ, in the pair approximation obtained by solving Eqs. (12) and (20) using Euler's method with step-size 0.01 for different initial conditions (bottom to top) $\rho_0 = 0.1, \ldots, 0.9$.

Carrying out the average over η_i and η_j (say $j = 1$) explicitly yields

$$\frac{d\phi}{dt} = -2\rho p_{1|1}(1 \mid 1) \sum_{m=0}^{3} B_m^3 \left[p_{1|1}(1 \mid 1)\right] \left|\Theta[m-1] - 1\right| \\ +2(1-\rho)\, p_{1|1}(1 \mid 0) \sum_{m=0}^{3} B_m^3 \left[p_{1|1}(1 \mid 0)\right] \Theta[m-2] \qquad (19)$$

which can be written more compactly as

$$\frac{d\phi}{dt} = -2\phi B_0^3\left(\frac{\phi}{\rho}\right) + 2(\rho-\phi)\left[B_2^3\left(\frac{\rho-\phi}{1-\rho}\right) + B_3^3\left(\frac{\rho-\phi}{1-\rho}\right)\right]. \qquad (20)$$

Equations (12) and (20) determine completely the time evolution of the ρ and ϕ and so are our final equations for the pair approximation for the extended majority-vote model. Through a somewhat cumbersome algebra, which involves expressing the binomial terms explicitly in terms of their arguments, we can show that Eq. (20) is indeed invariant to the symmetry $0 \leftrightarrow 1$. Figure 1 which shows the time evolution of the density of sites in state 1 confirms this claim. Most surprisingly, this figure uncovers an unexpected dependence on the choice of the initial condition ρ_0 and

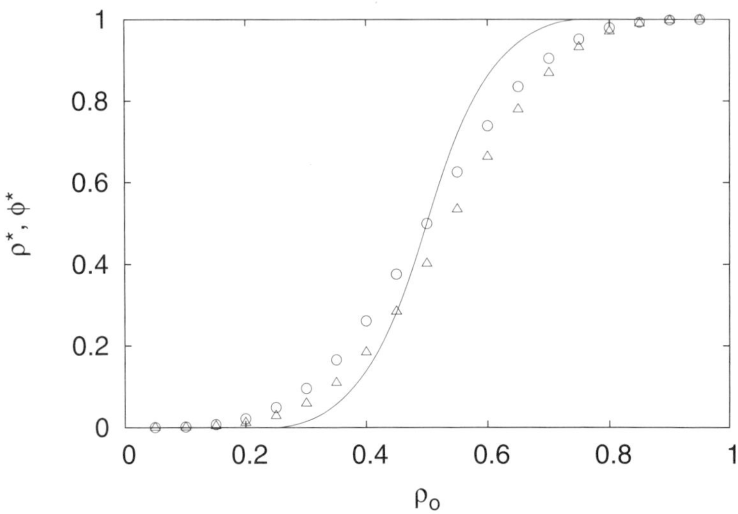

Figure 2. The fraction of sites in state 1 at equilibrium, ρ^* (represented by $\circ$), and the probability that two neighbors are in state 1 at equilibrium, ϕ^* (represented by $\triangle$) as functions of the initial fraction of sites in state 1, ρ_0. The solid line is the result of the pair approximation for which $\rho^* = \phi^*$. The initial condition is $\rho = \rho_0$ and $\phi = \rho_0^2$. The symbols show the results of the Monte Carlo simulations for a square lattice of linear size $L = 200$.

$\phi_0 = \rho_0^2$: indeed in the range $\rho_0 \in (\rho_m, 1 - \rho_m)$, where $\rho_m \approx 0.25$, the equilibrium solution ρ^* is a smooth function of ρ_0. The reason this happens is because $\rho = \phi$ solves the two equations (12) and (20) at equilibrium, i.e., $d\rho/dt = d\phi/dt = 0$, and so one of the unknowns is free to take any arbitrary value set by the dynamics.

In Fig. 2 we present the fixed point solutions of these equations for the usual situation in which the states of the sites of the initial configuration are set 1 with probability ρ_0 and 0 with probability $1 - \rho_0$. For this setting we have $\phi_0 = \rho_0^2$. The comparison with the results of the Monte Carlo simulations shows a good qualitative agreement between the pair approximation and the simulations. This is nevertheless an enormous progress compared with the single-site approximation which predicts homogeneous equilibrium states only.

Figure 3. The left panel exhibits a typical absorbing configuration of the extended majority-vote model for a lattice of linear size $L = 300$. Agents with feature 1 are painted black and agents with feature 0 are painted white. The fraction of black sites is 0.504. There are a total of $M = 822$ clusters and the largest one comprises $S_m = 10852$ agents. The right panel exhibits a random configuration where features 1 and 0 appear with the same frequency used to draw the left panel. For this configuration we have $M = 12256$ and $S_m = 396$.

4. Monte Carlo Simulations

Our aim in this section is to understand the nature of the cultural domains (clusters) that fragment the absorbing configurations of the extended majority-vote model in the thermodynamic limit. We recall that a cluster is simply a bounded region of uniform culture. In Fig. 3 we present a typical absorbing configuration of that model together with a random configuration. In order to quantify the evident differences between these two configurations we measure a few relevant statistical quantities, namely, the average number of clusters $\langle M \rangle$, the average size of the largest cluster $\langle S_m \rangle$, and the distribution of cluster sizes P_S. We must evaluate these quantities for different lattice sizes (typically we use 10^3 samples for each L) and then take an appropriate extrapolation procedure to infinite lattices ($L \to \infty$). This procedure depends on how those quantities scale with the lattice size.

To simulate efficiently the extended majority-vote model for large lattices we first make a list of the active agents. An active agent is an agent whose cultural feature is not the most frequent feature among its extended neighborhood. Clearly, only active agents can change their cultural features and so it is more efficient to select the target agent randomly from the list of active agents rather than from the entire lattice. In the case that the cultural feature of the target agent is modified by the majority-vote rule, we need to re-examine the active/inactive status of the target agent as well as of all its neighbors so as to update the list of active agents. The dynamics

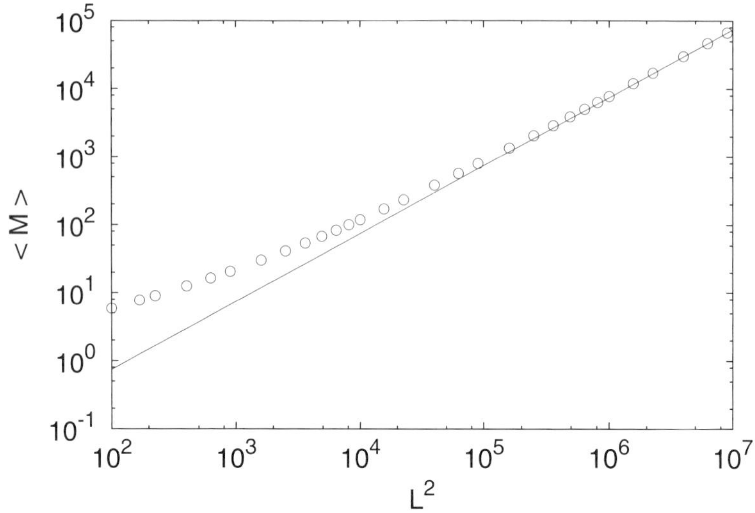

Figure 4. Logarithmic plot of the average number of clusters $\langle M \rangle$ as function of the the lattice area L^2. The solid straight line is the fitting $\langle M \rangle = 0.00742 L^2$ for large L. Each symbol represents the average over 10^4 samples so the error bar sizes are smaller than the symbol sizes.

is frozen when the list of active agents is empty. The implementation of a similar procedure allowed the simulation of Axelrod's model for very large lattices and the clarification of several issues regarding the stability of the homogeneous configuration[15,20].

We begin our analysis presenting the dependence of the average number of clusters $\langle M \rangle$ on the lattice size, Fig. 4. The increase of $\langle M \rangle$ with increasing L is the evidence that confirms that the extended majority-vote model exhibits a multicultural stationary regime in the thermodynamic limit, $L \rightarrow \infty$. More to the point, we find

$$\lim_{L \rightarrow \infty} \frac{\langle M \rangle}{L^2} = 0.00742 \pm 10^{-5}. \tag{21}$$

For the purpose of comparison, for random configurations in which sites are set to 1 or 0 with the same probability, this rate yields 0.1342 ± 0.0001. We note that the reciprocal of this rate is the average size of the clusters: $\langle S \rangle \approx 134.8$ for the extended majority-vote model and $\langle S \rangle \approx 7.45$ for random configurations.

The average size of the largest cluster $\langle S_m \rangle$ is shown in Fig. 5 which indicates that this quantity scales with $\ln L^2$ for large L. In the thermody-

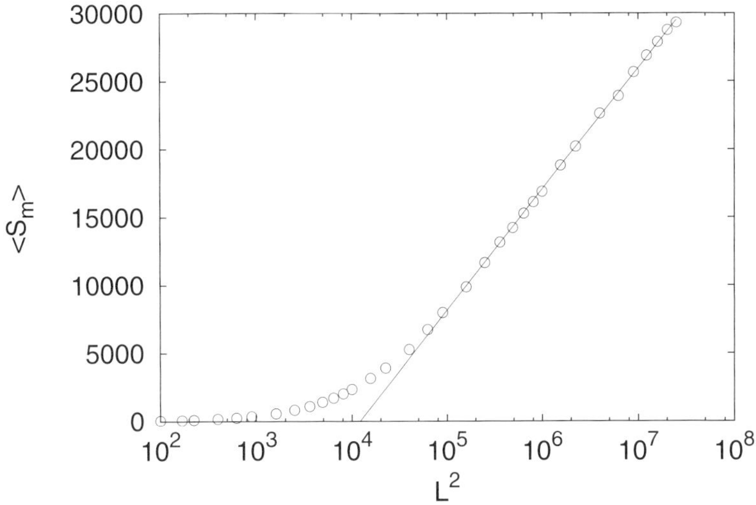

Figure 5. The average size of the largest cluster $\langle S_m \rangle$ as function of the the lattice area L^2. The solid line is the fitting $\langle S_m \rangle = 1933 \ln L^2 - 36363$. Note the logarithmic scale in the x-axis.

namic limit the relevant quantity is then

$$\lim_{L\to\infty} \frac{\langle S_m \rangle}{\ln L^2} = 1933 \pm 10. \tag{22}$$

We found that $\langle S_m \rangle$ scales with $\ln L^2$ for random configurations also, though the ratio $\lim_{L\to\infty} \langle S_m \rangle / \ln L^2 = 53.0 \pm 0.2$ is considerably smaller.

It is interesting to note that the standard deviation $\left[\left\langle S^2 \right\rangle - \langle S \rangle^2\right]^{1/2}$ tends to the constant value 695.5 ± 0.5 in the thermodynamic limit. This amounts to a large but finite variance and so in order to satisfy the Tchebycheff inequality the set of clusters whose size grows like $\ln L^2$ must have zero measure. The probability that a randomly chosen site belongs to one such a cluster vanishes like $(\ln L)/L^2$ and so the measure of that set will be vanishingly small if its cardinality is finite.

Up to now we found no qualitative differences between the organization of clusters produced by the extended majority-vote rule or by assigning randomly the digits 1 and 0 to the sites of the square lattice. In fact, for both types of configurations the quantities $\langle M \rangle$ and $\langle S_m \rangle$ exhibit the same scaling behavior with the lattice size L. However, a more detailed view of the cluster organization is given by the distribution P_S of cluster sizes S,

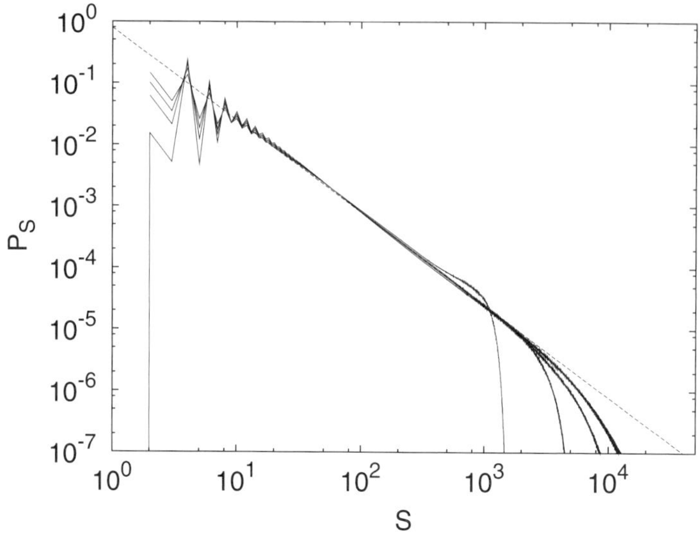

Figure 6. Logarithmic plot of the distribution of cluster sizes S for (solid lines from left to right at $P_S = 10^{-6}$) $L = 50, 100, 200$ and 1000. The dashed straight line is the fitting $P_s = 0.79S^{-1.5}$. The distribution was obtained through the analysis of 10^7 absorbing configurations.

which is shown in Fig. 6 for different lattice sizes. The data is well fitted by the power-law $P_S \sim S^{-1.5}$ over more than three decades. In addition, the figure indicates that the region where the power-law fitting holds increases with increasing L. There is a difficulty, however, to mingle this power-law distribution with our previous findings of finite values for the mean and variance of P_S, since the distribution $P_S \sim S^{-1.5}$ has infinite mean and variance. To answer this conundrum we present in Fig. 7 the same data plotted in a semi-logarithmic scale. This figure shows that for large lattice sizes the distribution P_S is in fact exponential, the power-law behavior being valid only in an intermediate regime of not too large clusters.

Finally, we note that the distribution of cluster sizes for the random configurations illustrated in Fig. 8 is an exponential distribution $P_s \approx 0.00085 \exp(-0.02S)$ for large S, which then differs qualitatively from the seemingly power-law behavior depicted in Fig. 6 for the extended majority-vote model.

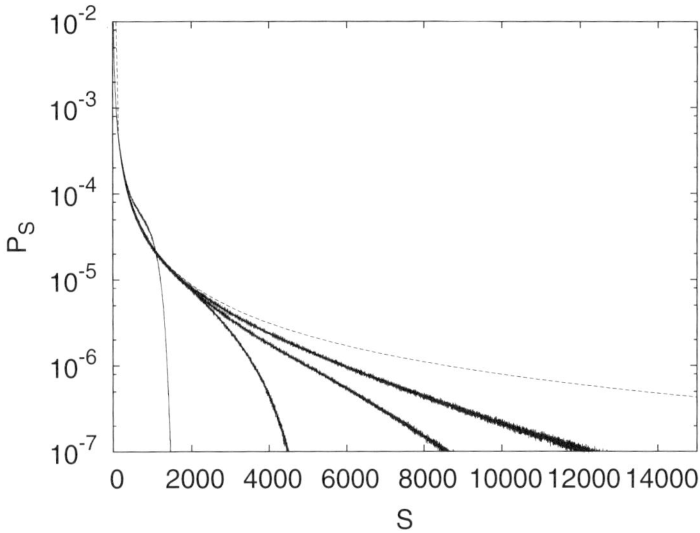

Figure 7. Same as Fig. 6 presented in a semi-logarithmic scale.

5. Conclusion

From the perspective of the statistical physics, we found the claim that the extended majority-vote model exhibits a nontrivial multicultural regime most intriguing[16]. In fact, we expected it to behave similarly to the majority-vote models commonly considered in the physics literature, which are basically Ising models with zero-temperature Glauber kinetics[19], and so exhibit only two homogeneous absorbing configurations in the thermodynamic limit. Our first impression was that the multicultural regime was an artifact of the small lattice size ($L = 20$) or of some 'non-physical' elements of the original model such as the nearest and next-nearest interactions (Moore neighborhood) and the parallel update of the sites.

Accordingly, we modified the original model proposed by Parisi *et al.* by introducing the usual nearest neighbors interactions (von Neumann neighborhood) and the random sequential update of sites. A careful finite size scaling analysis of a few relevant measures that characterize the absorbing configurations demonstrated that the non-homogeneous absorbing configurations not only persist in the thermodynamic limit but seem to produce a non-trivial distribution of cluster sizes P_S that decays as a power law $P_S \sim S^{-1.5}$ for large but not too large cluster sizes S. For very large clus-

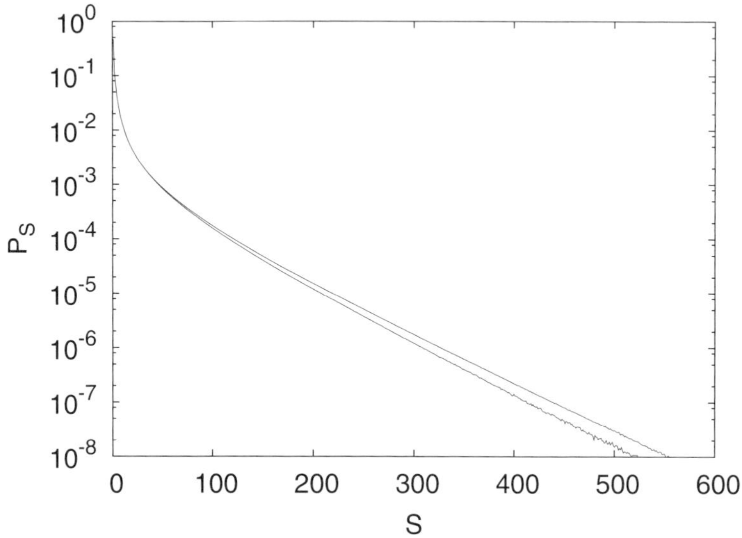

Figure 8. Semi-logarithmic plot of the distribution of cluster sizes S for random configurations with (solid lines from bottom to top) $L = 100$ and 300. The distribution was obtained through the sampling of 10^8 random configurations.

ter sizes, the distribution is exponential (see Figs. 6 and 7). Essentially, the reason for this somewhat unexpected outcome is the criterion of update in case of a tie for a site in the lattice bulk: the site remains unchanged rather than flipping to another state with probability 1/2 as usually done in statistical mechanics models[18].

This finding shows that, although the majority-vote rule actually biases the agents to become more similar to each other, the model does exhibit a stable multicultural regime. A similar conclusion holds for Axelrod's model as well[6,14,15], except that in Axelrod's model the similarity is a prerequisite for interactions - the 'birds of a feather flock together' hypothesis which states that individuals who are similar to each other are more likely to interact and then become even more similar[2]. (A similar assumption has been used to model the interspecies interactions in spin-glass like model ecosystem[21].) The majority-vote model is considerably simpler and converges to the absorbing configurations much faster than Axelrod's. However, the (short) inventory of advantages stops here: in disagreement with the claim of Parisi *et al.*[16] we found that the absorbing configurations of the extended majority-vote model are vulnerable to the noisy effect of flipping the cultural traits of the agents with some small probability (results not shown) It is well-known that this type of noise destabilizes the heterogeneous configurations of Axelrod's model too[9]. Of course, the extended majority-vote model lacks the main appealing feature of Axelrod's model, namely, the existence of a non-equilibrium phase transition that separates the homogeneous and the polarized regimes in the thermodynamic limit. In that sense it would be interesting to find out how a similar transition could be induced in the zero-temperature majority-vote model.

Acknowledgments

This research was supported by The Southern Office of Aerospace Research and Development (SOARD), grant FA9550-10-1-0006, and Conselho Nacional de Desenvolvimento Científico e Tecnológico (CNPq).

References

1. B. Latané, *American Psychologist* **36**, 343 (1981).
2. S. Moscovici, *Handbook of Social Psychology* **2**, 347 (1985).
3. R. Boyd and P. J. Richerson, *Culture and the evolutionary process* (University of Chicago Press, Chicago, 1985).
4. A. Nowak, J. Szamrej and B. Latané, *Psychological Review* **97**, 362 (1990).
5. C. Castellano, S. Fortunato and V. Loreto, *Rev. Mod. Phys.* **81**, 591 (2009).
6. R. Axelrod, *J. Conflict Res.* **41**, 203 (1997).
7. R. L. Goldstone and M. A. Janssen, *Trends Cog. Sci.* **9**, 424 (2005).
8. J. Kennedy, *J. Conflict Res.* **42**, 56 (1998).
9. K. Klemm, V. M. Eguíluz, R. Toral, M. San Miguel, *Phys. Rev. E* **67**, 045101R (2003).
10. J. M. Greig, *Conflict Res.* **46**, 225 (2002).
11. K. Klemm, V. M. Eguíluz, R. Toral, M. San Miguel, *Physica A* **327**, 1 (2003).
12. K. Klemm, V. M. Eguíluz, R. Toral, M. San Miguel, *Phys. Rev. E* **67**, 026120 (2003).
13. J. Marro and R. Dickman, *Nonequilibrium Phase Transitions in Lattice Models* (Cambridge University Press, Cambridge, UK, 1999).
14. C. Castellano, M. Marsili and A. Vespignani, *Phys. Rev. Lett.* **85**, 3536 (2000).
15. L. A. Barbosa and J. F. Fontanari, *Theory Biosci.* **128**, 205 (2009).
16. D. Parisi, F. Cecconi, F. Natale, *J. Conflict Res.* **47**, 163 (2003).
17. T. M. Liggett, *Interacting Particle Systems* (Springer, New York, 1985).
18. M. J. de Oliveira, *J. Stat. Phys.* **66**, 273 (1992).
19. R. J. Glauber, *J. Math. Phys.* **4**, 294 (1963).
20. L. R. Peres and J. F. Fontanari, *J. Phys. A: Math. Theor.* **43**, 055003 (2010).
21. V. M. de Oliveira and J. F. Fontanari, *Phys. Rev. Lett.* **89**, 148101 (2002).

ESTIMATING THE PHOTOSYNTHETIC INHIBITION BY ULTRA VIOLET RADIATION ON THE ANTARCTIC PHYTOPLANKTON ALGAE*

CARLOS MARIO OCAMPO MARTÍNEZ†

Maestra en Biomatemáticas, Universidad del Quindío,
Cra. 15 calle 12 N, AA 460.

FERNANDO ROBERTO MOMO

Instituto de Ciências, Universidad Nacional de General Sarmiento,
Los Polvorines (Malvinas Argentinas) Buenos Aires, B1613GSX, Argentina

GLADYS ELENA SALCEDO ECHEVERRY

Departamento de Matemáticas, Universidad del Quindío,
Cra. 15 calle 12 N, AA 460, Armenia (Quindío), Colombia.

The thinning of the ozone layer has resulted into an increase in the UVR (Ultra Violet Radiation) doses entering the water column in the Antarctic Sea; such that together with several physical chemical influential factors in the vertical mixing are affecting photoinhibition to the phytoplankton. Starting with a single mathematical model among other factors the algal cell self-repairing during a vertical mixing cycle, we obtained on 1-hr basis estimations of photosynthetic inhibition of the Antarctic algae. We obtained such estimations by using data from Melchior scientific station, a research facility located in the Archipielago Melchior, in the Billingshausen Sea. Samplings were registered between February 11th and March 10th, 2002. We calculated two photosynthetic inhibition time series: The first one by using the irradiance series measured at these wavelengths: 305, 320, 340 and 380 nm. A second series was obtained by using the irradiance series calculated after verifying the statistical hypothesis related to the proportionality of the sampled series.

Introduction

The so-called ozone holes are those zones in the atmosphere where an abnormal reduction or a thinning is experienced by the ozone layer. This event

*This work is supported by grants conacyt 0125851, p51739-r, 105844.
†Tcorresponding autor e-mail:cmom24@gmail.com.

annually occurs in Polar Regions during spring time and it is followed by the ozone layer recovery taking place on summer time. In the Antarctic the ozone reduction attains a 70% during spring. This phenomenon, which is still going on, was documented and demonstrated by Sir Gordon Dobson (G.M.B. Dobson) in 1960.

He regarded this occurrence to the extreme climatological conditions reigning in the Antarctic continent.

Ultraviolet radiation (UVR, 280-400 nm) is known to strongly affect chemical and biological processes in the aquatic medium [3, 6], and it is still more important if considering its effect on the loss of extratospheric ozone due to the chlorofluorocarbon compounds (CFC) emitted by human activities as well, which deteriorated the ozone layer, leading to an increase in the UVR penetrating the atmosphere.

Phytoplankton algae suffer from damages at several levels because of the increased UVR levels, including the one by photosynthetic inhibition [1, 4, 10, 11, 12], which in turns alters the algae numbers and size and the algae oxygen supply to the oceans. The UVR reaching each depth within the ocean, and thereby its phytoplankton, depends on the physical [7] and chemical properties affecting the vertical mixing in the water column [9] and on the wavelength of the irradiance penetrating at different depths. In addition, the fact that when there is a higher temperature in the air mass nearby the surface, the water surface becomes warmer than the layers below, and so the water column gets strongly stratified implying that algae do not circulate and remain for long time at certain water depth.

In this study we used data collected from February 11th 2002 through March 9th 2002 obtained in the Melchior Scientific Research facility located in Archipelago Melchior of the Bellingshausen Sea. An effective damage or cumulative percentage of photoinhibition was calculated for the Antarctic phytoplankton algae by considering the incident irradiance, chemical and physical factors and the vertical mixing of the water column. These factors were used in the BWF-PI model (BWF-Photoinhibition) [8] which combines a proper BWF (Biological Weighting Function) [10] and the phytoplankton response to its RUV exposure in a single model that implicitly entails the algae self-repairing during the vertical mixing cycle. On the other hand, we also estimated a percentage of RUV cumulative photoinhibition damage by using estimated irradiance series at 320, 340 and 380 nm-wavelengths. These irradiance series were obtained from the irradiance series measured at 305 nm-wavelengths.

1. The data analysis

The variables of study were: Incidence irradiance, air temperature, wind speed and tidal range which were registered every hour from February 11th to March 10th, 2002. We also have the extinction coefficient profiles and the incident irradiance on the water surface at 7:00, 13:00 and 19:00 hours. In what follows we call these hourly samples as time series.

1.1. *Irradiance time series analysis*

We can observe in Figure 1 that the irradiance time series for the wavelengths 305, 320, 340 and 380 nm denoted by i305, i320, i340 and i380, respectively, present two well defined characteristics: the presence of a periodic component that explains its seasonal behavior and a strong similarity among them. In order to verify statistically the presence of hidden periodicities we apply the test of Fisher [16], and to verify the similarity of the four irradiance series, we apply a test that compares the similarity among the respective periodograms as given in Salcedo [15]. The general idea consists in to compare the spectral density function of Series i305 versus the spectral density functions of Series i320, i340 and i380, respectively. We can say that the behavior of the four irradiance time series is very similar except by a scale variation on horizontal axes due to the different wavelengths. Hence, before applying the Fisher test we must transform the series to put them at the same scale. The estimated proportionality coefficients were $\hat{\alpha}_1 = 13,9$, $\hat{\alpha}_2 = 31$ and $\hat{\alpha}_3 = 43,6$ which satisfy that $i320 = \hat{\alpha}_1 i305$, $i340 = \hat{\alpha}_2 i305$ and $i380 = \hat{\alpha}_3 i305$. After correcting the scale, we can test the presence of hidden periodic components through a periodogram analysis. In Figure 2 appear the periodograms of every irradiance series and we can observe clearly that they are dominated by a very large peak at frequency $w = 28/671$. This frequency corresponds to a period of $P = 1/w = 23.9642 = 24$hours. It indicates that the data exhibit an approximated 24-hour cycle that is expected due to the solar cycle. This periodic component can be removed to avoid biases in the test of similarities among the four irradiance time series. The transformation $Y_t = (1 - B^{24})X_t = X_t - X_{t-24}$ is proper to remove the seasonal deterministic components, where X corresponds to every irradiance time series. The periodograms for the new time series Y appear in Figure 3. The comparison procedure consists in to verify the equality of these new periodograms that represent the series Y, respectively. Since the *p-values* are higher than the level of significance 0.05, we cannot reject the null hypothesis of equality of periodograms; hence we can conclude that the irradiance time series are really proportional.

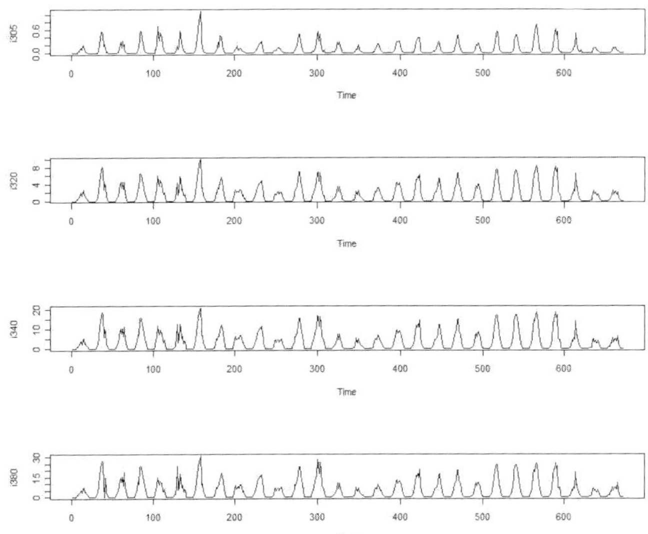

Figure 1. Irradiance time series i305, i320, i340 and i380 nm.

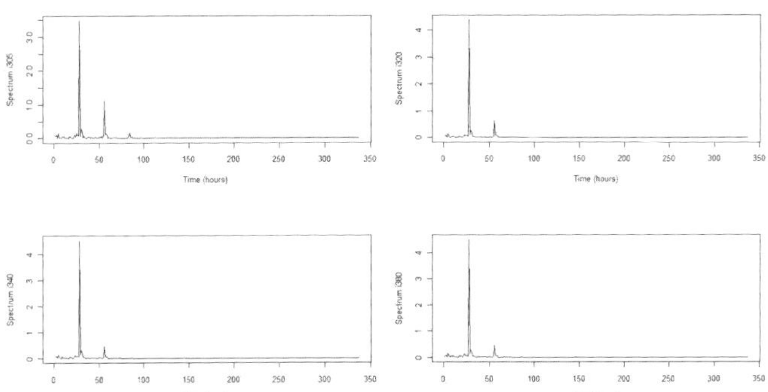

Figure 2. Periodograms of the times series i305, i320, i340 and i380 nm.

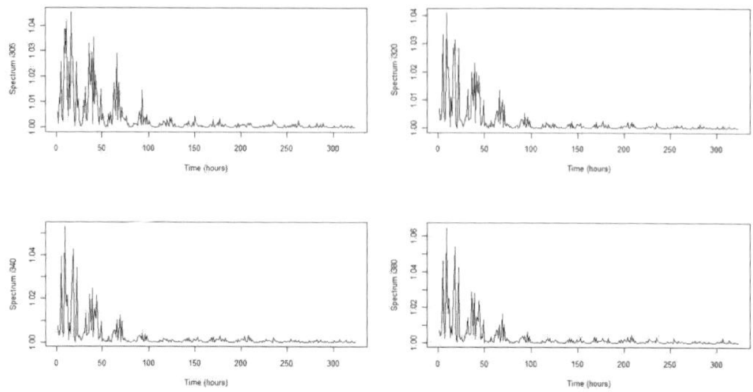

Figure 3. Periodograms of the times series i305, i320, i340 and i380 nm after removing the seasonal deterministic components.

2. Photoinhibition estimated based on the sampled irradiances

The cumulative rate of photosynthetic effective damage on the algae of phytoplankton due to the irradiance is estimated considering several physical, chemical and biological factors. From an integrated model of photosynthetic response to the UVR, i.e. a BWF-PI (BWF-Photoinhibition) model [8], given by the expression

$$P^B = \frac{{P_{opt}}^B}{1 + E^*_{inh}} \tag{1}$$

where P^B_{opt} represents the optimal rate of photosynthesis in absence of photoinhibition and the achieved rate of photosynthesis; E^*_{inh}represents the weighted inhibitory irradiance or the total damage which is dimensionless. From Eq. (1) can be determined the expression for the rate of photoinhibition given by

$$P^B = \frac{{P_{opt}}^B}{1 + E^*_{inh}} \tag{2}$$

The weighted inhibitory irradiance is a dimensionless measure of the inhibitory dose of irradiance in the water column calculated as function of the vertical mixing at depth Z, the time t, the spectral irradiance on the surface of the seawater I and the attenuation coefficient k. E^*_{inh} is a modification of the expression given by Neale P. J. *et al.* [8] and is defined as

$$E_{inh}^{*} = \sum_{305nm}^{380nm} \varepsilon(\lambda) IM(\lambda) \Delta\lambda \tag{3}$$

where $\varepsilon(\lambda)$ is the BWF (Biological Weighting Function) $\left(\mathrm{mWm}^{-2}\right)^{-1}$ at the wavelength λ (nm) and $IM(\lambda)$ $\left(\mathrm{mWm}^{-2}\mathrm{nm}^{-1}\right)$ [7] is the light intensity in the water column at λ. There exist different expressions to estimate the BWF [5, 10, 14] depending on the best fit to the study problem. In this case, we use a function whose curve-shaped line falls to zero and depends on the incident wavelength λ[10]. This function can be simplified if we suppose it almost linear on the logarithmic scale, and is expressed as

$$\log \varepsilon(\lambda) = b(\lambda - 280) + c \tag{4}$$

where the factor $(\lambda - 280)$ will be the starting point or reference point, $b = -0.025$ and $c = -2.5$. Accordingly to Momo et al. [8], in a coastal Antarctic setting with a depth less than 50 m, as is the case of our data, primary production (photosynthesis) is influenced by physical and biological factors, such as the depth and the vertical mixing due mainly to the wind effect, the attenuation of light intensity in the water column and the seawater density. Thus, the chlorophyll quantity that can be found in the water column depends on the quantity of light $(IM(\lambda))$ in it, which depends on the light extinction coefficient k , the irradiance on the water surface I, the depth and water column mixing speed Z. Then, the expression for the light intensity in the water column is given by

$$IM(\lambda) = \frac{I_0}{2}\left(1 + e^{-k_d Z_t}\right) \tag{5}$$

Values for the light extinction coefficient k_d, can be obtained from the profiles determined at 7:00, 13:00 and 17:00 hours in the Melchior Scientific Research Station. The mixing depth Z is calculated from a modification of expression proposed by Momo *et al.* [8] and is given by

$$Z_t = \sqrt{\frac{0,0018K_w}{9800\frac{d\rho}{dz}} U^{\frac{3}{2}}} \tag{6}$$

which is proportional to the wind speed U and duration, and inversely proportional to the extent of water stratification $d\rho/dz$. This value represents the depth at which phytoplankton algae gets stirred up in the water column. Values for $\rho(s,t,p)$ and k_w were determined using the international equations of UNESCO [2]. Seawater density depends on the variables Salinity (s), Temperature (t) and Pressing (p) that were taken from the profiles accomplished at the study field at 1:00, 7:00, 13:00 and 19:00 hours.

The time series U in Eq. (6) was estimated every six hours, at 1:00, 7:00, 13:00 and 19:00 hours, since the mixing depth to a given time depends on the wind speed at previous hour. I_0, corresponds to the irradiance series i305, i320, i340 and i380, respectively.

From Eq. (2) we obtained a hourly time series for the photosynthetic inhibition rate by UVR on the algae of Antarctic phytoplacton, which represents the cumulative effective damage. This cumulative photoinhibition by UVR appears in Figure 4 for every study day between 7:00 and 20:00 hours. We can observe photoinhibition values up to 7.6% where the extreme values for the photoinhibition correspond to the first days of the study. Significant values for the photoinhibition occur between 9:00 and 18:00 hours with some isolated peaks up to 20:00 hours attaining its maxima values between 12:00 and 14:00 hours. From Figure 4, we also can observe that in general, the photoinhibition rate decreases according to the change of the end of summer to the beginning of fall.

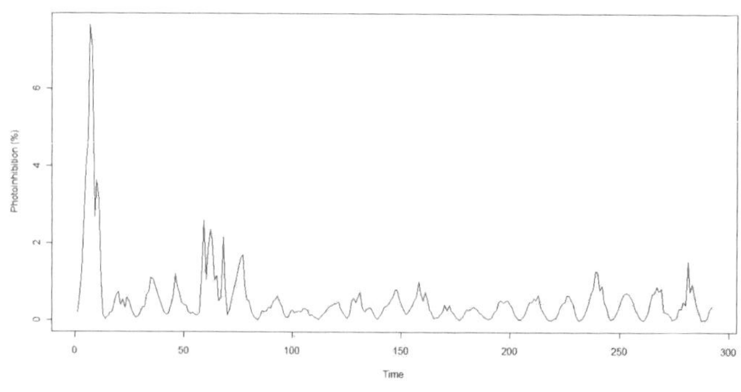

Figure 4. Hourly time series for the cumulative photosynthetic inhibition rate by UVR on the algae of Antarctic phytoplankton between February 11th and March 9^{th}9, 2002.

Figure 5 exhibits the weighted inhibitory irradiance E^*_{inh} for the differ-

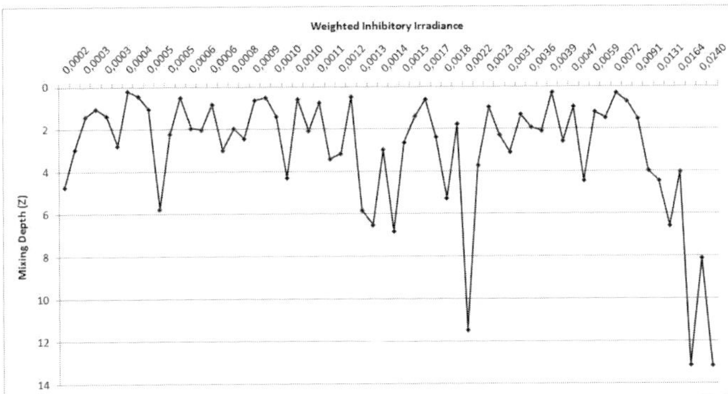

Figure 5. Weighted inhibitory irradiance E^*_{inh}

ent mixing depths at 7:00, 13:00 and 19:00 hours for the study days. In general, E^*_{inh} can penetrate the water column up to 4 m of depth with some extreme cases wherein can penetrates up to 14 m of depth.

In particular, Figure 6 shows the calculated photoinhibition for the days with the highest values. Those days were February 12th, 16th and 17th, corresponding to the end of summer in the Antarctic Ocean. In order to compare this situation with those days for which photoinhibition levels were lowest, in Figure 7 we give the calculated photoinhibition for February 14th and 24th. Notice that while in Figure 6 the photoinhibition attains values up to 8% approximately, in Figure 7 it only attains values up to 1.1%.

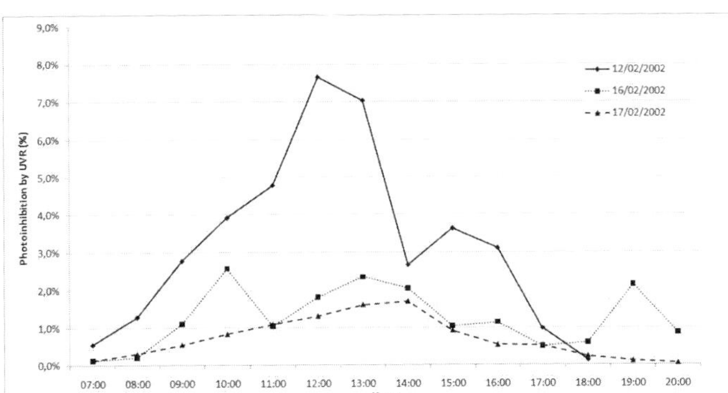

Figure 6. Photosynthetic inhibition rate by UVR on the algae of Antarctic phytoplankton for February 12th, 16th and 17th

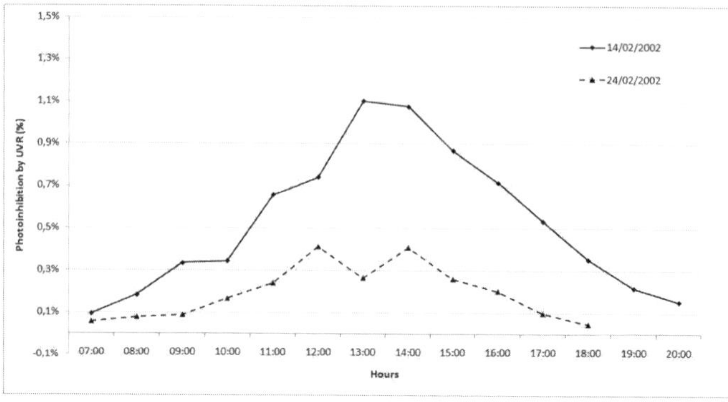

Figure 7. Photosynthetic inhibition rate by UVR on the algae of Antarctic phytoplankton for February 14th and 24th.

3. Estimated photoinhibition based on the estimated irradiance

In Section 1.1 we verified the null hypothesis of proportionality among the sampled irradiance time series. Based on that conclusion, in this section we again estimate the cumulative photosynthetic inhibition rate by UVR on the algae of Antarctic phytoplankton, from Series i305. Figure 8 exhibits the couple of time series of cumulative photoinhibition rate; the solid line represents the photoinhibition based on the sampled irradiances and the dashed line represents the photoinhibition based on the estimated irradiances. The two time series are very approximated, this fact suggests that the researches can calculate the photoinhibition from an only set of irradiance data; in our case, we used the Series i305. We also calculate the photoinhibition for days February 14th and 24th which appear in Figure 9. Notice again the similarity with the curves in Figure 7, indicating again the viability of calculate the photoinhibition daily from Series i305.

Conclusions

(1) The irradiance time series i305, i320, i340 and i380 at different wavelengths after correcting the scale and removing the seasonal component are statistically similar. Then, it is possible to use the sampled irradiance at wavelength 305nm to estimate the irradiance at wavelengths 320, 340 and 380 nm.

(2) The cumulative photosynthetic inhibition rate by UVR on the al-

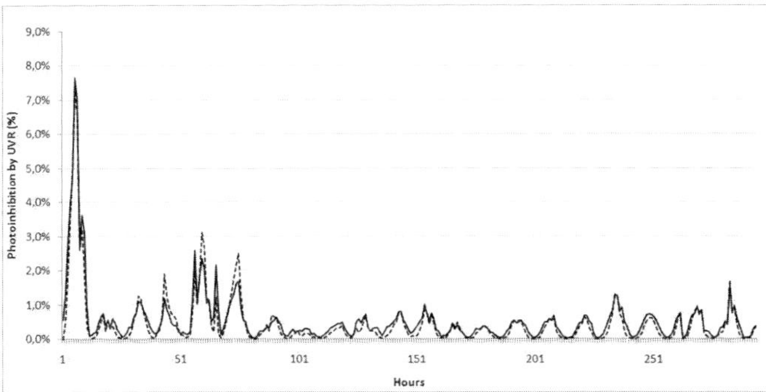

Figure 8. Time series of cumulative photoinhibition rate by UVR based on the sampled irradiance (solid line) and the estimated irradiance (dashed line).

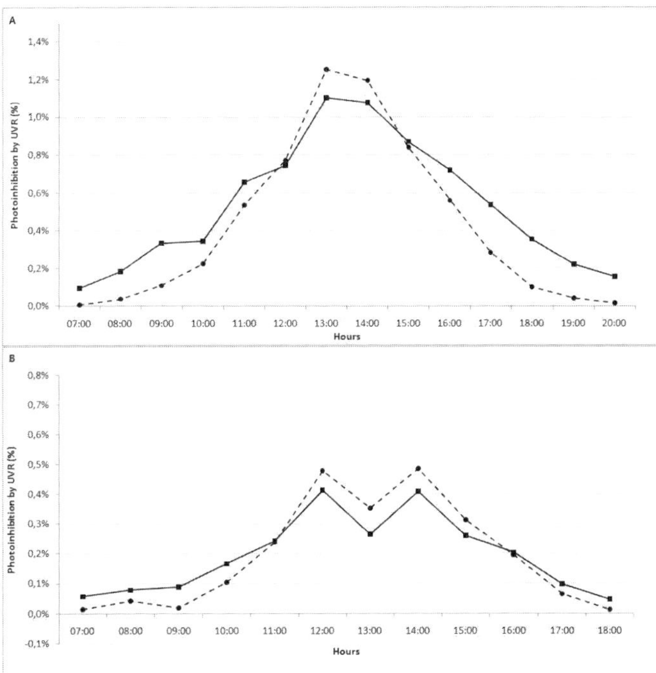

Figure 9. Time series of Photosynthetic inhibition rate based on the sampled irradiance (solid line) and the estimated irradiance (dashed line) for (A) February 14th and (B) February 24th.

gae of Antarctic phytoplankton attains values of photoinhibition up to 7.6% between 12:00 to 14:00 hours, for some days in particular.

(3) The irradiance that most has influenced the photosynthetic inhibition on the algae of Antarctic phytoplankton is the irradiance at wavelength 320 nm.

(4) From sampled irradiance at wavelength 305nm it is possible to obtain good approximations for both the cumulative photosynthetic inhibition and the daily photoinhibition.

Acknowledgments

The authors acknowledge the support provided by of University of Quindío and National University of General Sarmiento. Authors also are thankful to Ph.D. Gustavo Ferreira for providing the data from Melchior Scientific Research Station in Argentina, and Ph.D. Pedro Pablo Cardona by his suggestions.

References

1. G. J. Anita, E. Buma, W. Helbling, M. Karin de Boer, V. E. Villafae, *Photochemistry and Photobiology: Biology* **62**, 9-18 (2001).
2. N. P. Fofonoff, R. C., Millar Jr., *UNESCO Technical Papers in Marine Science* **44**, (1983).
3. Instituto de Ciencias del Mar. Caractersticas generales de la Antrtida [on line], *http://www.icm.csic.es/bio/outreach_s.htm*, cited in October, 2009.
4. J. J. Cullen, P. J. Neale, M. P. Lesser *Science* **258**, 645-650 (1992).
5. J. J. Cullen, P. J. Neale, M. P. Lesser, *The effects of Ozone Depletion on Aquatic Ecosystems* **R.G. Landes Company**, 97-118 (1997).
6. F. R. Momo, E. Ferrero, M. Eöry, M. Esusy, J. Iribarren, G. Ferreira, I. Schloss, B. Mostajir, S. Demers, *Modeling Community-Level Effects of UVR in Marine Ecosystems Photochemistry and Photobiology* **82**, 903-908 (2006).
7. F. R. Momo, I. Scholss, *Congreso Latinoamericano de Biomatemática*, Guanajuato, México, Noviembre 2002.
8. P. J. Neale, *Cambridge University Press* **72**, 100 (2000).
9. P. J. Neale, R. F. Davis, J. J. Cullen, *Letters to Nature* **392**, 585-589 (1998).
10. P. J. Neale, M. P. Lesser, J. J. Cullen, *Antarctic Research Series* **62**, 125-142 (1993).
11. P. J. Neale, *Photochemistry and Photobiology: Biology* **62**, 1-8 (2001).
12. P. J. Neale, *Acuatic Sci.* **63**, (2001).
13. K. E. Rasmus, W. Granéli, S. A. Wangberg, *Deep-Sea Research II* **51**, 2583-2597 (2004).

14. R. D. Rundel, *Physiol. Plant.* **58**, 360-366 (1983).
15. G. E. Salcedo, *Msc Thesis (in portuguese). Univ. of S. Paulo, Brazil* (1999).
16. W. S. Wei, *Time Series Analysis: Univariate and Multivariate Methods*, Addison-Wesley Publishing Company Inc., (1990).

THE SPATIOTEMPORAL DYNAMICS OF AFRICAN CASSAVA MOSAIC DISEASE

Z. LAWRENCE and D. I. WALLACE
Department of Mathematics
Dartmouth College, HB 6188
Hanover, NH, 05755, USA
E-mail: Dorothy.Wallace@Dartmouth.edu

African Cassava Mosaic Disease, a vector-borne plant disease, causes massive food shortages throughout sub-Saharan Africa. A system of ordinary differential equations is used to find the equilibrium values of the whitefly vector and the cassava plants it affects. The temporal ODE system is modified to incorporate the spatial dynamic. The resulting system of advection-diffusion equations is analyzed using finite differencing in MATLAB to assess the spatiotemporal spread of ACMD. The partial differential equations system is systematically altered and solutions are assessed in terms of the relative cassava yield they predict. Simulations include parameter sensitivity analysis, spatial modifications, analysis of the impact of a source term, and initial condition variance. Results are compared with field data. Practical implications of these simulations for controlling ACMD are explored. Data suggests that the use of windbreaks and ACMD resistant strains of cassava will have the most beneficial impact on cassava yield.

1. Introduction

The average person in Africa eats about 80 kg of cassava per year [1] and total annual production is approximately 85 million tons, more than any other crop in the African continent [2]. A starchy, potato-like source of carbohydrates, cassava root is a versatile food source that can be prepared and consumed according to a variety of different methods. Cassava is a major part of the diets of many of the world's citizens; the health and stability of this staple sub-Saharan African food crop is therefore of the utmost importance. Cassava production has been severely affected by a devastating virus known as the African Cassava Mosaic Disease (ACMD). First discovered in 1928, this virus is transmitted via the whitefly *Bemisia tabaci Gennadius* and can result in substantial losses to cassava harvests. [3] An affected plant has a chlorotic mosaic on infected leaves. Although the stem and roots of the cassava plant do not show symptoms of the disease, chlorosis of the leaves interferes with photosynthesis and affected plants thus experience reduced tuberization [4]. Significant disease spread can therefore result in drastically reduced yield of cassava root, the starchy, edible part of the plant.

Bemisia tabaci Gennadius are commonly referred to as whiteflies due to the bright white shade of their wings. Whiteflies are vectors for ACMD and if a whitefly is carrying ACMD, it may inject the virus into the plant while feeding [5]. Additionally, female whiteflies may deposit up to 300 eggs into the mesophyll of the plant leaves. As the site of development of the eggs, the cassava plant plays a vital role in the lifecycle of this pest. Without plants on which to feed and lay eggs, the whiteflies cannot survive. Whiteflies have two different flight patterns. Short distance flights are less than 15 feet in diameter and are generally shaped like a loop. Long distance flights, in which the insects passively drift in air currents, are dependent on wind and can carry the insect significant distances depending on the properties of the air currents [6]. Adult whiteflies have limited ability to direct their flight [7] but will choose to stay on or leave a plant host based on its suitability for feeding and breeding [6]. Movement of the insects and the disease they carry is thus highly dependent on the wind.

So what is being done to solve one of African agriculture's biggest problems? Several different approaches have already been analyzed in attempts to hinder the spread of this virulent virus. Phytosanitation is defined as the use of virus-free stem cuttings as planting material matched with the removal, or roguing, of infected plants from within the field. This approach focuses on systematically removing infected plant material [8]. But infected plants do not always display clear symptoms and diseased material can be introduced into new areas through the use of infected cuttings [3]. The existence of ACMD resistant varieties of cassava has been a major breakthrough in the development of strategies to combat this disease. The known resistant varieties show lower rates of ACMD, develop inconspicuous symptoms, and exhibit infection that is generally "less completely systematic than in more susceptible plants" [3]. Previous mathematical analysis on ACMD has often relied on geostatistics to make sense of field experiments and explain the spatial spread of the whitefly vector [9, 10]. Such analysis relies on the extrapolation of collected data and has shown that the spatial dynamics of ACMD and other such diseases are influenced by wind patterns. Additionally, various plant diseases have been studied using stochastic methods. Such analysis incorporates the probabilistic nature of disease to model transmission through non-deterministic methods. [11, 12,

13, 14].

Previous studies modeling the spatiotemporal dynamics of disease have relied on partial differential equations to demonstrate diffusion. Such research focusing on disease patterns in mammals has shed light on the mathematics governing vector-transmitted epidemics and the importance of spatial dynamics in understanding and analyzing disease spread [15, 16, 17]. An application of these methods to ACMD will provide valuable insight into the spread of the whitefly vector in space. The use of a spatiotemporal model of ACMD will enable analysis on possible methods of limiting disease incidence with the ultimate goal of maximizing harvest and preventing deadly food shortages.

In section 2 we present a temporal model for ACMD from the literature. Section 3 extends the model to the spacial domain. In section 4 we describe the numerical methods used to analyze the system, as well as describing convergence issues. Section 5 contains two analyses of sensitivity, on for the temporal model and one for the extend spacial model, and compares these. Sections 6,7, 8, and 9 contain numerical experiments on ACMD control via windbreaks and alternative spacial arrangements of plantings. Section 10 summarizes the results.

2. Holt's temporal model

Holt *et al* [3] conducted a study to model the temporal dynamics of ACMD with the goal of understanding how different variables impact the kinetics of the disease. Since contact between healthy and diseased plants occurs by means of the whitefly vector, a system of four differential equations representing healthy plants, diseased plants, non-infective whiteflies, and infective whiteflies is developed to explain the progression of the disease [3]. Holt's analysis considers the effects of the susceptibility of the plant to ACMD, the extent of the use of healthy as opposed to diseased plants for cutting, and the extent of roguing of diseased plants. Other factors that are likely to effect disease dynamics such as the intensity of cropping, the rate of crop turnover, the extent of reversion, the virulence of the disease, and the dynamics of the vector population and disease transmission are included in the model. Parameters are estimated to best represent real-world transmission of ACMD [3].

In this model, X represents the number of healthy plants, Y represents the number of diseased plants, U represents the non-infective whitefly population, and V represents the infective whitefly population. The following equations are defined:

$$dX/dt = rX(1 - (X + Y)/K) - dXV - gX \tag{1}$$

$$dY/dt = dXV - aY - gY \tag{2}$$

$$dU/dt = b(U + V)(1 - (U + V)/m(X + Y)) - eYU - cU \tag{3}$$

$$dV/dt = eYU - cV \tag{4}$$

Holt [3] gives parameter values reproduced in Table 1. Note that the parameter range for growth rate has been corrected from the original paper with the author's permission.

Table 1. Parameter values and ranges.

Parameter	$StandardValue$	$Range$
K, plant density	$0.5\ m^{-2}$	0.01-1
r, growth rate	$0.05\ day^{-1}$	0.025-0.2
a, plant loss/roguing rate	$0.003day^{-1}$	0-0.033
g, harvesting rate	$0.003day^{-1}$	0.002-0.04
m, maximum vector abundance	$500plant^{-1}$	0-2500
b, maximum vector birth rate	$0.2day^{-1}$	0.1-0.3
c, vector mortality	$0.12\%day^{-1}$	0.06-0.18
d, infection rate	$0.008vector^{-1}day^{-1}$	0.002-0.32
e, acquisition rate	$0.008vector^{-1}day^{-1}$	0.002-0.0032

Setting all equations equal to zero and solving for the equilibrium yields the following values :

$$X_{equ} = 0.07557 \tag{5}$$

$$Y_{equ} = 0.30382 \tag{6}$$

$$U_{equ} = 74.434 \tag{7}$$

$$V_{equ} = 1.5076 \tag{8}$$

Initial conditions for all variables in the spaciotemporal simulations are set near these equilibrium values.

3. A spaciotemporal model

The system of ordinary differential equations, in equations 1 to 4, represents the temporal dynamics of the disease. In actuality, however, the perpetuation of ACMD is highly dependent on spatial variables. Thus, we use Holt's ODE model as a basis and modify it to incorporate the spatial dynamics of the spread of the disease.

The whitefly vector should be equally likely to fly in any direction. This movement, the characteristic short distance flight patterns of whiteflies, can be represented mathematically by diffusion.

Cassava plants are stationary objects that do not diffuse. Only the whiteflies carrying the disease can be thought of as diffusing bodies. Assuming that diffusion along the z-axis, or depth-wise in space, will not affect disease transmission, we incorporate only North-South diffusion along the y-axis and East-West diffusion along the x-axis.

Applying diffusion to the model, we modify equations 3 and 4 to include the Laplacians of U and V. In accordance with the principals of diffusion, the equations for the infective and non-infective whitefly populations can be expressed as follows, where the $c_{i,j}$'s represent the diffusion coefficients specific to whitefly type (infective or non-infective) and direction (along the x axis or the y axis).

$$U_t = b(U+V)(1-(U+V)/m(X+Y)) - eYU - cU + c_{11}U_{xx} + c_{12}U_{yy} \tag{9}$$

$$dV/dt = eYU - cV + c_{21}V_{xx} + c_{22}V_{yy} \tag{10}$$

The movement patterns of the whitefly are also highly affected by wind. These tiny insects cannot fly long distances when depending solely on their own power. But, the wind can transport a whitefly much farther than it would normally travel on its own in a short period of time. In the cassava fields in Ivory Coast studied by Lecoustre *et al* [9] and Fargette *et al* [18], there is a prevailing southwest wind. Whiteflies are thus more likely to travel in a northeasterly direction than in any other direction. This process of being transported by wind can be modeled using advection, where the $a_{i,j}$'s represent the advection coefficients specific to whitefly type (infective or non-infective) and direction (along the x axis or the y axis).

Combining these two processes in an advection-diffusion equation yields the following equations:

$$U_t = b(U+V)\frac{(1-(U+V)}{m(X+Y))} - eYU - cU + c_{11}U_{xx}$$
$$+ c_{12}U_{yy} + a_{11}U_x + a_{12}U_y \tag{11}$$

$$dV/dt = eYU - cV + c_{21}V_{xx} + c_{22}V_{yy} + a_{21}V_x + a_{22}V_y \tag{12}$$

Diffusion now represents the short distance flight of the whiteflies in all directions and advection represents the long and short distance flights of the whiteflies as influenced by wind direction and strength. Equations 1 and 2, which represent the healthy and unhealthy plant populations, remain unchanged.

4. Numerical methods

We apply the finite difference method to solve and analyze this system of partial differential equations in MATLAB. Let G be a 50x50 numerical grid to represent the plane in space where the cassava field is located.

For each entry in G, there can, theoretically, be movement to any other point on the grid. Thus, our diffusion and advection matrices must have n=2304 entries in each row to represent the 2304 non-zero entries in grid G; so, the diffusion and advection matrices are both of the size 2304 x 2304. MATLAB has a function, "delsq(G)", which constructs the finite difference Laplacian on grid G and forms the diffusion matrix.

Mathematically, this diffusive activity is expressed by the following equation:

$$U_{xx} + U_{yy} = (u_{i+1,j} - 2u_{i,j} + u_{i-1,j} + u_{i,j+1} - 2u_{i,j} + u_{i,j-1})h^{-2} \tag{13}$$

where the error is of the order of h^2. [19]

Using the same grid, G, as above, we construct the advection matrix, C, which is stored in MATLAB as a sparse matrix. Discrete advection is expressed by the following equation:

$$U_x + U_y = (u_{i-1,j} - u_{i+1,j} + u_{i,j-1} - u_{i,j+1})(2h)^{-1} \tag{14}$$

In this approximation, the error is of the order of 2h [19].

In the field, wind does not blow equally in all four cardinal directions. Wind also does not have the same velocity in all locations within the field. A southwestern wind pattern predominates in the cassava fields studied by Fargette *et al* [18] in Ivory Coast and the advection matrix must be weighted to represent differing wind velocities.

A coefficient column, referred to as adx in the MATLAB code, represents the vector that dictates advection in the x direction. Similarly, the vector ady represents the wind velocity in the y direction. Each entry in the advection matrix is weighted by its corresponding entry in these column vectors; adx represents the relative velocity of the wind along the x axis, but when it is paired with ady, which represents the relative velocity of the wind along the y axis, a 2 dimensional vector field of wind velocity in the 2-dimensional cassava field is created.

In the standard MATLAB code created for analysis, higher values are given to those entries in adx and ady that correspond to the southern and western edges because these areas are most exposed to the wind. Lower values are given to the parts of the field that are more shielded from the wind, such as the middle and the northeastern sections. This assumption stems from the field data collected by Fargette *et al.* They found that the highest concentration of whiteflies occurred in the southwestern sections of the fields due to an "upwind edge effect" that indicates higher levels of flight activity along the upwind edges. This increased flight activity is here attributed to greater wind velocity. [18] Thus, we use a model in which wind strength decreases as we move from south to north and from west to east.

Wind and diffusion move the whiteflies among the cassava plants within the field, but they also bring new whiteflies into the field. To model the movement of insects into the field, we add a source term to the partial differential equations so that at each time-step, a small amount of whiteflies, both infective and noninfective, are entering the field from the southwest.

The creation of grid G, with zeros along the boundary, facilitates the implementation of the finite difference method. However, the methods described above allow for whiteflies to be moved only among the non-zero entries of G. Thus, when whiteflies move beyond the numbered entries of G and into the boundaries due to advection or diffusion, they are lost from the system. In real fields, when a whitefly flies outside the boundaries of a field, it may still reenter the field by flying back towards the direction from which it came. This movement is prohibited by the model and thus whiteflies enter the field only due to the source terms established in the southwest. Figure 1 shows a typical simulation.

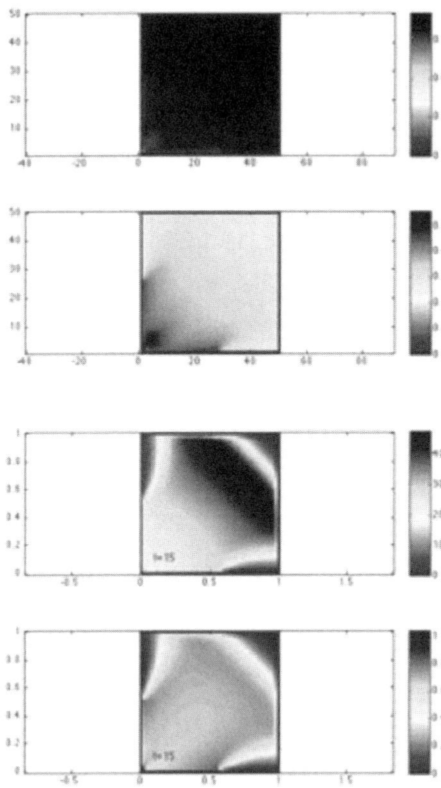

Figure 1. Simulation of the standard spatiotemporal dynamics of the ACMD system on a one-unit square with resolution of 50 by 50. a) distribution of healthy plants b) distribution of unhealthy plants c) distribution of non-infective whiteflies d) distribution of infective whiteflies

The solution calculated via the finite difference method is only an approximation of the true value of the solution of the system of equations. Further refining the grid spacing would allow for a more accurate solution; however, refinement ad infinitum is not feasible. Additionally, as the change in time between iterations, dt, tends to zero, the solution of the PDE system should converge to the analytical solution. To evaluate the error in the approximation used for the purposes of our analysis, we evaluate how the solution changes at a particular time, T=15, as dt is decreased.

The original time step, dt, is 0.003; dividing this value in half. As dt is halved from 0.003 to 0.0015, the difference between the two sets of X values is on the order of 10-4, as is the difference between the two sets of Y values. The error is thus less than 0.1%. The values of V and U have converged and the difference is zero. This code has very small error with respect to dt.

The grid resolution in the simulations performed in this paper is 50x50. To test the error with respect to h, we increase this resolution and compare the results. As the grid is refined, the finite difference method solution should converge to the analytical solution. We found that scaling of $1/h^2$ applied to the source terms is needed for convergence. The difference between the values of X and Y calculated at a resolution of 50x50 and those calculated at a resolution of 100x100 is less than 0.1. The error with respect to grid spacing is greater than the error with respect to time. However, our approximations have an overall low level of error.

5. Sensitivity to parameters

Systematically changing the value of each parameter in the ordinary differential equations model by 50% of the standard value suggested by Holt *et al* and examining the effect these changes have on the ratio of healthy plants to unhealthy plants provides insight into which parameters have the greatest impact on cassava yield. Ideally, the healthy cassava population would increase as the unhealthy cassava population decreases. Maintaining a large presence of unhealthy cassava is a waste of land and resources even if it is accompanied by a large amount of healthy cassava.

Simple deduction will show that if the desired effect of any change in the system is an increase in the ratio of healthy plants to unhealthy plants, certain variables, namely r, maximum replanting rate, a, plant loss/roguing rate, and c, vector mortality should be increased by 50% while others, namely m, maximum vector abundance, b, maximum vector birth rate, d, infection rate, and e, acquisition rate, should be decreased by 50%.

A sensitivity analysis on the system of ordinary differential equations proposed by Holt *et al* shows that reducing maximum vector birth rate to 50% of its standard value has the largest positive effect on the outcome of the temporal model. Solving the ODE system with this parameter alteration resulted in the highest X value and lowest Y value of all trials in which one variable was altered by 50% while all others were held constant. Second only to the effect of reducing maximum vector birth rate was the effect of increasing vector mortality to 150% of its standard value. Increasing the portion of whiteflies dying appears to have a significant positive effect on the cassava yield.

This analysis implies that the population dynamics of the whitefly are vital to the propagation of this disease in time. Model analysis predicts that controlling and limiting the vector population should lead to increased cassava yield. Experimentation with pesticides and whitefly traps are thus recommended as control method, based on the sensitivity of the temporal model.

With the goal of minimizing disease and maximizing cassava yield, we systematically alter different aspects of the spaciotemporal model and compare the results. In the standard advection diffusion program for ACMD, the total value of healthy plants, as calculated by summing the entries in the vector, X, representing health cassava plants, is 1640.5 and the total value of unhealthy plants, calculated similarly, is 745.9. Thus the ratio of healthy plants to total plants is 0.6874. Any increase in this ratio would be beneficial to the overall cassava crop, but an optimal increase in this ratio would be accompanied by an increase in total healthy cassava as this increase represents a higher yield of edible root. Under these circumstances, there would be not only a greater proportion of cassava that is healthy, but also a greater abundance of cassava in general. Therefore the food supply would increase. Unless otherwise specified, simulations have 100 time steps, corresponding to 15 time units of growth.

Performing the same parameter sensitivity analysis on the updated PDE system serves to assess the impact of adding the spatial dynamics to the system and to further examine the effects of the different parameters. All parameters were held at standard value except the one parameter under consideration in each MATLAB run. Figure 2 shows the results.

In all of the trials except the first one, in which maximum replanting rate was altered, the expected effects occur. The total number of healthy plants and the percent of plants that are healthy both increased. Increasing r, the maximum replanting rate, had a significantly negative impact on the total number of healthy plants.

Decreasing the maximum vector abundance to 50% of its standard value has only a very small impact on the abundance of both healthy and unhealthy cassava. The whitefly population remains under this maximum

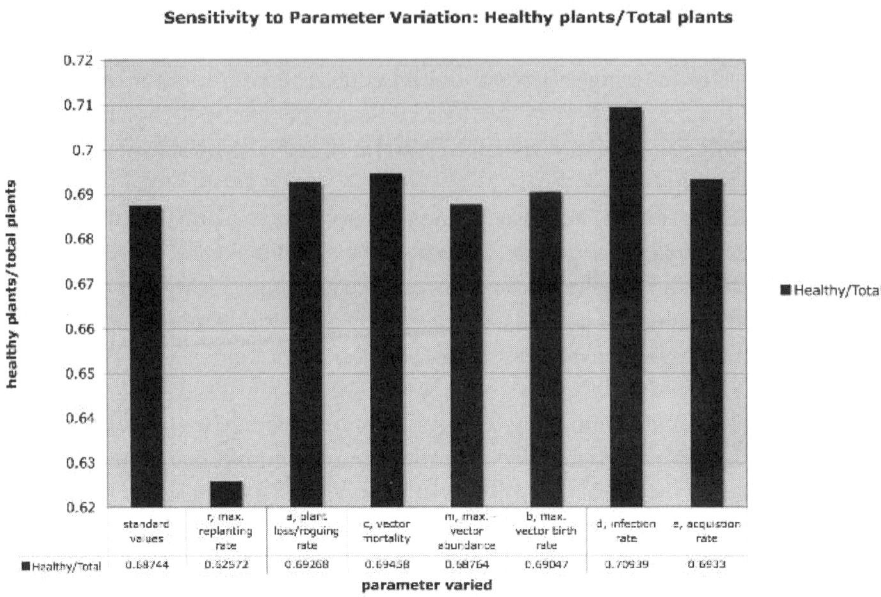

Figure 2. Sensitivity to parameter variation

abundance in the experiment and the system is therefore not affected by a change in this value so long as the maximum vector abundance is greater than the peak whitefly population.

The greatest positive impact on cassava yield occurs with a reduction in infection rate. In this scenario, the increase in total healthy plants is accompanied by a decrease in total unhealthy plants. The coupling of these two changes is the ideal result as it not only increases cassava yield from healthy plants but also decreases unhealthy plant numbers thereby reducing wasted resources and space.

Increasing vector mortality also raises cassava yield significantly. This result is concurrent with the analysis on the ordinary differential equations system. However, the parameter sensitivity analysis in the PDE system places more importance on the infection rate than does the analogous analysis performed on the ODE system. Incorporating the spatial dynamics of ACMD affects the predicted response of the disease to intervention.

6. The effects of wind

Previous work on ACMD has included observation of the distribution of disease in fields in Ivory Coast. This analysis shows an increased rate of disease incidence among plants in the southwest corner of experimental fields. The wind-exposed borders on the south and west also showed high disease incidence and a general trend of decreasing disease with increasing distance from the southwest border. The lowest incidence was found in the middle of the field and in the northeast sections [9].

The simulations created for our analysis of the spatiotemporal dynamics of ACMD in many ways reflect these field findings. Highest disease incidence is found in the southwest corner and along the south and west borders of the simulation. Our model does display low disease rate in the center of the field; however, the lowest rate of incidence is predicted in the north, east, and northeast sections of the field as opposed to the center.

These differences may be due to differing strength of wind across the field. It is possible that the cassava plants themselves act as a sort of windbreak such that lower wind speeds are present in the center of the field. The whiteflies would be less likely to be advected into the center of the field and a lower disease incidence could thereby be observed near the middle. This wind pattern is not represented by our standard simulation.

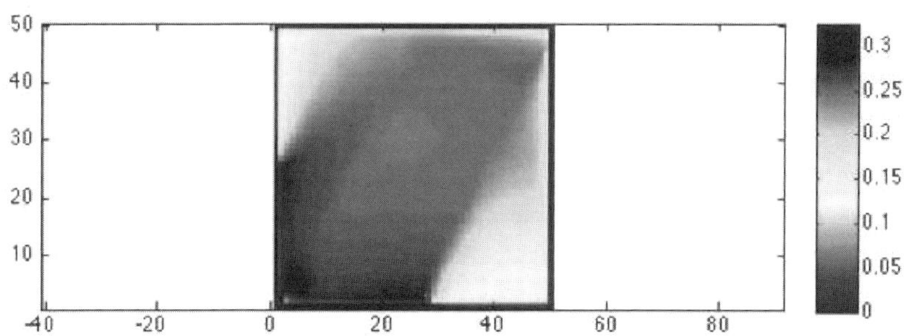

Figure 3. Simulation of the standard spatiotemporal dynamics of the unhealthy plants of the ACMD system on a one-unit square with resolution of 50 by 50 over 1000 time steps.

If our model is run for 1000 time steps instead of the 100 time steps used for most of the analysis in this paper, we see the emergence of an area in the middle of the field with lower disease incidence than the regions directly surrounding it, as shown in Figure 3. While this area fits the pattern recorded by Lecoustre *et al* [9], the lowest disease incidence in the simulation remains the corners where medium level disease incidence was observed in the field. Our model, while accurate in many regards, is not a perfect fit for this field data.

7. Whitefly "firebreaks": some numerical experiments

With a spatial model that gives a fairly accurate representation of the collected data, we can analyze the impact of different spatial modifications on the cassava yield of the fields. We modified the spatial dynamics of the system to search for ways of increasing cassava yield and limiting disease. By creating a hole in the cassava field represented by grid G in MATLAB, we hoped to establish a trap for the whiteflies. Female whiteflies lay their eggs on the leaves of cassava plants. These leaves also serve as the main source of food for the insects; the whiteflies are thus dependent upon the cassava. By creating a section of the field that is free from plants, we hypothesize that whiteflies that reach this area will die off, unable to survive without access to cassava plants on which to feed and reproduce. Thus, there should be a reduced number of infective whiteflies, especially downwind of the plant-free plot. Because the whiteflies carry ACMD and transmit it to previously healthy plants, a reduction in whitefly population may lead to a decrease in ACMD prevalence among plants.

As predicted, this hole in the field reduces the total number of whiteflies. As the area of the plant-free square increases, the populations of both infective and non-infective whiteflies continue to decrease.

Following the removal of a 3x3 plot of cassava plants from the middle of the field, there is a 2.02% reduction in the population of infective whiteflies. Increasing the size of the plant-free area to a 5x5 square, more than doubling the plant-free area, causes a further 2.54% reduction in this population. This trend continues as the size of the cassava-free zone is increased.

As the size of the plant-free square is increased, the total healthy cassava population increases. However, the simulation with a 5x5 dead zone has only a 0.129% increase in healthy cassava, representing a rise in the percentage of total cassava that is healthy from a standard value of $67.9\%A0toanimproved68\%$. While this spatial modification does increase cassava yield, it does so minimally. If the labor required to enact such a situation in a real field in Ivory Coast would be significant, then the implementation of a centered cassava-free zone would not be implicated from a cost-benefit standpoint.

The visualization of the distribution of whiteflies in a field with the above specifications illustrates a shadow effect in which the area directly northeast of the zone with no plants has a reduced whitefly population. Since whiteflies are unable to survive for any extended period of time without cassava plants, there are very few whiteflies living in the area with no cassava. Thus, the area directly downwind of this dead zone has a very limited number of insects being advected into it and the result is the observed shadow.

To evaluate the power of this shadow in limiting disease spread to the downwind regions of the field, we reoriented the cassava-free zone and ran the program with no plants in a region closer to the southwestern borders of the field. As expected, running the MATLAB program with a 3x3 cassava free area in the southwest region of the field yields a decreased number of whiteflies, both infective and non-infective, as compared to not only the control case, but also the case with a 3x3 dead zone in the center of the field. The unhealthy plant population is also further decreased and the healthy plant population increased such that the percentage of plant material that is healthy rises to 68.1%. However, as before, this increase is quite small and the subsequent gain in cassava yield is most likely not enough to account for the added labor costs of creating this abnormal spatial distribution in planting.

We also modified the layout of the field by creating plant-free zones that divide the field into 4 smaller fields. By placing these dead zones so that the northeastern plot is the largest of the four subsections of the cassava field, we hoped to minimize the infective whitefly population and maximize healthy plants. The numerical results are in Figure 4.

As hypothesized, the subplot located in the southwest of the field has the highest concentration of infective whiteflies and of unhealthy plants. In fact, the dead zone creates such an effective barrier that relatively few whiteflies pass into the northeastern subplot and the total number of whiteflies, both infective and non-infective, is significantly reduced from the standard simulation. The total amount of unhealthy plant material is decreased and the total amount of healthy plant material is increased, raising the percentage of total healthy plants to 69.9% from an original 67.9%. Thus, total cassava yield is increased at the same time that the percentage of plant material that is healthy is increased. The coupling of these changes represents an ideal change in the dynamics of the field; an increased harvest is accompanied by a reduction in resources dedicated to the maintenance of

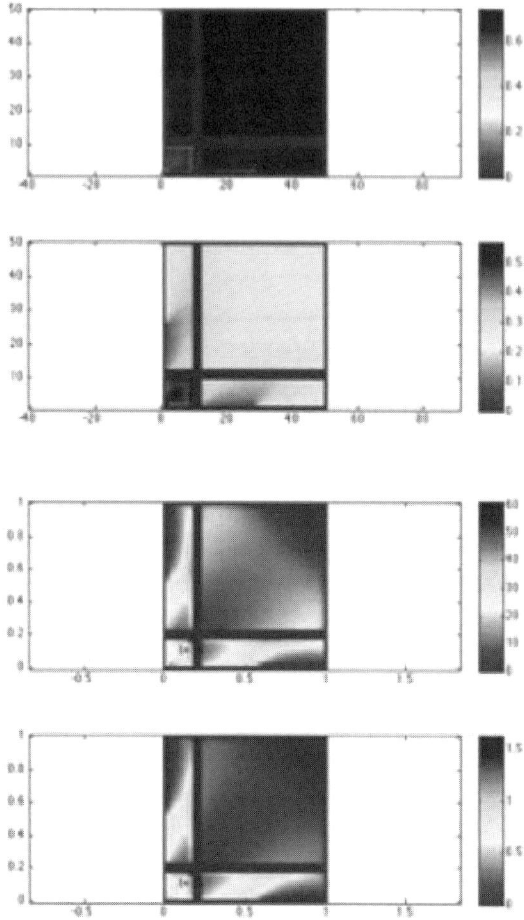

Figure 4. Simulation of the standard spatiotemporal dynamics of the unhealthy plants of the ACMD system on a one-unit square with resolution of 50 by 50 with plant-free zones dividing the field into four subfields. a)distribution of healthy plants b) distribution of unhealthy plants c) distribution of non-infective whiteflies d) distribution of infective whiteflies

unhealthy plants. Implementation of cassava-free zones organized in this manner is thus recommended for trial in Ivory Coast.

8. Windbreaks: more numerical experiments

In the field, cassava growers have experimented with building windbreaks to protect against ACMD [18]. The installation of a windbreak is simulated in MATLAB by limiting the number of whiteflies that enter the field. Presumably, a windbreak does not prevent absolutely all movement of whiteflies into the field, so to represent this barrier the source terms for whiteflies are altered. The new source terms are 10% of their initial value. This demonstrates a significantly reduced, but still existent, stream of whiteflies into the field due to natural flight patterns.

This windbreak scenario yields an increase in the percentage of plants that are healthy from the standard value of 68.7% to an improved value of 72.5%. Along with this increase, there is also an increase in the total healthy plants from 1640.2 to 1710. Thus the installation of this windbreak appears to be beneficial in that it limits the percentage of the field that is wasted due to the presence of unhealthy plants at the same time that it increases the total cassava yield. The positive effect of a windbreak is greater than the positive effect of any of the previously analyzed initiatives implying that it would be beneficial to develop real world applications of this simulation.

Next, we evaluated the dynamics of the wind. The standard model developed in section 4 assumes that wind strength decreases from south to north and from west to east due to the cassava plants themselves acting as windbreaks or buffers. What if, instead, the wind blows with constant speed at all points of the field? The wind in the area of Ivory Coast where Lecoustre *et al* [9] and Fargette *et al* [18] studied comes predominantly from the southwest, but the varying speeds within the spatial domain of the cassava fields were not recorded. The MATLAB program was altered to reflect constant wind in the southwest direction. The new value of the advection coefficient is chosen as the mid-point of the previous advection coefficient vector field. We avoid assigning wind speeds that are extremely powerful or extremely weak.

This alteration in the program yields very similar results in terms of plant population numbers as the initial program. The sum of all healthy plant material is now 1634.5 which represents 68.5% of total plants as opposed to 1640.2 plants representing 68.7% of total plants in the original simulation. This similarity is to be expected given the choice of advection constant as the median of the advection coefficient column from the earlier model. However, with constant advection, the number of non-infective flies rose to 76401 from 52791, while the infective whitefly population similarly saw a rise to 1449.9 from 1068.7. This increase in the population of the vector, especially the infective population, is not ideal and can lead to increased levels of ACMD among plants in the long run.

9. Planting density

The initial conditions for the cassava populations used in the above simulations are based on the equilibrium values of these populations as calculated from the ordinary differential equations used by Holt *et al* [3] [equations 1-4]. Clearly though, altering the equations to incorporate the spatial dynamics of ACMD greatly influences the outcome of the system. Therefore, we experiment with these initial conditions in hopes of discovering an ideal density for cassava planting.

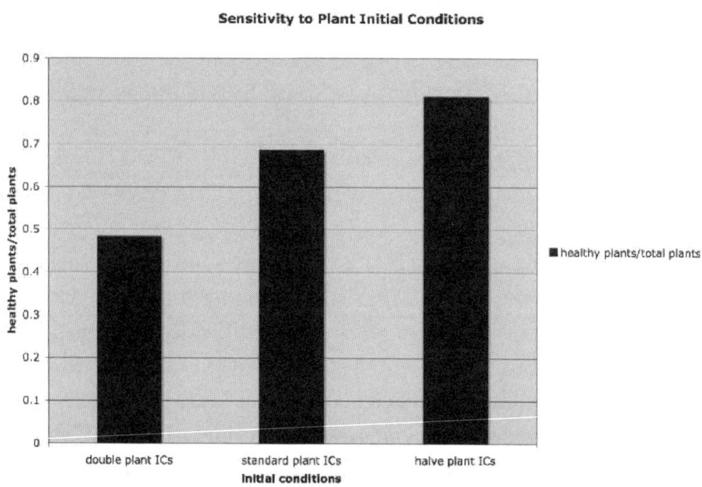

Figure 5. Sensitivity of the proportion of healthy plants to changes in the initial concentrations of healthy and unhealthy plant populations after 100 time steps in the standard simulation

Figure 5 clearly indicates that a less crowded field is beneficial to the cassava yield. Beginning the simulation with one half the number of plants, both healthy and unhealthy, results in significantly higher yield of cassava root after 100 time steps. This result may be due to the whitefly's dependence on the cassava. A reduction in cassava plants imposes a natural limitation on the whitefly population that can be sustained. With fewer total whiteflies, the vector population experiences a limited ability to spread disease among the plants and the plants therefore prosper. This rise in percentage of plants that are healthy is accompanied by an increase in the total number of healthy plants and a decrease in the total number of unhealthy plants. This ideal scenario represents a beneficial increase in cassava yield. While it may be counterintuitive for farmers to decrease the concentration of plants in their fields in order to increase the yield, this strategy is recommended for trial in fields in Ivory Coast.

10. Discussion

African Cassava Mosaic Disease causes severe food shortages throughout the African continent. A spatiotemporal model incorporating advection and diffusion based on the ordinary differential equations suggested by Holt *et al* [3] allows for the simulation of the dynamics of this disease and the vector that carries it. Systematically altering the model to represent modifications to cassava fields provides insight into the best methods of controlling ACMD and increasing harvest.

Previous analysis on parameter variation in the nonlinear interactions of the ordinary differential equations suggests that controlling the whitefly population would lead to increased cassava yield. Use of pesticides and other measures to increase vector mortality would therefore be implicated.

Similar analysis on the partial differential equations supports the data suggesting that reducing vector birth rate and increasing vector mortality would increase harvest. However, the transition from a temporal model to a spatiotemporal model fundamentally changes the system and the sensitivity analysis does not yield identical results. Such analysis on the partial differential equations suggests that reducing the infection rate would be most beneficial. Perhaps use of a naturally ACMD resistant strain of cassava could replicate this reduction in infection rate in real world applications. Such tactics are currently being employed in some cassava growing regions of Kenya [20] and Uganda. Through the National Network of cassava workers (NANEC), more than 250,000 ha of improved, resistant varieties of cassava have been planted in Uganda. The success of such programs is "explicitly illustrated by high yields of new cassava varieties that bridge the yield-gap caused by the devastating [A]CMD epidemic" [21].

The evidence that reduction in infection rate through the use of ACMD resistant varieties of cassava increases cassava yield validates the results of our simulation and further establishes the need to increase the availability and use of such strains of cassava all across Africa.

Spatial modifications relying on removal of subsections of the field appear to be minimally beneficial. The implementation of these plant-free zones are likely to create a net monetary loss because the resulting increase in cassava yield is so small that the added revenue from this additional crop may be less than the labor costs of creating the spatial modifications to the field.

However, the division of the cassava field into four unequal subfields through the elimination of all plant material in strips such that the subplot to the southwest is the smallest and the subplot to the northeast is the largest, results in a 2.32% increase in cassava yield. We believe that this represents a significant percentage and we therefore suggest the implementation of such spatial modifications.

Significantly reducing the source term representing entry of new whiteflies into the field resulted in an increase in the percentage of total cassava that is healthy as well as an increase in total healthy plant material. The model thus suggests that preventing the entry of whiteflies into the field would increase cassava yield. Installing a windbreak along the upwind edges of the field would hopefully accomplish this trap. Experimentation with the use of windbreaks is suggested as a disease control method.

Additionally, decreasing the initial concentration of cassava in the field resulted in the greatest increase in the percentage of healthy plant material of all of the simulations used for analysis. Of course, the application of this result depends upon the field conditions currently being employed by cassava farmers, but this analysis suggests that overcrowding of the field at the beginning of the growing season can have detrimental effects on the amount of cassava harvested. Increased roguing, or the systematic removal of infected plant material, is beneficial to cassava yield; but, selective removal of all plant material, both healthy and unhealthy, also ultimately increases the amount of healthy plants.

Acknowledgments

The authors wish to acknowledge the generosity of the Neukom Institute, the National Science Foundation Epscor Program, the local chapter of the Association for Women in Mathematics and the Dartmouth Mathematics Department for supporting Zoe Lawrence to present this paper at the Society for Mathematical Biology Annual Meeting 2010.

References

1. M. Roest, http://www.fao.org/news/story/en/item/8490/icode/, *Food and Agriculture Organization*
2. J. P. Legg, and J. M. Thresh. Cassava mosaic virus disease in East Africa: a dynamic disease in a changing environment. *Virus Research* **71**, 135-149, (2000).
3. J. Holt, M. J. Jeger, J. M. Thresh and G. W.-Nape *J. Appl. Ecology.* **34 3**, 793-806, (1997).
4. International Institute of Tropical Agriculture. www.iita.org. 11 Oct. 2009
5. B. James, J. Yaninek, P. Neuenschwander, A. Cudjoe, W. Modder, N. Echendu and M. Toko, *Pest Control in Cassava Farms.* International Institute of Tropical Agriculture. Nigeria: Wordsmithes Printers, Lagos, (2000).
6. R. F. L. Mau and J.L. Martin Kessing, Department of Entomology, Honolulu, Hawaii, *http://www.extento.hawaii.edu/kbase/crop/Type/b_tabaci.htm* Apr. 2007. retrieved 25 Oct. 2009.
7. D. N. Byrne, T. B. Bellows, Jr. and M. P. Parrella. 1990. Whiteflies in agricultural systems. In: Whiteflies: Their Bionomics, Pest Status and Management, D. Gerling (ed.). Intercept, Hants, United Kingdom, pp. 227- 261.
8. Thresh, J. M. and R. J. Cooter. *Plant Pathology* **54 5**, 587-614, (2005).
9. R. Lecoustre, D. Fargett, C. Fauquet and P. de Reffye, *Ecology and Epidemiology* **79 9**, 913-920 (1989).
10. D. N. Byrne, R. J. Rathman, T. V. Orum and J. C. Palumbo *Oecologia* **105 320**, (1996).
11. A. J. Diggle, M. U. Salam, G. J. Thomas, H. A. Yang, M. O'Connell and M. W. Sweetingham, *Phytopathology* **92 10**, 1110-1121, (2002).
12. G. J. Gibson *Phytopathology* **87 2** 139, (1997)
13. Krone, S. A Spatial model of range-dependent succession. Journal of Applied Probability. 37, 1044(200).
14. B. Szymanski, T. Caraco, *Evolutionary Ecology* **8 3**, 299-314, (1994).
15. T. Caraco, S. Glavanakov, G. Chen, J. E. Flaherty, T. K. Ohsumi, and B. K. Szymanski, *The American Naturalist* **160 3**, 348-359, (2002).
16. P. Marcati and M. A. Pozio. *J. Math. Bio.* **9**, 179- 187(1980).
17. J. D. Murray, E. A. Stanley and D. L. Brown, *Proc. R. Soc. Lond. B. Biol. Sci.* **229 1255**, 111-150, (1986).
18. D. Fargette, and J. C. Thouvenel, *Ann. Appl. Biol*, **106 2**, 285-295, (1985)
19. W. H. Press, S. A. Teukolosky, W. T. Vetterling and B. P. Flannery, Numerical Recipes. New York: Press Syndicate of the University of Cambridge, (1986).
20. Kenya Agricultural Research Institute. www.kari.org. 22 May 2010.
21. National Agricultural Research Organisation. www.naro.go.ug. 8 Oct. 2009.

THE BIOACCUMULATION OF METHYLMERCURY IN AN AQUATIC ECOSYSTEM

N. JOHNS, J. KURTZMAN, Z. SHTASEL-GOTTLIEB,
S. RAUCH and D. I. WALLACE*

Department of Mathematics,
Dartmouth College, HB 6188
Hanover, NH 03755, USA
E-mail: Dorothy.Wallace@Dartmouth.edu

A model for the bioaccumulation of methyl-mercury in an aquatic ecosystem is described. This model combines predator-prey equations for interactions across three trophic levels with pharmacokinetic equations for toxin elimination at each level. The model considers the inflow and outflow of mercury via tributaries, precipitation, deposition and bacterial methylation to determine the concentration of toxin in the aquatic system. A sensitivity analysis shows that the model is most sensitive to the rate of energy transfer from the first trophic level to the second. Using known elimination constants for methyl mercury in various fish species and known sources of input of methyl mercury for Lake Erie, the model predicts toxin levels at the three trophic levels that are reasonably close to those measured in the lake. The model predicts that eliminating methyl mercury input to the Lake from two of its tributary rivers would result in a 44 percent decrease in toxin at each trophic level.

1. Introduction

Toxins that enter an ecosystem are generally observed to be more concentrated per unit of biomass at higher trophic levels. This phenomenon is known as "bioaccumulation". Organisms can take up a toxic substance through lungs, gills, skin or other direct points of transfer to the environment. Predators, however, have a major source of toxin in their prey. The mechanism by which predators gain toxin (catching prey) is offset somewhat by mechanism of elimination (toxicokinetics). In this paper we model these two processes to see how they, together, produce the phenomenon of bioaccumulation. The model we develop is a simplified food web with

*Corresponding author

three trophic levels interacting with damped Lotka-Volterra dynamics. The model developed here could represent any toxin passed through food. As an application we consider the case of methylmercury in Lake Erie.

Section 2 gives some background information on methylmercury. Section 3 presents the basic model, analyzed in section 4. Numerical results are given for the basic model in section 5. Section 6 refines the portion of the model concerned with elimination of methylmercury from a trophic level through first order toxicokinetics. Section 7 refines the model describing input of methylmercury to the lake and the resulting ambient concentration in Lake Erie water. Section 8 describes the results of the refined model. In section 9 we investigate the sensitivity of the model to changes in parameters, varying each parameter 20% from its default value. Section 10 describes the effect of reducing methylmercury concentrations in the water on toxin concentrations in the trophic levels, using the elimination of toxin sources from the Detroit River and smaller tributaries as an example. Section 11 summarizes all results in a short discussion.

2. Background

Mercury poisoning is a significant health risk for people of all ages and is particularly severe for fetuses, infants and young children. While mercury exposure can occur in a number of ways, people in the United States are most commonly exposed to mercury through the consumption of fish or shellfish containing methylmercury.

Mercury can be emitted into the air as a byproduct of manufacturing or coal burning activities, and is released from volcanoes. Once in the atmosphere, this mercury falls in the form of precipitation and can pollute water sources. Bacteria living in the soils and sediments in and around these water sources convert the mercury into its toxic form, methylmercury (MeHg). In this paper we will ignore sediment exchange processes, and look only at toxins entering from tributaries and air, and toxins leaving via photodemethylation. In the case of Lake Erie, polluted tributaries are a significant source of methylmercury. These sources are included in the system.

Methylmercury makes its way up the aquatic food chain, becoming more concentrated with each trophic level. The model for bioaccumulation of toxin presented here is based on organisms in three trophic levels. The system of trophic levels describes the position that a species occupies in the food chain- essentially what that species eats and what eats them. Inherent in this system is a transfer of energy, nutrients and toxins embedded in tissues between organisms. The concentrations of such toxins, including methylmercury, can be modeled using a simple pharmacokinetics model, in which a trophic level uptakes a toxin at a some rate and expels the toxin, through excretion, at a second rate of elimination. Since some amount of the toxin remains in the organism tissue thereafter, it is presumably transferred in its entirety to the next trophic level when the organism is eaten. In this sense, a modification of a simple mathematical model for trophic dynamics in combination with a pharmacokinetic model for chemical accumulation in animal tissue provides a theoretical model for bioaccumulation of methylmercury across trophic levels.

3. Basic Model for Bioaccumulation

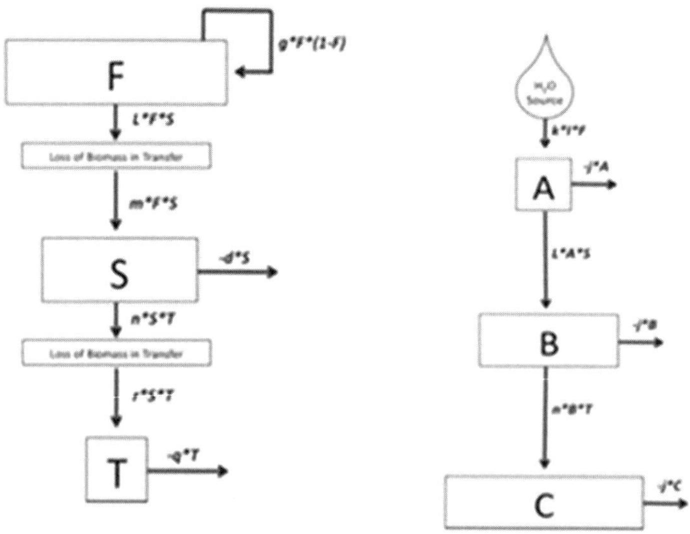

Figure 1. Box model for trophic dynamics and parallel pharmacokinetics

In any given ecosystem a natural food web exists. The lowest level of the web generally corresponds to the smallest organisms with the largest population, while the highest level is generally occupied by the largest organisms with the smallest population. While most isolated ecosystems can be broken down into hundreds of trophic levels, for the sake of simplicity, this paper will invoke a model for a tripartite trophic system.

3.1. *Predator prey*

The model defined in this paper defines three trophic levels:

(1) The photosynthetic/asexual producing level, which derives its nutrients from abiotic sources. This population will be defined by the variable F, for First trophic level.
(2) The second trophic level, which derives it nutrients for sustenance and growth from population F. This population will be defined by the variable S, for Second trophic level.
(3) The third trophic level, which derives its nutrients for sustenance and growth from population S. This population will be defined by the variable T, for Third trophic level.

For the sake of simplicity, implicit in the model are the following assumptions:

(i) There is a natural carrying capacity for the population of F.
(ii) The only predator of F is S; the only predator of S is T.
(iii) T does not have any predators.
(iv) Only a fraction of biomass lost due to predation of populations F and S is transferred to the populations of S and T respectively.
(v) F grows at a relative rate, defined as g.
(vi) S dies from natural causes, unrelated to predation by T, at a relative rate of d.
(vii) T has a relative natural death rate of q.

Given these assumptions and variables, we use a standard damped predator prey model to describe the dynamics of the trophic levels. All units are scaled to a total carrying capacity of 1 unit for F. The prime notation refers to the derivative with respect to time. As these equations are standard in the literature we omit further description.

$$F' = gF(1 - F) - LFS \tag{1}$$

$$S' = mFS - dS - nST \tag{2}$$

$$T' = nST - qT \tag{3}$$

3.2. *Bioaccumulation*

The full bioaccumulation model is an amalgamation of the aforementioned predator-prey model and a pharmacokinetics model. We use a first order pharmacokinetics model which assumes that the organisms of population F have a relative rate of uptake of methylmercury from the environment and a relative rate of elimination of methylmercury from their systems. In this section the rate of elimination is assumed to be the same across trophic levels, an assumption to be refined in Section 6.

F, the first trophic level, is the only population with a an uptake rate that depends only on its own biomass. The higher levels of the food chain are assumed to retain all of the toxin that is preserved in the tissues of their prey. Therefore, the bioaccumulation model in some sense parallels the predator-prey model. Thus, for populations S and T, the relative rate of methylmercury uptake is equivalent to the relative rate of biomass uptake in the population growth model for each predator group.

The model uses the following assumptions and variables.

(i) Variables A,B,C represent the absolute toxin concentrations in each trophic level F, S, and T, respectively.
(ii) The relative rate of uptake of toxin by F is defined as variable k.
(iii) The relative rate of elimination is defined as j.
(iv) The rates of toxin uptake are proportional to the predation rates for T and S, making them LAS and nBT respectively.
(v) L and n also describe the population lost from each trophic level by predation, and therefore serves as a second means by which methylmercury is eliminated from a trophic level.

With these assumptions the following equations describe the amount of toxin in each trophic level.

Rate of bioaccumulation for population F
= uptake due to consumption of I- first order pharmacokinetic elimination - loss due to predation by the next trophic level

$$A' = kFI - jA - LAS \tag{4}$$

Rate of bioaccumulation for population S

= gain due to predation on A - first order pharmacokinetic elimination - loss due to predation by the next trophic level

$$B' = LAS - jB - nBT \tag{5}$$

Rate of bioaccumulation for population T
= gain due to predation on S - first order pharmacokinetic elimination

$$C' = nBT - jC \tag{6}$$

Before we describe the basic results of this model it is worth mentioning the units of measure of various quantities. Biomass of the various levels is scaled to whatever units describe the lowest level. The carrying capacity could represent one gram of F per liter of water, or one total unit of biomass in the system. Which units are chosen depends on the units in which the input, I, is described. In this paper we usually use a per volume basis for all units.

4. Analysis of Basic Model

Now that we have determined that our models work separately, we must put them together in order to really understand the bioaccumulation process. The graphs of A, B and C only tell us the total amount of methylmercury present in each trophic level. We need to find the amount of methylmercury per unit biomass to determine whether a certain type of fish is safe to eat with respect to methylmercury levels. In order to do this, we solve for the equilibrium values of F, S, T, A, B and C by setting the derivative equations equal to zero:

$$F_{equil} = 1 - \frac{(Lq)}{(rg)} \tag{7}$$

$$S_{equil} = \frac{q}{r} \tag{8}$$

$$T_{equil} = -\frac{d}{n} + \frac{m}{n}(1 - \frac{Lq}{rg}) \tag{9}$$

$$A_{equil} = \frac{F_{equil}kI}{(LS_{equil} + j)} \tag{10}$$

$$B_{equil} = \frac{LA_{equil}S_{equil}}{(j + nT_{equil})} \tag{11}$$

$$C_{equil} = \frac{nB_{equil}T_{equil}}{j} \tag{12}$$

The Jacobian for this system was computed. For all parameters given in subsequent parts of this paper, the eigenvalues were computed numerically in Matlab and the equilibrium values were found to be stable.

5. Results of Basic Model

Our goal is to combine these two models and apply them using literature values in order to understand the process of bioaccumulation in this ecosystem. Because of the complexity of the population model and limitations in available data, some assumptions were made in this process.

We chose three species as proxies for our three trophic levels as follows, assuming that data found on these species is applicable to the Lake Erie model:

(i) The zooplankton species *Daphnia magna* as species F
(ii) The yellow perch as species S
(iii) The large mouthed bass as species T

Research by Tsui and Wang [1] gives the rate of uptake, k, and the rate of elimination, j, of methylmercury by our proxy species F as $k = 0.46$ and $j = 0.056$ per hour which were converted to a per day basis.

We assume that the values of k and j are consistent for all species in this basic model. We will see that the model behaves reasonably. In Section 6 we refine the assumptions on excretion parameters, which depend on fish species, size, temperature of water and other factors but certainly differ greatly among trophic levels.

We define the growth rate g of population F to be 0.3% per day, based on [2]. The concentration of mercury in the Lake Erie system, I is taken to be 1.9010^{-9} L/g. Sources vary for this number and we will refine it in Section 6.

As approximately 10 % of the biomass lost due to predation is gained by the subsequent trophic level, we define 0.1 as the value of constants m and r. These parameters are the most difficult to measure in the field. We will see in Section 8 that the model is sensitive to them. Unfortunately, we have only rule-of-thumb estimates to guide choice of these parameters. We define the values of L and n, the constants representing loss in biomass due to predation, to be 0.5. This estimate was based on the assumption that roughly half of the overall mortality rate of each species was due to predation. The mortality rates for species S and T were estimated as well. Because the model is in terms of mortality per day, we took the constants d and q to be 0.035, which implies that each species loses roughly 3.5 % of its population daily due to mortality outside of predation.

Initial values of A, B and C are all defined as zero, as our model assumes that the concentration of methylmercury in each trophic level is zero at time t=0. Determination of initial values for F, S and T was more complicated and published values were unavailable. However, literature tells us that population size per trophic levels varies such that S should be roughly 15-20 % as large as F, and that T should be roughly half as large as S, [3]. We therefore set starting populations at $F = 0.95$, $S = 0.14$, $T = 0.075$. Units are in terms of the percent of the total carrying capacity, which is equal to 1, which could be taken as biomass total or per unit of lake volume. Numerical runs used BGODEM software (Reid, 2008), which uses a Runge-Kutta algorithm to numerical integrate systems of ordinary differential equations.

First, we consider our predator-prey model. We see in Figure 2 that our model presents a classic predator-prey relationship. On the left we see the population sizes in each trophic level will change slightly until reaching equilibrium. On the right we can see that the amount of toxin in each trophic level will peak and then reach a point of equilibrium.

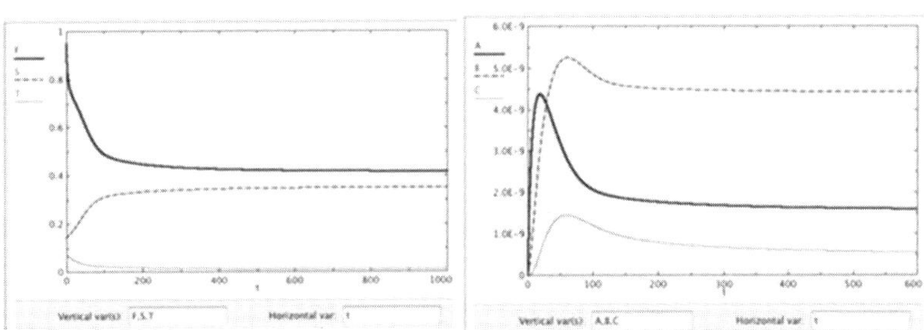

Figure 2. Typical output with biomass on the left and toxin on the right.

Using the equilibrium calculations from Section 4 we find that:

$$F_{equil} = 0.4174 \tag{13}$$

$$S_{equil} = 0.3496 \tag{14}$$

$$T_{equil} = 0.1347 \tag{15}$$

$$A_{equil} = 1.581 * 10^{-9} \tag{16}$$

$$B_{equil} = 4.404 * 10^{-9} \tag{17}$$

$$C_{equil} = 5.304 * 10^{-10} \tag{18}$$

If the units for F, S, and T are biomass per gram of carrying capacity of F (either total or per unit lake volume) then the units for A, B, and C are grams of toxin in respective trophic levels (either total or per unit lake volume). In order to find the concentration of methylmercury per unit biomass, we simply calculate the ratio of bioaccumulation to population size for each trophic level:

$$\frac{A_{equil}}{F_{equil}} = 3.78 * 10^{-9} \tag{19}$$

$$\frac{B_{equil}}{S_{equil}} = 1.259 * 10^{-8} \tag{20}$$

$$\frac{C_{equil}}{T_{equil}} = 3.9 * 10^{-8} \tag{21}$$

So if one unit of lake volume can support one gram of biomass of F in the absence of predators, then the units of A, B, and C become grams of toxin per gram of biomass in the respective trophic level. These results show that although the actual amount of methylmercury in each trophic level may be low, the ratio of methylmercury concentration to population size is relatively high. Furthermore, there is a significant increase in the ratio as you move up the food chain; the concentration of methylmercury per unit biomass in the third trophic level is 10.3 times higher than in the first. This is significant because it demonstrates that our model actually does duplicate a system of bioaccumulation.

Equations (10), (11), and (12) show that equilibrium values for toxin levels are all constant multiples of the input level, I. Figure 3 illustrates how the concentration of methylmercury per unit biomass in each trophic level increases with the level of contamination of the water.

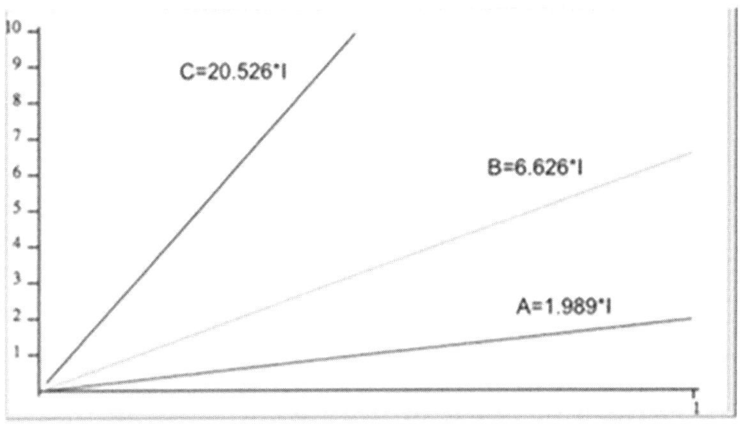

Figure 3. Equilibrium toxin concentration rises linearly with input values, with the highest trophic level having the greatest concentration of toxin per unit biomass.

Note that higher predators increase toxin per unit biomass much more quickly than the lower levels, in accordance with observation.

6. Refinement of Excretion Parameters for Lake Erie

Although a great deal of ingested MeHg is retained in bodily tissues for a long period of time, organisms do have the ability to remove it from their bodies. The rate at which this occurs is dependent on numerous parameters, especially body mass. The MeHg excreted by the organisms constituting the trophic levels associated with mercury concentrations A, B, and C is reintroduced into the water of the lake. It is assumed that all such excretions are composed entirely of MeHg. The input from this process is modeled by three different equations since each population has a unique rate of excretion.

$$I_1 = J_1 A \tag{22}$$

$$I_2 = J_2 B \tag{23}$$

$$I_3 = J_3 C \tag{24}$$

J_1, J_2, and J_3 are the rates of excretion (d^{-1}) for the populations of F, S, and T, respectively. They are multiplied by the total concentration of methylmercury per trophic level, as given by A, B, and C. When the original bioaccumulation model is adjusted for the elimination rate differences among the observed species, the overall behavior of the model is greatly affected.

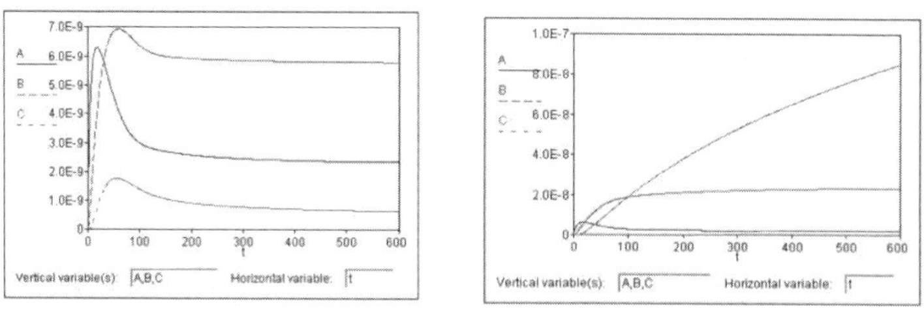

Figure 4. Typical runs for the system. On the left, $j_1 = j_2 = j_3$. On the right are the adjusted pharmacokinetic parameters as in Table 1.

The graph on the left is computed as in the basic model in section 3, with each species eliminating MeHg at the same rate. The graph on the right assumes that different organisms have different such rates. It is immediately seen that the overall character of the graph is changed. C, which has the lowest levels on the left, now has the highest methylmercury concentration and takes much longer to reach a state of equilibrium. In both studies the toxin per biomass goes up with trophic levels but the effect is much more pronounced in the refined model. Values used for the elimination rates are given in Table 1 along with sources for these.

7. Refinement of Inputs for Lake Erie

The concentration of mercury in a given lake is subject to a wide variety of processes and ecological events. In the case of Lake Erie, inflow from contributing rivers and streams increases the mercury concentration in the lake at a rate relative to the concentration in the tributaries. Mercury also enters the Lake Erie ecosystem by way of the atmosphere, primarily through wet precipitation and dry deposition. Once in the lake ecosystem, elemental mercury is methylated by bacteria, then absorbed by microorganisms such as plankton, thereby entering the food chain. Thus, this section will examine the concentration of MeHg within the Lake Erie ecosystem and use this information to improve upon the bioaccumulation model put forward in Section 3.

The processes contributing methylmercury to the Lake Erie ecosystem will first be modeled individually and then will be combined into a single differential equation governing I, the total concentration of MeHg in the lake. In doing so, we will be able to see how changes in individual components of the equation may influence the overall outcome of the model. Figure 5 summarizes the processes that affect the concentration of mercury in the lake, as well as the bioaccumulation of MeHg among trophic levels.

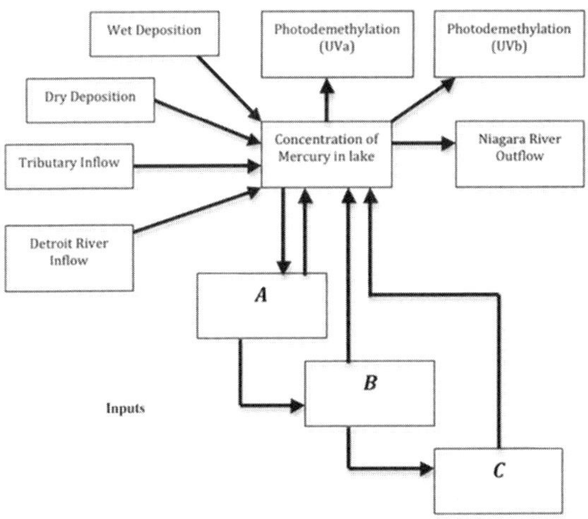

Figure 5. Box model for sources of methyl mercury input to Lake Erie.

7.1. *Inflow of the Detroit River*

The Detroit River is the primary tributary to the lake and its contribution to the total mass of MeHg in the lake is modeled as a first-order differential equation:

$$I_4 = F_D M_D \tag{25}$$

where F_D is the flow rate of the river (L/day) and M_D is the concentration of mercury in the river water (g/L).

7.2. *Inflow of Subsidiary Tributaries*

Tributaries in both the United States and Canada flow into Lake Erie. For simplicity, their input to lake mercury concentration will be analyzed collectively. The structure of the mathematical model is similar to the previous inflow.

$$I_5 = F_T M_T \tag{26}$$

where F_T is the flow rate of the tributaries (L/day) and M_T is the mercury concentration (g/L).

7.3. *Wet Deposition of Mercury*

Mercury can enter an ecosystem through rain and snow. A model for this process is found in the literature [4].

$$I_6 = C_w Ps \tag{27}$$

Here C_w is the concentration of mercury in the precipitation (g/L), P is the depth of precipitation falling on Lake Erie (m), and s is the surface area of the lake (m^2).

7.4. *Dry Deposition of Mercury*

Dry deposition, especially industrial output into the atmosphere, provides another avenue of entry for mercury. The literature, [4], again provides a model.

$$I_7 = C_p V_d s(.9) \tag{28}$$

Here, C_p is the average atmospheric concentration of mercury, V_d is the particle deposition velocity, s is the lake surface area, and the value of .9 is meant to correct for the 10% of the time that it is raining.

7.5. *Outflow from the Niagara River*

The Niagara River is the primary route by which water leaves the Lake Erie ecosystem. It is modeled in a similar fashion as the inputs from the Detroit River and the various tributaries:

$$I_8 = -F_N I \tag{29}$$

Where F_N is the flow rate constant of the river as a ratio to the overall lake volume. It is essentially what percent of the lake's water leaves through the Niagara River per day and hence and thus has units of d^{-1}. I is the concentration of MeHg in the lake (g/L).

7.6. *Photodemethylation of MeHg*

It has been shown that ultraviolet light from the sun can break down MeHg in the lake ecosystem into products that are easily evaporated from the lake's surface. For simplicity, it is assumed that products of MeHg that are evaporated by this process do not return. The model for this process is found in the literature, [5]. As indicated by the authors, UVA and UVB radiation must be considered separately.

$$I_9 = \int_0^3 .5k_{pA} I H e^{x k_A} dx \tag{30}$$

Where k_{pA} is a first-order rate constant $(m^2 E^{-1})$, I is the concentration of mercury in the lake (g/L),H is the average incident light strength $(E * m^{-2} * d^{-1})$, k_A is the attenuation coefficient for UVA radiation in Lake Erie water (m^{-1}), and x is the depth. The integral is calculated from 0 to 3.5 meters because at a depth of 3.5 m, the lake water has attenuated 99% of all UVA radiation.

$$I_1 0 = \int_0^1 .8k_{pB} I H e^{x k_B} dx \tag{31}$$

All quantities are as above and any change in subscript is simply an indication that UVB radiation is now being considered. As in the previous equation, the integral is calculated from 0 to 1.8 meters because at a depth of 1.8 m, the lake water has attenuated 99% of all UVB radiation.

7.7. *Values of Constants*

The relevant literature provides values for the constants used in the model. These are described in Table 1 and sources are given.

Table 1. Constants and parameters for the system

parameter	value	units	source
J_1	.056	d^{-1}	[6]
J_2	.01	d^{-1}	[6]
J_3	.00095	d^{-1}	[6]
F_D	5.61610^{11}	L/d	[4]
M_D	1.010^{-8}	g/L	[4]
F_T	4.18710^{10}	L/d	[4]
M_T	3.510^{-8}	g/L	[4]
C_W	2.010^{-8}	g/L	[4]
P	.85	m	[4]
s	2.5710^{10}	m^2	[4]
C_p	2.210^{-9}	g/L	[4]
V_d	.2	cm/sec	[4]
g	.3	d^{-1}	[1]
L	.5	%	basic model
F_N	.001	d^{-1}	Wikipedia
k_pA	2.1610^{-3}	m^2E^{-1}	[5]
k_pB	1.2510^{-3}	m^2E^{-1}	[5]
k_A	1.3	m^{-1}	[7]
k_B	2.5	m^{-1}	[7]
H	46.1	$Em^{-2}d^{-1}$	[5]
V	4.810^{14}	L	Wikipedia
d	.035	%	basic model
q	.035	%	basic model
m	.1	%	basic model
r	.1	%	basic model
n	.5	%	basic model

7.8. *Description of the concentration model for input of methylmercury to Lake Erie*

The constants allow an estimate of I_9 and I_{10}. Calculating the integral over the specified values results in the following:

$$I_9 = .0758 * I \tag{32}$$

$$I_{10} = .0228 * I \tag{33}$$

For practical purposes, we will define two new constants: $R_A = .0758d^{-1}$ and $R_B = .0228d^{-1}$. The differential equation for the concentration of MeHg in Lake Erie can now be combined. Since the input terms of the differential equation describe the mass of mercury entering the lake, the terms I1 through I7 must be divided by the volume, V, of the lake (L) in order to determine the total concentration of mercury in the lake. Additionally, since roughly 22% percent of the mercury that enters the lake by way of precipitation, inflow, and deposition is MeHg, we multiply these terms (I_4 through I_7) by (.22). We assume that the return of toxin to the water from elimination by organisms is negligible compared to the volume of water, and ignore the contribution of those terms. Our final equation is

$$I' = .22(F_D M_D + F_T M_T + C_w PS + C_p V_d S(.9)) - (F_N + R_A + R_B)I \tag{34}$$

The quantity I reaches a stable equilibrium, I_{equil}, easily calculated from this equation, which will be incorporated as a constant in the basic model.

8. Analysis and Results of Enhanced Model

We can now refine the parameters of the original model.

$$F' = gF(1 - F) - LFS \tag{35}$$

$$S' = mFS - dS - nST \tag{36}$$

$$T' = nST - qT \tag{37}$$

$$A' = kFI_{equil} - j_1 A - LAS \tag{38}$$

$$B' = LAS - j_2 B - nBT \tag{39}$$

$$C' = nBT - j_3 C \tag{40}$$

With these modified equations and a constant input I calculated as the equilibrium value of equation (34), we have the following equilibrium values for our quantities.

$$F_{equil} = 1 - \frac{(Lq)}{(rg)} \tag{41}$$

$$S_{equil} = \frac{q}{r} \tag{42}$$

$$T_{equil} = -\frac{d}{n} + \frac{m}{n}(1 - \frac{Lq}{rg}) \tag{43}$$

$$A_{equil} = \frac{F_{equil} k I_{equil}}{(LS_{equil} + j_1)} \tag{44}$$

$$B_{equil} = \frac{LA_{equil} S_{equil}}{(j_2 + nT_{equil})} \tag{45}$$

$$C_{equil} = \frac{nB_{equil} T_{equil}}{j_3} \tag{46}$$

It is also worth considering the ratios of toxin concentration at succeeding levels. The ratio of the second to first levels, for example, is given by equation (47).

$$\frac{B_{equil}}{S_{equil}} \frac{F_{equil}}{A_{equil}} = \frac{LA_{equil}}{(j_2 + nT_{equil})} \frac{(LS_{equil} + j_1)}{kI_{equil}} \tag{47}$$

This ratio is roughly on the order of S/T, the ratio of biomass from second to third trophic levels, no matter what the various constants are. A similar result holds for the ratio from third to second levels. Also it on the order of L^2, where L represents the predation rate, which is another measure of biomass transfer, this time from the first level to the second.

9. Sensitivity of the model to parameter values

Many of the parameters in the model are fairly rough estimates. We tested the sensitivity to model parameters by varying each parameter 20% from the default value given in Table 1. Figure 6 gives a visual key to the sensitivity of the equilibrium values of all quantities to the parameters listed. The horizontal scale is percent change from the default value, denoted by the vertical lines at 100%, 200% etc.

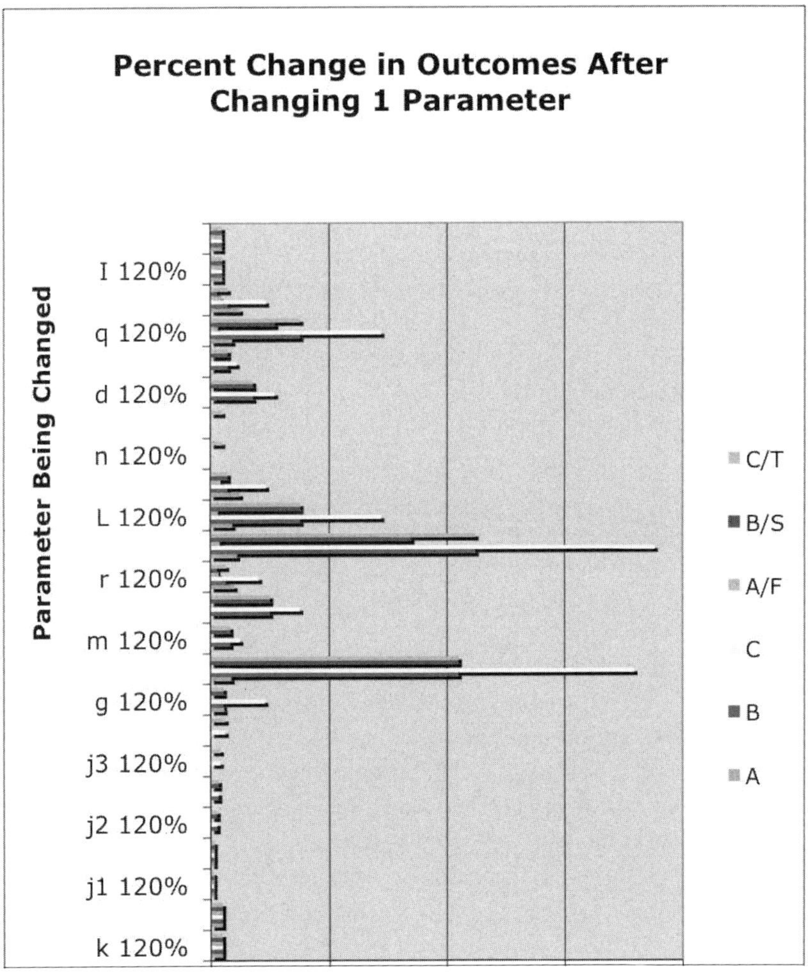

Figure 6. Sensitivity of equilibrium values to parameters of the model.

The model was extremely sensitive to constants controlling biomass transfer, exactly those constants most difficult to measure. This suggests two things. First, the process of biomass transfer is the cause of the phenomenon of bioaccumulation. The outputs of the model were far more sensitive to these parameters than to the adjustment of elimination constants j_1, j_2, j_3. Bioaccumulation occurs in this model even when all are equal (as in the basic model of section 3). Second, the actual predicted values of toxin per gram cannot possibly be reliable. However we also note that experimentally determined toxin levels also show wide variability, as in [8] for example.

10. The Effect of Reducing River-borne Contaminants

The model allows us to estimate changes in toxins at all trophic levels as a result of potential interventions. As an example, we can model what occurs to the observed species should the Detroit River and the lake's various smaller tributaries cease contributing to the influx of mercury. Though cleaning the river and streams of mercury would in reality be an extremely difficult task, it would nevertheless be considerably easier than stopping the atmospheric contributions to the lake's overall MeHg concentration. Additionally, doing so would allow us to gauge the relative importance of atmospheric mercury contribution against direct contribution via adjoined water ways.

If we assume no change in toxin inputs to Lake Erie, parameters in Table 1 lead to an equilibrium methylmercury concentration of 9.0210^{-11} g/L. Setting I_4 and I_5 equal to zero in equation (34) results in a lower equilibrium value for I, reducing it by about 44%. In the more toxic environment, the equilibrium value C in the in a value of 1.09210^{-7} (g/g), whereas in the environment with reduced toxin it reaches an equlibrium of 6.2610^{-8} (g/g). It is easy to see that, at equilibrium, all quotients A/F, B/S, and C/T are just multiples of the input I. Thus any percent reduction in input level will have a corresponding percent reduction in toxin concentration (grams toxin per gram biomass) at all trophic levels.

Although the system is nonlinear, this relationship is linear and independent of any parameters tested in Section 9. As a statement about proportionality, it is also independent of the constants determining I_{equil}. The relationship would change, however, if we took into account the return loop of toxins from the trophic levels to I, which we assumed was negligible.

11. Discussion

The model constructed in this paper relies on many assumptions, starting with an oversimplification of the food web into three trophic levels related by simple damped Lotka-Volterra dynamics, with elimination through first order kinetics. it uses some estimates for parameters that have been measured carefully and other estimates that are just rough guidelines. It assumes that toxins returned to water via elimination can be ignored. It assumes complete mixing of toxins entering Lake Erie via tributaries, which really should be considered point sources. In order to get actual predicted concentrations out of the model, we must assume that the average biomass of the lowest trophic level is about 1 gram per liter. In spite of all this the model yields some useful results.

11.1. *Qualitative results*

Standard predator prey relations coupled with first order elimination kinetics are enough to guarantee bioaccumulation will occur as the trophic level rises. Figure 3 shows this relation for one set of parameters. Equation (47) illustrates that the ratio of toxin concentrations from one level to the next rises as the transfer of biomass rises. This relationship holds for all parameters. The predator prey relations are the main "cause" of bioaccumulation in this model, in the sense that the toxin amounts are very sensitive to changes in parameters governing the predator prey relations, as seen in Figure 6.

11.2. *Quantitative results*

In this model, any percent reduction in ambient toxin levels in the environment (I_{equil}) results in an equal percent reduction in toxin concentration in all trophic levels. That is, the relationship between toxin concentration in any trophic level and toxin concentration in the surrounding water is linear, even though the underlying model is nonlinear. This relationship holds across all choices of constants and is therefore fairly reliable.

We can also use the model to predict actual toxin levels. As discussed in Section 10, the value of C, the predicted amount of methylmercury in the highest trophic level, is 1.09210^{-7} (g/g). From equation (15) T_{equil} is about .135 and so the toxin concentration for that level is about 4.610^{-7} g/g or .46 mg/kg. Weis [8] estimates tissue concentrations from field data for a variety of species of fish in the Canadian Great Lakes area. The highest estimated concentrations are for Northern pike and range between .397 and .603 mg/kg. So the toxin levels predicted by this model are actually within known ranges. Because numerical values predicted by this model are reasonably close to reality, the model also provides some weak confirmation that rough estimates of biomass transfer are likely to be close to correct.

Acknowledgments

The authors wish to acknowledge the generosity of the Neukom Institute, the National Science Foundation Epscor Program, the local chapter of the Association for Women in Mathematics and the Dartmouth Mathematics Department for supporting Nicole Johns to present this paper at the Society for Mathematical Biology Annual Meeting 2010.

References

1. M. T. K. Tsui and W-X. Wang, *Environmental Science and Technology* **38**, 808-816, (2004).
2. L. Wu, and D. A. Culver, *Journal of Great Lakes Res.* **20 3**, 537-545, (1994).
3. J. W. Kimball, *Biology*, online text, retrieved 12 Feb 2009. *http://users.rcn.com/jkimball.ma.ultranet/BiologyPages/F/FoodChains.html*
4. T. J. Kelly, J.M. Czucza, P.R. Sticksel, G.M. Sverdrup, P.J. Koval, and R.F. Hodanbosi, *J. Great Lakes Res.* **17 4**, 504-516, (1991).
5. I. Lehnherr and V. L. St. Louis, *Environ. Sci. Tehcnol.* **43**, 5692-5698, (2009).
6. M. Trudel and J. B. Rasmussen, *Environ. Sci. Technol.* **31**, 1716-1722, (1997).
7. M. Amyot, G. Mierle, D. Lean, and D. J. McQueen, *Geochimica et Cosmochimica Acta* **61 5**, 975-987, (1997).
8. I. M. Weis, *Environ. Res.* **95**, 341-350, (2004).

THE HUMORAL IMMUNE RESPONSE: COMPLEXITY AND THEORETICAL CHALLENGES

GITIT SHAHAF, MICHAL BARAK,
NETA ZUCKERMAN and RAMIT MEHR
*The Mina and Everard Goodman Faculty of Life Sciences,
Bar-Ilan University, Ramat-Gan, Israel*

The immune response involves cells of various types, including B, T and Natural Killer (NK) lymphocytes expressing a large diversity of receptors which recognize foreign antigens and self-molecules. The various cell types interact through a complicated network of communication and regulation mechanisms. These interactions enable the immune system to perform the functions of danger recognition, decision, action, memory and learning. As a result, the dynamics of lymphocyte repertoires are highly complex and non-linear. The humoral (antibody-generating) immune response is one of the most complex responses, as it involves somatic hypermutation of the B cell receptor (BCR) genes and subsequent antigen-driven selection of the resulting mutants. This process has been and still is extensively studied using a variety of experimental methods - ranging from intravital imaging to studying the mutations in BCR genes - and has also been one of the most often modeled phenomena in the theoretical immunology community. The problem for modelers, however, is that until recently kinetic data on the humoral immune response were so limited that all models could fit those data. We have addressed this - and the challenge of following individual clones - by combining modeling with a novel immuno-informatical method of generation and quantification of lineage trees from B cell clones undergoing somatic hypermutation. We applied these new analyses to the study of humoral response changes in aging, chronic or autoimmune diseases and B cell malignancies. Finally, we used simulations to answer some theoretical questions regarding the evolution of BCR genes.

1. Introduction

1.1. *Imumune system complexity*

One of the most intriguing challenges to theoretical biologists is presented by the adaptive immune system - one of the only two biological systems capable of continuously learning and memorizing its experiences.

The cells responsible for identifying, recognizing and remembering any foreign antigen are B and T lymphocytes, which have a special receptor for the antigen on their surface. Every clone of lymphocytes expresses a different receptor, creating a large repertoire of lymphocytes capable of identifying any existing antigen. B cells express the immunoglobulin molecule, either as a cell surface receptor (BCR[a]) for the antigen, or as a secreted antibody.

The formation of the lymphocyte repertoire is a multi-step process. In the beginning of the development of a B or T lymphocyte, a random rearrangement of antigen receptor variable region gene segments occurs at the DNA level [1]. Then selection takes place: only functional cells that contain a non-self-reactive receptor continue in the maturation process. In this way, a vast repertoire of naive cells is formed. These cells now survey the body, searching for a foreign antigen. Gene rearrangement on the DNA level is unique to the immune system, and is the main factor in generating the diversity and variability of antigen receptors in the vertebrate body.

The study of lymphocyte repertoires - the generation of antigen receptor repertoire diversity, the dynamics of lymphocyte development under normal or immune-deficient conditions, the forces of selection and interaction with antigen that shape lymphocyte repertoires - is particularly challenging for theoreticians. The difficulty in modeling lymphocyte repertoires does not result from the non-linear behavior of each component in the system, nor from the astronomic numbers of lymphocyte clones in the human patient or the experimental animal. Even though actual modeling of comparable number of clones would require a much higher computing power than is available to the average theoretician, studies on a smaller number of clones can often give sufficient insight into the behavior of the whole system. The main difficulty stems from the need to formulate a description of the system on several levels: genetic, molecular, cellular and systemic. The various aspects of immune system cell behavior are not tractable by the classical mathematical models of cell populations which have traditionally been used. For example, when studying the selection of developing B lymphocytes in the Bone marrow (BM) and spleen [2- 6], there is only so much one can do with a model that is limited to the level of cell populations. Before long, the rearrangement of B cell receptor (antibody) genes [7-8], and the interaction with the antigen that led to B cell activation, hypermutation and affinity maturation [9-10], must be taken into account. Hence it is not only the population dynamics within each clone (i.e., cell division, differentiation and death processes) that must be modeled, but also the meta-dynamics:

[a]Abbreviations used in this paper: B cell receptor (BCR); complementarity determining regions (CDRs); germinal center (GC); immunoglobulin (Ig).

creation, selection and elimination of whole clones in the population, and the molecular interactions or genetic changes that form the basis for these meta-dynamics.

The solution to the demand for increasing complexity depends on the problem at hand. No model can integrate all aspects of the real system - nor should it attempt to do so, because models must be simple enough to be tractable. Hence it is up to the researcher to decide which aspects of the system to focus on, and which aspects to neglect, in each part of the study. A modeler may choose to construct a simulation of a number of individual cells, and follow them in time, as we have done in our studies of isotype switch [11], but this is feasible only if the processes are very simple and the number of cells that must be simulated in order to receive statistically significant results is not large. Alternative solutions may be to follow one cell at a time, or follow whole clones of cells rather than individual cells in time, as demonstrated below. In any case, computational challenges abound, not only in formulating models and running simulations thereof, but also in creating novel methods of analysis of the exponentially growing amounts of data generated daily by both models and experiments.

1.2. *The B cell receptor (BCR)*

The immune system recognizes the great variety of antigens by specific receptors expressed on B and T lymphocytes, called B cell and T cell receptors (BCR and TCR, respectively); BCRs are also called immunoglobulins (Ig). The Ig is made up of a constant region (responsible for function - Fc) and a variable region (responsible for antigen binding - Fab). The Ig is composed of two copies of a heavy chain (IgH) and two copies of a light chain (IgL). The variable region is composed of both IgH and IgL, while the Fc is composed of parts of the IgH chains. The genes that encode the BCR are assembled in the process of V(D)J recombination during B cell development, from dispersed variable (V), diversity (D) and joining (J) minigene elements in the IgH, and V and J segments in the IgL. Each B cell expresses only one type of BCR, achieved by allelic exclusion and also isotype exclusion at the light chain [12].

The antigen binding site of Igs is formed by six hypervariable regions that are also called complementarity determining regions (CDRs): three from the VL and three from the VH (13). In an antibody, the major determinants of the specificity and affinity of these six regions for an antigen are their structure, the size, shape and chemical characteristics of their surface residues, and the positions of those regions relative to each other [14-16].

1.3. *The humoral immune response*

The antigen recognition signal, along with activation signals from T helper cells, activates naive B cells into dividing. Most activated B cells differentiate into plasma cells, which secrete their Ig molecules as antibodies. In the first few days following an infection, a subset of activated B cells in secondary lymphoid organs (spleen, lymph nodes and gut-associated lymphatic tissues) form clusters of rapidly dividing cells called germinal centers (GC) [17]. There, they undergo rapid somatic hypermutation (SHM) of the Ig variable region (IgV) gene, so the mutations change its structure and in effect change the receptor's affinity to the antigen. Most mutations reduce the affinity of the antibody to the antigen, or render the protein or the gene nonfunctional, but a small fraction of the mutations improve the affinity. A subsequent selection process ensures that only B cells whose receptors best match the antigen survive: all GC B cells undergo programmed cell death (apoptosis), unless their receptors bind the antigen with sufficient affinity to capture, process and present the antigen to helper T cells [17-19]. B cells which successfully passed the selection process differentiate to form the memory B cell pool, which largely mediates the response during repeated encounters with the same antigen, or become antibody-secreting plasma cells in the current immune response [20]. The resulting progressive increase in affinity for the antigen over time is called affinity maturation [18, 19, 21].

1.4. *Somatic hypermutation (SHM)*

SHM of Ig genes is several orders of magnitude faster than normal somatic mutation, and is generated by a different mechanism [22-24]. The SHM process is initiated by AID (activation induced cytidine deaminase), which deaminates C nucleotides to Uridine (U), causing a U-G mismatch. Subsequently, this mismatch undergoes DNA repair by one of the cell's several DNA repair error-prone mechanisms, which may lead to various types of mutations. It is not fully clear yet how SHM is triggered and regulated, however it is known that GC initiation and structural organization [25-27] require both cognate and co-stimulatory interactions with T cells [28-30].

Most of the mutations occur in RGYW/WRCY (R=A/G, Y=C/T, W=A/T) hotspot motifs [24] and are characterized by transition mutations occurring at about a 3:2 ratio of frequency relative to transversion mutations. The mutations are most concentrated at the CDRs, which are the areas of the BCR that are exposed to the antigen. Accordingly, a replacement mutation in a CDR has a better chance to improve the affinity of the BCR than a mutation in the framework regions (FRs), which are responsible for the structural integrity of the BCR.

1.5. *Theoretical challenges in the study of B cell repertoire generation and function*

Many questions regarding the complex processes of B cell diversification, mutation and antigen-driven selection are still open, such as how SHM is triggered and regulated; how SHM and selection interact dynamically to shape the memory B cell repertoire - whether they occur simultaneously, sequentially or alternately; does selection operate continuously or otherwise; is there recycling in GCs; when is somatic hypermutation initiated and when is it terminated; is selection based on affinity alone; whether there is negative selection in GCs; how B cells decide which heavy chain isotype to switch to; and how GC cells decide whether to differentiate into memory or plasma cells. Many studies by various theoretical groups (reviewed below) have addressed these questions, as well as questions which were raised by theoreticians rather than experimentalists, such as how the GC reaction is terminated [31-32], and how the immune system monitors its performance [33]. Understanding these natural mechanisms thoroughly is crucial for finding the impairments of B cell generation and function that are involved in the etiology and pathogenicity of various diseases.

In this paper we review work in our group aimed at addressing some of the above questions. To characterize B cell repertoires and their function, we developed novel algorithms for the construction of lineage trees from Ig genes of the responding B cell clones [34] and quantifying their shape properties [35-36]. Bioinformatical tools which include identification of motifs around mutated nucleotides, selection assessment through analysis of replacement and silent mutations, and AA substitution frequency analysis [37-38] are then applied to the constructed lineage trees, followed by statistical analysis which correlated tree characteristics with B cell response dynamic parameters [10]. Furthermore, we used modeling to understand the dynamics of the immune response, as well as address questions related to the evolution of antibody genes, as reviewed below.

2. Following Clonal B cell dynamics

2.1. *Following clonal dynamic using lineage trees*

To study SHM and follow the dynamics of the responding B cell clones, the clonal relationships of B cells should be assigned based on their Ig gene sequences, and mutations in Ig genes must be identified by comparing the mutated sequence to the pre-mutation, rearranged germline sequence. Usually, this sequence is not available, and thus we must reconstruct it by identifying the original gene segments used - based on the highest homology to the mutated sequence - and the junction region, for which there is no germline equivalent, so it must be deduced from the consensus of all mutated sequences in the same clone. There are several programs that researchers use for this purpose: the IMGT/V-QUEST [39]; SoDA [40], which uses a dynamic programming sequence alignment algorithm that takes into account the variation around the IGHV-IGHD and IGHD-IGHJ junctions resulting from the competing effects of nucleotide addition and exonuclease action; and iHMMune-align, which uses explicit hidden Markov models (HMM) of the various BCR generation processes, and was particularly designed to improve IGHD gene identification [41]. Once the germline sequences has been deduced, clonally-related sequences are identified as having the same germline segments.

Clonal B cell dynamics can be explored by examining the lineage trees that can be generated, in the same way as phylogenetic trees, from the Ig gene sequences of the responding B cells [9,18,42-43]. The original sequence ("root") is the known or deduced germline sequence identified as described above. Lineage trees trace the most probable evolution from the known root to the experimentally identified sequences (which may include both 'leaves' and internal nodes in the tree, as diversification is still ongoing at the time of sampling - see Figure 1. Note that the procedure does not intend to re-construct the particular sequence of events that actually took place. Rather, the procedure aims to find the minimal sequence of events that could have led to the observed sequences assuming that the minimal sequence of events is the most probable natural scenario. Such trees can be used to explore the dynamical properties of Ig affinity maturation. We have found that available phylogenetic tree creation softwares are not suitable for Ig gene lineage tree creation, not only because they assume all sequences are terminal sequences in the tree (or "leaves"), but also because most assume that the tree is binary, and that the root is unknown, which is not the case here. A cell population sharing the same Ig gene sequence, and hence

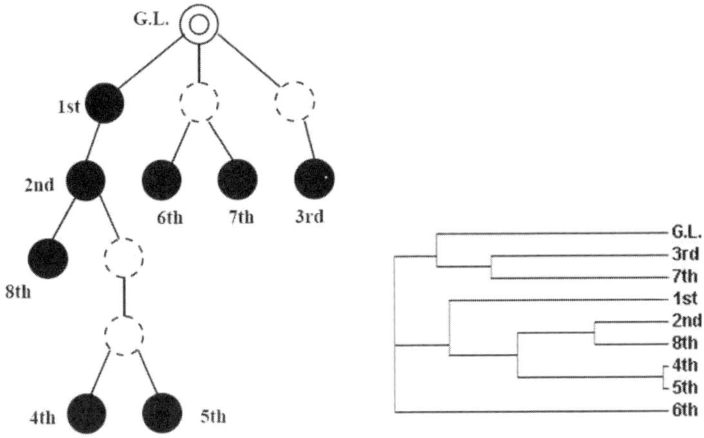

Figure 1. Differences between phylogenetic trees and Ig lineage trees. Left: A lineage tree is composed of all internal nodes, where every node represents a single mutation. The root of the tree (double circle) is known, and sampled data (filled nodes) may represent internal nodes in the tree. Deduced split and pass-through nodes (dashed circles) are added to the tree if needed. Right: A phylogenetic tree constructed by ClustalW (www.ebi.ac.uk/Tools/clostalw) from the same alignment as the Ig lineage tree in the left, using the default parameters of ClustalW. In such trees, the root may be known or unknown, sampled data can only be represented as leaves in the tree, and internal nodes are not calculated nor shown on the tree. In the lineage tree, the 6th and 7th sequences are closer than in the phylogenetic tree, as an intermediate mutation was added, causing the 7th sequence to be closer to the new node than to the 3rd sequence. Thus, using phylogenetic tree algorithms such as ClustalW for constructing Ig gene lineage trees requires much manual correction to get to the correct lineage tree.

represented by one node in the tree, can generate many different mutants. Hence we wrote our own program for the purpose, called IgTree©.

2.2. *The IgTree©program*

We developed a program using a heuristic algorithm that was tailored to handle the construction of Ig gene lineage trees, where all the mutations are represented as individual nodes, and observed sequences are not necessarily leaves. Trees that genuinely represent Ig gene evolution are not necessarily binary, include all internal nodes, and allow representation of sampled data as internal nodes (Figure 1). The algorithm we developed complies with all the requirements for Ig gene lineage trees. In addition, the algorithm can handle a succession of mutations (gaps or adjacent point mutations) as one mutational event, thus enabling it to also handle gene conversion events.

The goal of the algorithm is to create a lineage tree with the minimal number of mutations (nodes in the tree), where every node is separated from its immediate ancestor by only one mutation.

The tree-constructing algorithm performs the following set of steps. 1. Read the data, including the root sequence, and marking duplicate sequences - sequences that were derived from different B-cells but are identical. 2. Calculate the distance between each pair of the distinct input sequences and find possible ancestor-progeny relationships. 3. Construct the preliminary tree which includes only the samples sequences. 4. Add internal nodes to the tree to represent all individual mutations, thus creating the full tree. 5. Check whether inclusion of reversions can improve the tree. 6. Change the internal nodes to follow the changes suggested in step 5 above. 7. Repeat the reversion correction until the first of the following two conditions is met: (a) No improvement in the tree can be gained by adding more reversions, and (b) An upper limit to the number of reversion cycles (typically 5) was reached. 8. Output the tree. 9. Perform mutation analyses and output the results.

The program IgTree©[34] implements this algorithm, generates lineage trees specifically for IgV gene clonal sequences, and enumerates mutation frequencies and sequence motifs on a per-tree basis. Then MTree©[35-36] measures lineage tree characteristics. Automating lineage tree construction from IgV sequences and identifying all mutations including gaps and reversions, has aided in facilitating and standardizing the usage of lineage trees, thus advancing the usage of lineage trees in more studies. The programs IgTree©and MTree©contribute to the creation of a fully automated process of lineage tree analysis. In addition, several mutation analyses procedures, such as replacement and silent mutation analysis, mutational motif counts and amino acid substitutions, were integrated into IgTree©, providing much information regarding diversification and selection of Ig genes in the GC.

2.3. *Application of lineage tree and mutation analyses to various diseases*

Studies implementing our above-described algorithms have demonstrated the usefulness of the lineage tree analysis method, resulting in new insights regarding GC reactions in different tissues [44], GC B cell lymphomas such as follicular lymphoma, diffuse large B-cell lymphoma and primary central nervous system lymphoma; chronic inflammation such as Ulcerative colitis

[37], Light chain amyloidosis [45-46], autoimmune diseases such as myasthenia gravis, multiple sclerosis, Sjgren's syndrome and Rheumatoid arthritis [38, 47-48], and additional processes of diversification, such as gene conversion in rabbits and chickens [49]. Thus, analyzing mutations in IgV genes via the construction and quantification of mutational lineage trees yielded many novel insights into human diseases.

3. Modeling Overall GC dynamics

Theoretical approaches utilized so far in the study of affinity maturation include mathematical models exploring the dynamical interactions between SHM and clonal selection [9, 31 50-57], their spatial segregation [58-63], and the follicular niches and GC size distributions [32]. We built a hybrid mathematical/agent-based model to address dynamical questions using a different approach - by analyzing the shapes of Ig gene mutational lineage trees [10]. Previous experimental studies used lineage trees only qualitatively, to illustrate the biological process of IgV gene diversification, but the biological meaning of the shape differences between lineage trees remained unclear. Therefore the theoretical challenge here was to identify the quantitative relationships between the parameters governing the GC immune response and the graphical properties of B lymphocyte mutational lineage trees, using our mathematical models and computer simulations of the GC reaction [10]. In particular, we aimed to find how graphical tree properties depend on the thresholds and schedule of GC selection.

Our clone-based model (Figure 2) is an extension of a previous model used to study repertoire shift [9]. To account for the complex dynamics of several B cell clones competing for the same antigen, the model consists of an array of B cell sub-clones, each characterized by its receptor's affinity to the antigen. These clones can proliferate, differentiate and mutate in response to antigenic stimulation. The simulation starts with several different clones that have different Ig genes (see below), where new mutants arising from each clone form new sub-clones. A simplified summary of the population dynamics within each clone is shown in Figure 2A.

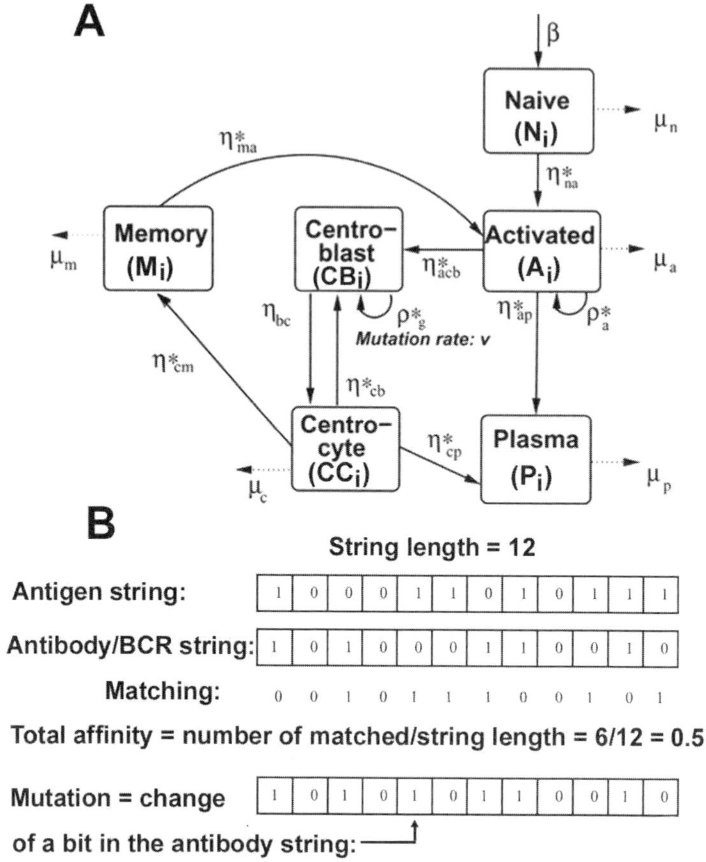

Figure 2. The main features of the GC response simulation. A: Population dynamics within each sub-clone (see text for the differential equations). The simulations integrate, in each "time step" and for each sub-clone, the equations representing the number of cells in each subset within this sub-clone. Transitions with antigen dependent rates are marked with asterisks. New cells arrive only into the original founder clones, but not their mutated descendants. B: A simple version of the antigen/BCR binding bit string model, where affinity is modeled by the number of bite-to-bit matches between the strings, divided by the string's length. Mutations are modeled by single bit flips.

Within each sub-clone i, the dynamics of B cell populations are described by the following set of differential equations.

$$dN_i/dt = \beta - (\mu_n + R\sigma_i\eta_{na})N_i \tag{1}$$

$$dA_i/dt = R\sigma_i(\eta_{na}N_i + \eta_{ma}M_i) + [R\rho_a - \mu_a - R\sigma_i\eta_{acb} - \eta_{ap}]A_i \tag{2}$$

$$dCB_i/dt = R\sigma_i\eta_{acb}A_i + R_G\sigma_i\eta_{cb}CC_i + (R_G\sigma_i\rho_b - \eta_{bc})CB_i \tag{3}$$

$$dCC_i/dt = \eta_{bc}CB_i - [R_G\sigma_i(\eta_{cb} + \eta_{cp} + \eta_{cm}) + \mu_c]CC_i \tag{4}$$

$$dM_i/dt = R_G\sigma_i\eta_{cm}CC_i - (\mu_m + R\sigma_i\eta_{ma})M_i \tag{5}$$

$$dP_i/dt = \eta_{ap}A_i + R_G\sigma_i\eta_{cp}CC_i - \mu_p P_i \tag{6}$$

Members of each sub-clone may be naive (with the number of naive cells in clone i represented by N_i), activated (A_i), centroblast (CB_i), centrocyte (CC_i), memory (M_i) or plasma (P_i) cells. Other than competition for antigen (R), no inter-clonal interactions (e.g. idiotypic regulation) are modeled here, for simplicity. R is measured relative to its initial concentration, so that the maximum concentration equals 1. The antigen concentration decreases with time due to consumption by activated cells and removal due to antibody production by plasma cells, according to:

$$dR/dt = -R[\sum_{i=1}^{S}(A_i + fP_i)\sigma_i]/B_{\max} \tag{7}$$

R_G, the antigen concentration inside the GCs (equation 8), is assumed to decrease due to consumption by GC cells, and varies with time, as antigen from outside the GCs flows inside and is lost with in proportion to the number of GC cells in the simulation:

$$dR_G/dt = \gamma R - \delta R_G \sum_{i=1}^{S}(CB_i + CC_i)\sigma_i \tag{8}$$

The model required a representation of the actual sequence of the Ig variable region gene and the mutations it undergoes, in order to model the mutational process in more detail and create lineage trees. A bit string model was used to represent the multi-dimensional interaction between the antigen and the antibody. One bit string was used to represent the antigen, and other bit strings represent the antibodies of B cell clones, as shown in Figure 2B. Mutations occur only in the centroblast compartment and are expressed as changes in the value of affinity of the cell's receptor to the antigen σ_i, which is based on the matching between the two strings.

Selection was implemented as follows. Following each cell division, each daughter cell attempts to mutate (flip) each bit of its receptor sequence, with the probability of a mutation attempt that is given by the mutation rate. The affinity of the new receptor bit-string to the antigen bit-string is then calculated. If the affinity has increased by an amount equal to or exceeding the selection threshold, then the cell forms a new sub-clone. If not, the cell dies. For more details on model parameters see the original paper [10] .

Lineage trees are generated throughout the simulation. At the peaks of the primary and secondary response of each clone, the corresponding tree was recorded, and measured using MTree©[35], a rigorous graph theoretical algorithm for quantifying the shape properties of the trees, developed by our lab. Its input is an adjacency list describing the tree created in the simulation. Its output is a list of variables that describe tree properties such as the number of leaves, the number of nodes, etc.

The 25 shape properties of the trees variables we measured supply much information on the lineage trees, yet there are considerable overlaps between the information supplied by different parameters. Thus, we decided to reduce the number of variables by selecting the most informative ones. We considered in the analysis the measurements that best correlated with the biological parameters, based on ordinal logistic regression and identified the best subgroup of biological parameters for explaining the variability of each of the most informative tree shape variables. We found inter-dependencies between dynamic parameter value ranges, and correlations between several tree properties and the main parameters that affect lineage tree shapes which, in both primary and secondary response trees, are the mutation rate, the initial clone affinity and the selection threshold. Finally, analysis of the impact of biological parameters on tree properties identified the 7 key properties, on which future analyses have focused [10].

4. The evolution of antibody genes

The extraordinary diversity characterizing the antibody repertoire is generated both during evolution and during lymphocyte development [64]. Much of this diversity is due to the existence of IgV gene segment libraries [65-67], which were diversified during evolution and, in higher vertebrates, are used in generating the combinatorial diversity of antibody genes. The human IgV gene locus has undergone extensive evolutionary editing. This can be seen by the division to families, where every family probably started from a single segment that was duplicated and mutated to form sets of similar but not identical sequences. Inactive pseudogenes exist, which might indicate that gene conversion was used and abandoned by some species.

There are many theoretical questions related to Ig gene evolution, such as: What evolutionary parameters affect the size and structure of gene segments libraries? Are the number of segments in libraries of contemporary species, and the corresponding gene locus structure, a random result of evolutionary history, or are these properties optimal with respect to individual or population fitness? If a larger number of segments or different genome structures do not increase the fitness, then the current structure is probably optimized. To address those questions we developed a computer simulation of Ig gene evolution. The simulation algorithm is shown in Figure 3A. The model's basic unit is the organism.. Every organism has a "genome", represented by a collection of fixed-length bit strings, which represent its antibody V gene segments.

In every generation, the organisms' fitness values are determined by exposing the organism's phenotype, created from a part of its genome, to antigen library. After the fitness is calculated for all organisms, the organisms are divided into pairs. The organism with the highest fitness is paired with the second highest, the third with the fourth, and so on.

To create a new organism, a random number of genes is taken from each parent, and combined by a "genetic crossover" to create the organism's new gene library (Figure 3B). After a new organism is created, its genome has a chance of undergoing one or more mutations. After creating all the children, the system is ready for the next cycle.

We measured the effect of different parameters on gene segment library size and diversity, and the corresponding fitness. We found compensating relationships between parameters, which optimized Ig library size and diversity. From our results we may conclude that contemporary species Ig gene segment libraries were optimized by evolution. Diversity of the Ig seg-

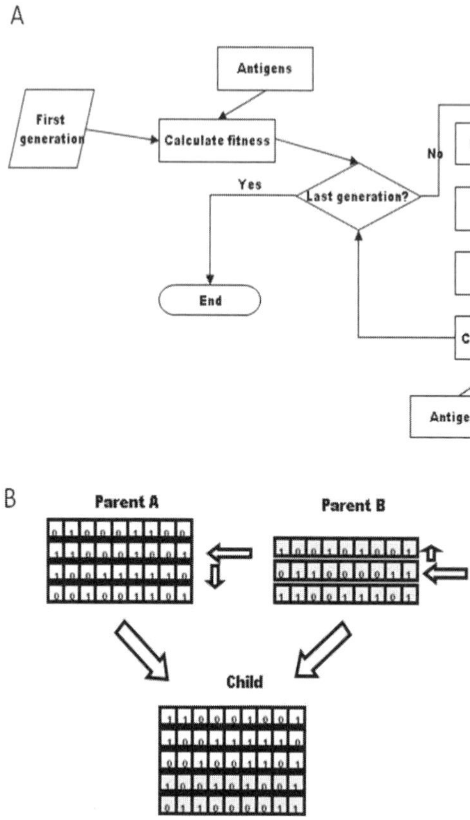

Figure 3. Ig gene evolution Simulation flow. Schematic representation of (A) the flow of the simulation, and (B) the crossover process used to create a child's genome from its parents' genomes.

ments, combined with the Ig library size, is thus sufficient to protect the organisms from almost all possible antigen.

5. Conclusion

In this review we have demonstrated the challenges facing researchers in theoretical immunology, focusing on the humoral immune response as an example of a complex, multi-scale system. We demonstrated how combining mathematical or agent-based modeling with immuno-informatical methods has helped in creating a study framework of this response, from the scale of individual mutations in Ig genes, through following single clones using Ig gene lineage trees, to modeling the whole response and even the evolution of the genetic system on which it is based. This review presented an exploration of several Ig gene diversification processes, which contribute to the complexity of the immune response of living organisms.

Acknowledgments

The work reviewed here was supported in parts by the following grants: Israel Science Foundation [grant numbers 759/01, 546/05 and 270/10]; A Human Frontiers Science Program Research Grant [to RM]; The Yeshaya Horowitz Association through a Center for Complexity Science PhD scholarship, a Ministry of Science and Technology PhD scholarship for advancing women in science, and a Bar Ilan University President's PhD scholarship [to NSZ]; and Dean's excellence PhD scholarships of The Mina and Everard Goodman Faculty of Life Sciences [to GS and NSZ].

References

1. F. Melchers, E. Boekel, T. Yamagami, J. Andersson, and A. Rolink, *Semin. Immunol.* **11**, 307 (1999).
2. R. Mehr, G. Shahaf, A. Sah, and M. Cancro, *Int. Immunol.* **15**, 301 (2003).
3. M. Gorfine, L. Freedman, G. Shahaf, and R. Mehr, *Bull. Math. Biol.* **65**, 1131 (2003).
4. G. Shahaf, K. Johnson, and R. Mehr, *Immunol.* **18**, 31 (2006).
5. G. Shahaf, D. Allman, M. P. Cancro, and R. Mehr, *Int. Immunol.* **16**, 1081 (2004).
6. G. Shahaf, M. P. Cancro, and R. Mehr, *PLoS One.* **5**, e9497 (2010).
7. R. Mehr, M. Shannon, and S. Litwin, *J. Immunol.* **163**, 1793 (1999).
8. G. Kalmanovich, and R. Mehr, *J. Immunol.* **170**, 182 (2003).
9. M. Shannon, and R. Mehr, *J. Immunol.* **162**, 3950 (1999).
10. G. Shahaf, M. Barak, N. Zuckerman, N. Swerdlin, M. Gorfine, and R. Mehr, *J. Theor. Biol.* **255**, 210 (2008)
11. B. Yaish, and R. Mehr, *Bull. Math. Biol.* **67**, 15 (2005).
12. C. Coleclough, R. P. Perry, K. Karjalainen, and M. Weigert, *Nature* **290**, 372 (1981).
13. T. T. Wu, and E. A. Kabat, *J. Exp. Med.* **132**, 211 (1970).
14. K. R. Abhinandan, and A. C. Martin, *Mol. Immunol.* **45**, 3832 (2008).
15. B. Al-Lazikani, A. M. Lesk, and C. Chothia, *J. Mol. Biol.* **273**, 927 (1997).
16. J. C. Almagro, *J. Mol. Recognit.* **17**, 132 (2004).
17. C. Berek, A. Berger, and M. Apel, *Cell.* **67**, 1121 (1991).
18. J. Jacob, G. Kelsoe, K. Rajewsky, and U. Weiss, *Nature* **354**, 389 (1991).
19. G. Kelsoe, *Immunol.* **8**, 179 (1996).
20. C. Berek, and C. Milstein, *Immunol. Rev.* **96**, 23 (1987).
21. H. N. Eisen, and G. W. Siskind, *Biochemistry* **3**, 996 (1964).

22. D. Bolland, and A. Corcoran, *Nat. Immunol.* **8**, 677-679 (2007).
23. N. S. Longo, and P. E. Lipsky, *Trends Immunol.* **27**, 374 (2006).
24. V. H. Odegard, and D. G. Schatz, *Nat. Rev. Immunol.* **6**, 573 (2006).
25. C. D. Allen, K. M. Ansel, C. Low, R. Lesley, H. Tamamura, N. Fujii, and J. G. Cyster, *Nat. Immunol.* **5**, 943 (2004).
26. S. Han, X. Zhang, G. Wang, H. Guan, G. Garcia, P. Li, L. Feng, and B. Zheng, *Blood.* **104**, 4129 (2004).
27. J. C. van Galen, D. F. Dukers, C. Giroth, R. G. Sewalt, A. P. Otte, C. J. Meijer, and F. M. Raaphorst, *Eur. J. Immunol.* **34**, 1870 (2004).
28. R. M. Bernstein, F. C. Mills, M. Mitchell, and E. E. Max, *Mol. Immunol.* **41**, 63 (2004).
29. H. Song, X. Nie, S. Basu, M. Singh, and J. Cerny, *Immunology.* **98**, 258 (1999).
30. Y. Xiao, J. Hendriks, P. Langerak, H. Jacobs, and J. Borst, *J. Immunol.* **172**, 7432 (2004).
31. C. Kesmir, and R. J. De Boer, *J. Immunol.* **163**, 2463 (1999).
32. N. Wittenbrink, T. S. Weber, A. Klein, A. A. Weiser, W. Zuschratter, M. Sibila, J. Schuchhardt, and M. Or-Guil, *J. Immunol.* **184**, 1339 (2010).
33. L.A. Segel, R.L. Bar-Or, *J. Immunol.* **163**, 1342 (1999).
34. M. Barak M, N.S. Zuckerman, H. Edelman, R. Unger, R. Mehr, *J. Immunol. Methods.* **338**, 67 (2008).
35. D.K. Dunn-Walters, H. Edelman, M. Banerjee, R. Mehr, *Dev. Immunol.* **9**, 233 (2002).
36. D. K. Dunn-Walters, H. Edelman, and R. Mehr, *Biosystems.* **76**, 141 (2004).
37. H. Tabibian-Keissar, N. S. Zuckerman, M. Barak, D. K. Dunn-Walters, A. Steiman-Shimony, Y. Chowers, E. Ofek, K. Rosenblatt, G. Schiby, R. Mehr, and I. Barshack, *Eur. J. Immunol.* **38** , 2600 (2008).
38. N. S. Zuckerman, W. A. Howard, J. Bismuth, K. Gibson, H. Edelman, S. Berrih-Aknin, D. Dunn-Walters, and R. Mehr, *Eur. J. Immunol.* **40**, 1150 (2010).
39. V. Giudicelli, D. Chaume, and M. Lefranc, *Nucleic Acids Res.* **32**, 435 (2004).
40. J. M. Volpe, L. G. Cowell, and T. B. Kepler, *Bioinformatics.* **22**, 438 (2006).
41. B. A. Gaëta, H. R. Malming, K. J. Jackson, M. E. Bain, P. Wilson, and A. M. Collins, *Bioinformatics.* **23**, 1580 (2007).
42. C. Kocks, and K. Rajewsky, *Proc. Natl. Acad. Sci. USA.* **85**, 8206 (1988).
43. J. Jacob, and G. Kelsoe, *J. Exp. Med.* **176**, 679 (1992).
44. M. Banerjee, R. Mehr, A. Belelovsky, J. Spencer, and D. K. Dunn-Walters, *Eur. J. Immunol.* **32**, 1947 (2002).
45. M. K. Manske, N. S. Zuckerman, M. M. Timm, S. Maiden, H. Edelman, G. Shahaf, M. Barak, A. Dispenzieri, M. A. Gertz, R. Mehr, and R. S. Abraham, *Clin. Immunol.* **120**, 106 (2006).
46. R. S. Abraham, M. K. Manske, N. S. Zuckerman, A. Sohni, H. Edelman, G. Shahaf, M. M. Timm, A. Dispenzieri, M. A. Gertz, and R. Mehr, *J. Clin. Immunol.* **27**, 69 (2006).
47. A. Steiman-Shimony, H. Edelman, M. Barak, G. Shahaf, D. Dunn-Walters, D. I. Stott, R. S. Abraham, and R. Mehr, *Autoimmun. Rev.* **5**, 242 (2006).

48. A. Steiman-Shimony, H. Edelman, A. Hutzler, M. Barak, N. S. Zuckerman, G. Shahaf, D. Dunn-Walters, D. I. Stott, R. S. Abraham, and R. Mehr, *Cell Immunol.* **244**, 130 (2006).
49. R. Mehr, H. Edelman, D. Sehgal, and R. Mage, *J. Immunol.* **172**, 4790 (2004).
50. B. Sulzer, J. L. van Hemmen, A. U. Neumann, and U. Behn, *Bull. Math Biol.* **55**, 1133 (1993).
51. T. B. Kepler, and A. S. Perelson, *Immunol.Today.* **14**, 412 (1993).
52. M. Oprea, and A. S. Perelson, *J. Immunol.* **158**, 5155 (1997).
53. S. M. Anderson, A. Khalil, M. Uduman, U. Hershberg, Y. Louzoun, A. M. Haberman, S. H. Kleinstein, and M. J. Shlomchik, *J. Immunol.* **183**, 7314 (2009).
54. M. Meyer-Hermann, Deutsch A, Or-Guil M, *J. Theor. Biol.* **210**, 265 (2001).
55. S. H. Kleinstein, Singh JP, *J. Theor. Biol.* **211**, 253 (2001).
56. D. Iber, Maini PK, *J. Theor. Biol.* **219**, 153 (2002).
57. S. H. Kleinstein, Singh JP, *Int. Immunol.* **15**, 871 (2003).
58. J. Mestas, and C. C. W. Hughes, *J. Immunol.* **172**, 2731 (2004).
59. M. Meyer-Hermann, *Immunol. Cell Biol.* **80**, 30 (2002).
60. M. Meyer-Hermann, *J. Theor. Biol.* **216**, 273 (2002).
61. M. Meyer-Hermann and T. Beyer, *Dev. Immunol.* **9**, 203 (2002).
62. C. Kesmir, and R. J. De Boer, *J. Theor. Biol.* **222**, 9 (2003).
63. T. Beyer, M. Meyer-Hermann, and G. Soff, *Int. Immunol.* **14**, 1369 (2002).
64. J. D. Hansen, and J. F. McBlane, *Curr Top Microbiol Immunol.* **248**, 111 (2000).
65. N. Pallarès, J. P. Frippiat, V. Giudicelli, and M. P. Lefranc, *Exp. Clin. Immunogenet.* **15**, 8 (1998).
66. V. Barbié, and M. P. Lefranc, *Exp. Clin. Immunogenct.* **15** , 171 (1998).
67. S. Das, M. Nozawa, J. Klein, and M. Nei, *Immunogenetics.* **60**, 47 (2008).

DNA LIBRARY SCREENING AND TRANSVERSAL DESIGNS

JUN GUO
Department of Mathematics, Langfang Normal College,
Langfang, 065000, China.
Email:guojun_lf@163.com

SUOGANG GAO
Math. and Inf. College, Hebei Normal University,
Shijiazhuang, 050016, China.
Email:sggao@hebtu.edu.cn.

WEILI WU
Department of Computer Science, University of Texas at Dallas,
Richardson, Texas 75080, USA.
Email: weiliwu@utdallas.edu.

DING-ZHU DU
Department of Computer Science, University of Texas at Dallas,
Richardson, Texas 75080, USA.
Email: dzdu@utdallas.edu.

Pooling design is an important mathematical tool to increase the efficiency of DNA library screening. The transversal design is a special type of pooling design, which is easy to implement and hence gets more attentions. In this paper, we give two new constructions of trasversal design.

1. Introduction

As more and more sequenced genome data are available to scientific research community, the study of gene functions has become a hot research direction. Such a study is supported by a high quality DNA library, usually obtained through a large amount of testing and screening. Thus, the efficiency of testing and screening becomes an important research topic. Pooling design is a mathematical tool to reduce the number of tests in

DNA library screening.

A pooling design is usually represented by a binary matrix with rows indexed with items and columns indexed with pools. A cell (i,j) contains a 1-entry if and only if the ith pool contains the jth item. This binary matrix is called the *incidence matrix* of the represented pooling design. By treating a column as a set of row indices each intersecting the column with a 1-entry, we can talk about the union of several columns. A binary matrix is *d-disjunct* if no column is contained in a union of other d columns. A binary matrix is *d^e-disjunct* if every column has at least $e+1$ 1-entries not contained in the union of some other d columns. Clearly, a d^0-disjunct matrix is d-disjunct. d^e-disjunctness have been studied extensively in the literature [1]-[15].

A pooling design is said to be *transversal* if it can be divided into disjoint families, each of which is a partition of all item such that pools in different parts are disjoint. A transversal design has a special matrix representation with rows indexed by families and columns indexed by items. A cell (i,j) contains entry k if and only if item j belongs to the kth pool in the ith family. This matrix representation is called a *transversal matrix* of the represented transversal design.

In [3], Du *et al.* extended the concept of d^e-disjunctness from the 0-1 matrix to the general matrix. For a general matrix, the union of d column vectors is defined to be a column vector each of whose components is the union of corresponding components of these d column vectors. A general matrix is *d^e_*-disjunct* if every column has at least $e+1$ components not contained in the union of some other d columns. A d^0_*-disjunct matrix is called d_*-disjunct. By [3, Lemma 6], A transversal design is d^e-disjunct if and only if its general matrix representation is d^e_*-disjunct.

In this paper, we present new constructions of transversal designs. These constructions are generalizations of [3].

2. Construction I

Let $\mathbf{F}_q$ be a finite field with q elements, where q is a prime power. Let $\mathbf{F}_q[\lambda]_k$ be the set of all polynomials of degree k with first coefficient 1 over $\mathbf{F}_q$. Then $|\mathbf{F}_q[\lambda]_k| = q^k$. Let m, s be two positive integers and let $P_{ms}(\mathbf{F}_q)$ be the set of all $m \times s$ matrices $A(\lambda) = (g_{ij}(\lambda))$ over $\mathbf{F}_q$, where $g_{ij}(\lambda) \in \mathbf{F}_q[\lambda]_k$. Clearly, $|P_{ms}(\mathbf{F}_q)| = q^{msk}$.

Let e be the upper bound for the number of possible errors in testing.

Suppose k satisfies

$$n \le q^{msk} \tag{1}$$

and

$$f = d(k-1) + 1 + e \le q. \tag{2}$$

We construct an $f \times n$ matrix $M(d,n,q,k,m,s,e)$ as follows: Its column indices are n distinct elements of $P_{ms}(\mathbf{F}_q)$. Its row indices are f distinct elements of $\mathbf{F}_q$. The cell $(x, (g_{ij}(\lambda)))$ contains element $(g_{ij}(x))$.

Theorem 2.1 *$M(d,n,q,k,m,s,e)$ is a general matrix representation of d^e-disjunct transversal design.*

Proof. Suppose $M(d,n,q,k,m,s,e)$ is not d^e_*-disjunct. Then it has a column $(g^0_{ij}(\lambda))$ which has at least $f-e$ components contained in the union of the other d columns $(g^1_{ij}(\lambda))$, $(g^2_{ij}(\lambda)), \ldots, (g^d_{ij}(\lambda))$. Thus, there exists a column $(g^u_{ij}(\lambda))\,(1 \le u \le d)$ containing at least k components of $(g^0_{ij}(\lambda))$. That is, there exists a $(g^u_{ij}(\lambda))\,(1 \le u \le d)$ such that $(g^0_{ij}(x_t)) = (g^u_{ij}(x_t))$ for at least k row indices x_t. It follows that $(g^0_{ij}(\lambda)) = (g^u_{ij}(\lambda))$, a contradiction. Hence, the desired result follows.

Remarks. If $m = s = 1$, then $M(d,n,q,k,m,s,e)$ is the $M(d,n,q,k,e)$ of [3]. Therefore, $M(d,n,q,k,m,s,e)$ is a generalization of $M(d,n,q,k,e)$.

By (1) and (2), k, q, m and s should be chosen to satisfy

$$\frac{1}{ms}\log_q n \le k \le \frac{q-1-e}{d} + 1. \tag{3}$$

There exists a positive integer k satisfying (3) if q satisfies

$$\frac{1}{ms}\log_q n \le \frac{q-1-e}{d}. \tag{4}$$

That is, it is sufficient to choose q satisfying

$$n^d \le q^{ms(q-1-e)}. \tag{5}$$

By $m, s \ge 1$, there exists a q satisfying (5) if q satisfies

$$n^d \le q^{q-1-e}. \tag{6}$$

Let q_e be the smallest number q satisfying (6). Then, we have following estimation on q_e.

Lemma 2.2

$$q_e \le e + 1 + (1 + h(d,n))\frac{d\log_2 n}{\log_2(d\log_2 n)},$$

where

$$h(d,n) = \frac{\log_2\log_2(d\log_2 n)}{\log_2(d\log_2 n) - \log_2\log_2(d\log_2 n)}.$$

Proof. Set

$$q_1 = e + 1 + (1 + h(d,n))\frac{d\log_2 n}{\log_2(d\log_2 n)}.$$

Note that $h(d,n) \ge 0$. Therefore,

$$\begin{aligned}(q_1 - 1 - e)\log_2 q_1 &> (q_1 - 1 - e)\log_2(q_1 - 1 - e)\\ &\ge \frac{(1+h(d,n))d\log_2 n}{\log_2(d\log_2 n)}\log_2\frac{(1+h(d,n))d\log_2 n}{\log_2(d\log_2 n)}\\ &> d\log_2 n.\end{aligned}$$

That is, q_1 satisfies (6). It follows that $q_e \le q_1$.

We need to find a prime power q satisfying $q \ge q_e$. Then, we can choose $k = \lceil \frac{1}{ms}\log_q n \rceil$. For such a choice of k, we have

$$\begin{aligned} f &= d(k-1) + 1 + e\\ &= d\left(\left\lceil \frac{1}{ms}\log_q n\right\rceil - 1\right) + 1 + e\\ &\le d\left(\left\lceil \frac{1}{ms}\log_{q_e} n\right\rceil - 1\right) + 1 + e\\ &\le d\left(\left\lceil \frac{q_e - 1 - e}{d}\right\rceil - 1\right) + 1 + e\\ &\le q_e.\end{aligned}$$

Since each family contains at most q^{ms} pools, the total number of tests is at most $q_e q^{ms}$.

Theorem 2.3 *There exist a prime power q and a positive integer k satisfying (1) and (2), such that $M(d,n,q,k,m,s,e)$ gives a transversal design with at most $q_e(2q_e)^{ms}$ tests.*

Proof. Let $q = 2^{\lceil \log_2 q_e \rceil}$. Then q is a prime power satisfying $q_e \le q < 2q_e$. Therefore, $q_e q^{ms} < q_e(2q_e)^{ms}$.

Remarks. We know that the smaller the value of f/n is, the better the pooling design is. Note that $M(d,n,q,k,e)$ of [3] is a $f_1 \times n_1$ matrix, where $f_1 = d(k-1)+1+e \le q$ and $n_1 \le q^k$. Therefore, we may take m, s in many cases such that $f/n < f_1/n_1$.

3. Construction II

Let $\mathbf{F}_q$ be a finite field with q elements. Let r be a positive integer and let $D_r(\mathbf{F}_q)$ be the set of all $r \times r$ diagonal matrices over $\mathbf{F}_q$. Clearly, $|D_r(\mathbf{F}_q)| = q^r$.

Let e be the upper bound for the number of possible errors in testing. Suppose k satisfies

$$n \le q^k \tag{7}$$

and

$$f = dr(k-1)+1+e \le q^r. \tag{8}$$

We construct an $f \times n$ matrix $M(d,n,q,k,r,e)$ as follows: Its column indices are polynomials of degree k with first coefficient 1 over $\mathbf{F}_q$. Its row indices are f distinct elements of $D_r(\mathbf{F}_q)$. The cell $(A, g(\lambda))$ contains element $g(A)$.

Theorem 3.1 $M(d,n,q,k,r,e)$ is a general matrix representation of d^e-disjunct transversal design.

Proof. Suppose $M(d,n,q,k,r,e)$ is not d^e_*-disjunct. Then it has a column $g_0(\lambda)$ which has at least $f-e$ components contained in the union of the other d columns $g_1(\lambda), g_2(\lambda), \ldots, g_d(\lambda)$. Thus, there exists a column $g_j(\lambda)$ $(1 \le j \le d)$ containing at least $r(k-1)+1$ components of $g_0(\lambda)$. That is, there exists a $g_j(\lambda)$ $(1 \le j \le d)$ such that $g_0(A_i) = g_j(A_i)$ for at least $r(k-1)+1$ row indices A_i. It follows that $g_0(\lambda) = g_j(\lambda)$, a contradiction. Hence, the desired result follows.

Remarks. If $r = 1$, then $M(d,n,q,k,r,e)$ is the $M(d,n,q,k,e)$ of [3]. Therefore, $M(d,n,q,k,r,e)$ is a generalization of $M(d,n,q,k,e)$.

By (7) and (8), k, q and r should be chosen to satisfy

$$\log_q n \le k \le \frac{q^r - 1 - e}{dr} + 1. \tag{9}$$

There exists a positive integer k satisfying (9) if q satisfies

$$\log_q n \le \frac{q^r - 1 - e}{dr}. \tag{10}$$

It is sufficient to choose q satisfying

$$n^{dr} \le q^{q^r - 1 - e}. \tag{11}$$

By $r(q-1-e) \le q^r - 1 - e$, there exists a q satisfying (11) if q satisfies (6). Let q_e be the smallest number q satisfying (6). We need to find a prime power q satisfying $q \ge q_e$. Then, we can choose $k = \lceil \log_q n \rceil$. For such a choice of k, we have

$$\begin{aligned} f &= dr(k-1) + 1 + e \\ &= dr(\lceil \log_q n \rceil - 1) + 1 + e \\ &\le dr(\lceil \log_{q_e} n \rceil - 1) + 1 + e \\ &\le dr\left(\left\lceil \frac{q_e^r - 1 - e}{dr} \right\rceil - 1\right) + 1 + e \\ &\le q_e^r. \end{aligned}$$

Since each family contains at most q^r pools, the total number of tests is at most $(q_e q)^r$.

Theorem 3.2 *There exist a prime power q and a positive integer k satisfying (7) and (8), such that $M(d, n, q, k, r, e)$ gives a transversal design with at most $(2q_e^2)^r$ tests.*

Proof. Similar to the proof of Theorem 2.3.

4. Generalization

Let m, s, r be three positive integers and let $P_{ms}(\mathbf{F}_q)$ and $D_r(\mathbf{F}_q)$ as the above.

Let e be the upper bound for the number of possible errors in testing. Suppose k satisfies

$$n \le q^{msk} \tag{12}$$

and

$$f = dr(k-1) + 1 + e \le q^r. \tag{13}$$

We construct an $f \times n$ matrix $M(d,n,q,k,m,s,r,e)$ as follows: Its column indices are n distinct elements of $P_{ms}(\mathbf{F}_q)$. Its row indices are f distinct elements of $D_r(\mathbf{F}_q)$. The cell $(A,(g_{ij}(\lambda)))$ contains element $(g_{ij}(A))$.

Theorem 4.1 $M(d,n,q,k,m,s,r,e)$ is a general matrix representation of d^e-disjunct transversal design.

Proof. Suppose $M(d,n,q,k,m,s,r,e)$ is not d^e_*-disjunct. Then it has a column $(g^0_{ij}(\lambda))$ which has at least $f-e$ components contained in the union of the other d columns $(g^1_{ij}(\lambda))$, $(g^2_{ij}(\lambda)),\ldots,(g^d_{ij}(\lambda))$. That is, for each row index A_t, $(g^0_{ij}(A_t)) = (g^u_{ij}(A_t))$ for some u. Thus, there exists a $(g^u_{ij}(\lambda))$ $(1 \le u \le d)$ such that $(g^0_{ij}(A_t)) = (g^u_{ij}(A_t))$ for at least $r(k-1)+1$ row indices A_t. It follows that $(g^0_{ij}(\lambda)) = (g^u_{ij}(\lambda))$, a contradiction. Hence, the desired result follows.

Remarks. If $m = s = 1$, then $M(d,n,q,k,m,s,r,e)$ is the $M(d,n,q,k,r,e)$ in Section 3. If $r=1$, then $M(d,n,q,k,m,s,r,e)$ is the $M(d,n,q,k,m,s,e)$ in Section 2.

By (12) and (13), k,q,m,s and r should be chosen to satisfy

$$\frac{1}{ms}\log_q n \le k \le \frac{q^r-1-e}{dr}+1. \tag{14}$$

There exists a positive integer k satisfying (14) if q satisfies

$$\frac{1}{ms}\log_q n \le \frac{q^r-1-e}{dr}. \tag{15}$$

That is, it is sufficient to choose q satisfying

$$n^{dr} \le q^{ms(q^r-1-e)}. \tag{16}$$

By $m,s \ge 1$ and $r(q-1-e) \le q^r-1-e$, there exists a q satisfying (16) if q satisfies (6). Let q_e be the smallest number q satisfying (6). We need to find a prime power q satisfying $q \ge q_e$. Then, we can choose $k = \lceil \frac{1}{ms}\log_q n \rceil$.

For such a choice of k, we have

$$\begin{aligned} f &= dr(k-1)+1+e \\ &= dr\left(\left\lceil \frac{1}{ms}\log_q n\right\rceil - 1\right)+1+e \\ &\leq dr\left(\left\lceil \frac{1}{ms}\log_{q_e} n\right\rceil - 1\right)+1+e \\ &\leq dr\left(\left\lceil \frac{q_e^r-1-e}{dr}\right\rceil - 1\right)+1+e \\ &\leq q_e^r. \end{aligned}$$

Since each family contains at most q^{msr} pools, the total number of tests is at most $q_e^r q^{msr}$.

Theorem 4.2 *There exist a prime power q and a positive integer k satisfying (12) and (13), such that $M(d,n,q,k,m,s,r,e)$ gives a transversal design with at most $q_e^r(2q_e)^{msr}$ tests.*

Acknowledgement

This research is supported in part by NSF of China under grant 10971052 and NSF of U.S.A. under grant CCF 0621829 and 0627233.

References

1. Y. Bai, T. Huang and K. Wang, *Discrete Applied Mathematics* **157** (2009), 3038-3045.
2. Y. Cheng and D. Du, *Journal of Computational Biology* **15**, (2008) 195-205.
3. D. Du, F. Hwang, W. Wu and T. Znati, *Journal of Computational Biology*, **13**,(2006) 990-995.
4. D. Du, F. Hwang, *Pooling designs and non-adaptive group testing: Important Tools for DNA Sequencing*, World Scientific, 2006.
5. A. G. D'yachkov, F. K. Hwang, A. J. Macula, P. A. Vilenkin and C. Weng, *Journal of Computational Biology* **12**, (2005) 1129-1136.
6. J. Guo, *Journal of Combinatorial Optimization* doi: 10.1007/s10878-008-9185-6.
7. J. Guo, Y. Wang, S. Gao, J. Yu and W. Wu, *Journal of Combinatorial Optimization* doi: 10.1007/s10878-009-9217-x.
8. T. Huang and C. Weng, *Discrete Math.* **282**, (2004) 163-169.
9. T. Huang, K. Wang and C. Weng, *European Journal of Combinatorics* **29**, (2008) 1483-1491.
10. A. J. Macula, *Discrete Math.* **162**, (1996) 311-312.

11. A. J. Macula, *Discrete Appl. Math.* **80** (1997) 217-222.
12. J. Nan and J. Guo, *Journal of Combinatorial Optimization* doi: 10.1007/s10878-008-9197-2.
13. H. Ngo and D. Du, *Discrete Math.* **243**, (2002) 167-170.
14. H. Ngo and D. Du, *DIMACS Series in Discrete Mathematics and Theoretical Computer Science* **55** (2000) 171-182.
15. X. Zhang, J. Guo and S. Gao, *Journal of Combinatorial Optimization* **17** (2009) 339-345.

MAPPING GENOTYPE DATA WITH MULTIDIMENSIONAL SCALING ALGORITHMS

S. E. LLERENA and C. D. MACIEL

Department of Electrical Engineering, São Paulo University
Av. Trabalhador São Carlense, 400, 13566-590 São Carlos, SP – Brazil
E-mails: sespezua@usp.br, maciel@sc.usp.br

Genotype data resulting of modern biomolecular techniques are characterized by having high dimensionality. Find patterns in this type of data is a complex and delayed work if it is performed solely by humans. A Multidimensional Scaling (MDS) technique was recently applied to map genotype data into helpful visual representations. Despite its announced success in helping to identify patterns, such conclusion was relative to the chosen MDS algorithm. There exist various MDS algorithms and it is unknown which of them would be more suitable in the mapping of genotype data. In this paper we present a comparative analysis of four popular MDS algorithms: classical MDS (CMDS), optimized MDS (SMACOF), Landmark MDS (LANDMARK) and FASTMAP. The analysis was performed using three comparison criteria: stress index, which measure the ability of the algorithms in accommodate the similarity information contained in the data into low dimensional spaces; clustering purity index, which measure the ability of the algorithms to preserve true group structures in the mapped space; and the computational time index, which measure the empirical computational costs of the algorithms. The results obtained in three well know datasets showed some differences in the measured criteria, with SMACOF presenting the better values of stress and clustering purity index, but an increased computational time. Additionally, SMACOF was used to map a genotype dataset that was previously mapped with the LANDMARK algorithm, resulting in a similar visual representation, with clusters more accurately recognizable.

1. Introduction

The rapid development of bio-molecular technologies observed in the last decades makes possible the study of the genetic diversity of a great variety of organisms. It is common to see genetic studies in which dense maps of genetic markers are genotyped in populations of considerable size[12].

Despite this growing ability in collecting genotype data, extract meaningful knowledge from such data has proved to be defiantly difficult to perform[11,17]. This difficulty was mainly due to the intrinsic high dimensionality (number of genetic markers) that this type of data possess, causing the problem known as *curse of dimensionality*[16,13]. This problem states that the data points tend to become equidistant from one another with the increment of the dimensionality, making the pattern identification difficult and computational expensive[6].

A common way to analyze the referred data is to represent them visually[19,12], taking advantage of the recognized human capability in discovering and interpreting patterns in low dimensional spaces. The representation of the data in low dimensional spaces requires the join interaction of dimensional reduction methods with data visualization methods. The appropriate selection of such methods is a subject of an active research[3]. Traditionally, hierarchical representations, such as dendrograms[6] were used to represent similarity relationships of genotype data. These representations have been useful in many cases. However, they showed to be inadequate when the number of dimensions and data items becomes large[8,15,18].

In an attempt to find a more intuitive visual representation, a Multidimensional Scaling (MDS) technique[2] were applied with some success in the visual mapping of genotype data[8]. MDS are dimensional reduction methods that take as input a matrix of item-item distance and assign to each item a coordinate in a reduced dimensional space (2 or 3 dimensions) trying to conserve the distance information. There exist several MDS algorithms[2], which differ in the way the input distance matrix is interpreted, the loss function used (stress) and the type of target space. In the previous work that applied MDS to mapping genotype data, the selection of the MDS algorithm was arbitrarily chosen. However, it is unknown how this selection can affect the visual representation of this type of data. This is important to know, since interpretations and conclusions drawn from the resulting visual representations can be biased by the selected MDS algorithm, which may not represent accurately the true similarity relations contained in the data.

With the intention to filling this gap, we present in this paper a comparative analysis of four popular MDS techniques: classical MDS (CMDS)[2], optimized MDS (SMACOF)[2], Landmark MDS (LANDMARK)[5] and FASTMAP[9]. Three comparison criteria were used for this purpose: 1) stress index, which measure how different is the distance matrix calculated in the mapped data from the original distance matrix; clustering purity index, which measure the ability of the algorithms to preserve true group structures in the mapped space; and the computational time index, which measure the processing time of the algorithms. The results of this analysis, carried out over three well know datasets (Iris plant, Wisconsin breast cancer, Image segmentation)[1], showed some differences in the measured criteria, with SMACOF presenting the lowest levels of stress and the greatest clustering purity index. On the other hand, FASTMAP showed the worst values of the measured criteria. For reasons of comparison, SMACOF was also used to map a genotype dataset that was previously represented by dendrograms[10] and by the LANDMARK algorithm[8]: a RFLP-PCR marker data obtained from a Brazilian collection of bacterial strains belonging to the genus Bradyrhizobium[10,14]. The resulting SMACOF 2D mapping of this dataset showed small visual differences with respect to the previous LANDMARK mapping. However, it was found that the clustering performed with K-means[6] algorithm over the SMACOF representation seems to be more natural than the corresponding clustering in the LANDMARK mapping.

Although SMACOF had the best mapping precision, it presented the worst processing time with respect to the other algorithms. This is due to the fact that it is an iterative optimization algorithm. If precision is not a fundamental issue, the LANDMARK algorithm would be more appropriate, since it presented the lowest computational time and an acceptable precision in the mapping.

The paper is organized as follows: Section 2 introduces the Multidimensional Scaling techniques. Section 3 presents a description of four MDS algorithms (CMDS, LANDMARK, FASTMAP and SMACOF). Materials and Methods are presented in Section 4, here is explained three comparative criteria (stress,computational time and clustering purity). Section 5 discusses the experimental results obtained with the proposed comparison criteria. Section 5.4 discusses the experimental results obtained with the Bradyrhizobium dataset. Finally, in Section 6 the conclusions are presented.

2. Multidimensional Scaling

MDS is a set of models that represent measures of proximity (a generic term for similarity or dissimilarity) among pairs of objects as distances between points in low-dimensional spaces. In MDS it is assumed that the measures of proximity, p_{ij}, for each pair (i, j) of n objects are given. MDS attempts to represent the proximities p_{ij} by distances d_{ij}, by assigning one point to each object in an m-dimensional space. The set of n assigned points X is also called the MDS space[3].

Formally, an MDS model is a function f that maps the proximities p_{ij} into the corresponding distances $d_{ij}(X)$ in the MDS space X^2. That is,

$$f : p_{ij}(X) \rightarrow d_{ij}(X). \tag{1}$$

A particular choice of f specifies the MDS model. The distances $d_{ij}(X)$ are unknown, and MDS finds a configuration X of predefined dimensionality m on which the distances are computed. The function f can be either explicitly specified or restricted to come from a particular class of functions. Usually, the function used to measure the distance between any two points i and j in the MDS space X is the Euclidean distance [2], which is defined as:

$$d_{ij}(X) = \left[\sum_{a=1}^{m}(x_{i,a} - x_{j,a})^2\right]^{1/2}. \tag{2}$$

Empirical proximities always contain noise due to measurement imprecision. Hence, one should not insist, in practice, that $f(p_{ij}) = d_{ij}(X)$, but rather that $f(p_{ij}) \approx d_{ij}(X)$, where $\approx$ can be read as "as equal as possible". Computerized procedures for finding an MDS representation usually start with some initial configuration and improve it by moving around its points in small steps (iteratively) to approximate the ideal model relation $f(p_{ij}) = d_{ij}(X)$ more and more closely. A squared error[2] of representation is defined by

$$e_{ij}^2 = [f(p_{ij}) - d_{ij}(X)]^2 \tag{3}$$

summing e_{ij}^2 over all pairs (i, j) yields a badness-of-fit measure for the entire MDS representation, `raw stress`[2],

$$\sigma_r(X) = \sum_{i<j}[f(p_{ij}) - d_{ij}(X)]^2 \tag{4}$$

the inequality $i < j$ in the summation mean that only is considered the sum over half of the data, since its assuming that the $f(p_{ij})$ is symmetric. When working with missing data[2], its common define positive weights w_{ij}, which are added into the stress function as follows:

$$\sigma_r(X) = \sum_{i<j} w_{ij}[f(p_{ij}) - d_{ij}(X)]^2 \tag{5}$$

the w_{ij} is set to 1 for all i, j when its values are not missing. However, if $x_{i,a}$ or $x_{j,a}$ is missing (or both), then w_{ij} should be set to 0 so that missing values do not influence the similarity. To avoid scale dependency[2], σ_r can be normalized as follows,

$$\sigma_1^2(X) = \frac{\sigma_r(X)}{\sum d_{ij}^2(X)} = \frac{\sum[f(p_{ij}) - d_{ij}(X)]^2}{\sum d_{ij}^2(X)}. \tag{6}$$

The term `proximity` is used in a generic way to denotes both similarity and dissimilarity values. For similarities, a high p_{ij} indicates that objects i and j are similar. Dissimilarity is a proximity that indicates how dissimilar two objects are. A small score indicates that the objects are similar, a high score that they are dissimilar. A dissimilarity is denoted by δ_{ij} and a similarity is denoted by (s_{ij}). Dissimilarities[2] can be obtained as a reciprocal of the similarity $1 - s_{ij}$.

The task of MDS was defined as finding a low-dimensional configuration of points representing objects such that the distance between any two points matches their dissimilarity "as closely as possible". We would prefer that each dissimilarity be mapped exactly into its corresponding distance in the MDS space. For those diferents measures of stress evaluates the quality of the mapping. In this paper we use two stress functions known as *Kruskal's stress*[2], defined as:

$$S1 = \sqrt{\frac{\sum_{i<j}[\delta_{ij} - d_{ij}(X)]^2}{\sum_{i<j} d_{ij}(X)}} \tag{7}$$

$$S2 = \sqrt{\frac{\sum_{i<j}[\delta_{ij} - d_{ij}(X)]^2}{\sum_{i<j} [d_{ij}(X) - \bar{d}]^2}} \tag{8}$$

where $\bar{d}$ is the mean measure of distances; d_{ij} is the distance computed from the mapped points.

3. MDS Algorithms

Here, we describe the four MDS algorithms used for the analysis. All the presented algorithms assume that the only available information is a $n \times n$ distance matrix $\Delta = (\delta_{ij})$, where each element δ_{ij} represents the dissimilarity between the objects i and j. It is also assumed that such matrix is non-negative, symmetric and obeys the triangular inequality. In the MDS space, it is used the Euclidian distance, because it is invariant under rotation, traslation and reflection. The general problem is defined as follows:

General problem
Given n objects and distance information about them (Matrix Δ)
Find n points in a m-dimensional space,
such that the distances are preserved as well as possible.

CMDS

Classical MDS (CMDS) was the first practical MDS method proposed by Torgerson and Gower[2]. Its basic idea is to assume that the dissimilarities are distances and then find coordinates that explain them. A matrix X of coordinates of the points, explain that distances. In CMDS is performed a process known as "double centering" which involves pre and post multiply the matrix of dissimilarities Δ by the matrix of centering $J = I - n^{-1}11'$ and by the factor $-\frac{1}{2}$, obtaining $-\frac{1}{2}J\Delta^2 J$. The next step is a process of eigen-decomposition to obtain the matrix solution of points $X = Q\Lambda^{1/2}$. Eigenvalues (Λ) equal to zero or negatives are ignored joint to the corresponding eigenvectors (Q).

LANDMARK

The LANDMARK algorithm was proposed by Silva and Tenenbaum[5]. This algorithm seeks to optimize the CMDS with respect to the calculation of eigen-decomposition in all the high-dimensional matrix of dissimilarities Δ. The first step of LANDMARK is calculate the CMDS procedure to map a subset of n chosen points of the dataset, referred as *landmark points* L, whose distance matrix is Δ_n. The second step is calculate the distance-based triangulation procedure, which uses the distances of the already-embedded landmark points to determine where the remaining points should be placed. Finally is carried out an analysis of coordinates by eigenvalues decomposition in the resulting point dataset. In this algorithm, the selection of *landmark set* can be a random choice or a MaxMin (greedy optimization) procedure, where the landmark points are chosen one at a time, and each new landmark maximizes the minimum distance to any of the existing landmarks.

SMACOF

The SMACOF (Scaling by MAjorizing COmplex Function) algorithm[2,4] also known as "Guttman Transformation" is based on the *iterative majorization method* for reducing a function of stress *(Stress Majorization)* using the Guttman transform[2]. Its major feature is that it guarantees lower Stress values in each iteration. The idea of the majorizatin method is to minimize the complex function $f(x)$ by using a more simple function $g(x, y)$ that can be easily minimized. The function g has to satisfy the inequality $f(x) \leq g(x, y)$, for a given y such that $f(y) = g(y, y)$. The method of minimizing f is iterative. At the beginning an initial value x_0 is given, and a funtion $g(x_1, x_0)$ is minimize to get x_1. Next is minimized $g(x, x_1)$ to get the next value (x), continuing in this manner until converge.

The SMACOF algorithm satisfies the requirements for minimizing a function using majorization, as described above. The main purpose is looking for a configuration X that minimizes the *raw stress* given in the Eq. (5).

To understand how SMACOF works, is necessary present some relationships and notation. Let Z be a possible configuration of points. The matrix V has elements given by the following $v_{ij} = -w_{ij}$, $i \neq j$ and $v_{ij} = \sum_{j=1, i \neq j}^{n} w_{ij}$. This matrix is not of full rank, and we will need the inverse in one of the update steps. So we turn to the *Moore-Penrose inverse*, which will be denoted by V^+.[a] Also is necessary define the matrix $B(Z)$ with elements

$$b_{ij} = \begin{cases} -\frac{w_{ij}\delta_{ij}}{d_{ij}(Z)} & \text{if } d_{ij}(Z) \neq 0 \\ \\ 0 & \text{if } d_{ij}(Z) = 0 \end{cases}, i \neq j \tag{9}$$

and

$$b_{ij} = - \sum_{j=1, i \neq j}^{n} b_{ij}. \tag{10}$$

Now is defined the *Guttman transform*[4], which is given by

$$X^k = V^+ B(Z) Z, \tag{11}$$

where the k represent the iteration number in the algorithm. If all of the weights are one (none of the dissimilarities are missing), then the transform is much simpler:

$$X^k = n^{-1} B(Z) Z, \tag{12}$$

here n represents the number of rows of the matrix of the distance matrix Δ. Summarizing, the main steps of SMACOF are:

[a]The Moore-Penrose inverse is also called the pseudo inverse and can be computed using the singular value decomposition.

(1) Find an initial configuration of points in R^d. This can either be random or nonrandom. Call this X^0
(2) Set $Z = X^0$ and the counter to $k = 0$
(3) Compute the raw stress $\sigma(X^0)$
(4) Increase the counter by $1 : k = k + 1$
(5) Obtain the Guttman transform X^k
(6) Compute the stress for this values between the two iterations. If this is less than some pre-specified tolerance or if the maximum number of iterations has been reached, then stop.
(7) Set $Z = X^k$, and go to step (4).

FASTMAP

The Fastmap algorithm was proposed by Faloustos and Lin[9]. This algorithm seeks to optimize the CMDS regarding the calculation of self-decomposition, where every solution vector of coordinates is obtained by iterations recursive. The Fastmap performs iterative protection orthogonal of points over lines called pivots, which are found by selecting the two most distant points of a hyperplane projection. The Figure 1 depicted this process:

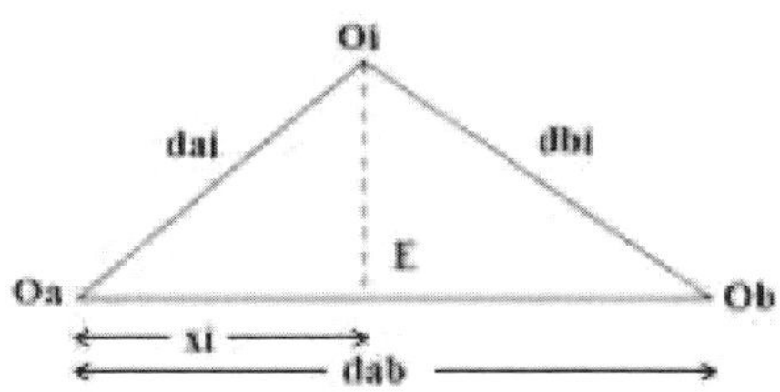

Figure 1. Ilustration of the "cosine law" projection on the line O_aO_b.

Using the *cosine law* Eq. (13), can be find the value of the projection x_i Eq. (14) from a point O_i over the pivot line E as follow:

$$d_{b,i}^2 = d_{a,i}^2 + d_{a,b}^2 - 2x_i d_{ab} \tag{13}$$

$$x_i = \frac{d_{a,i}^2 + d_{a,b}^2 - d_{b,i}^2}{2d_{ab}}. \tag{14}$$

In LANDMARK this procedure is repeated until obtain all the distances between points over one hyperplane perpendicular to the pivot. By means of this step a new projection is found from the subtraction of two points already projected, this is used for the Pythagorean Theorem. The new projection would represent the Euclidean distance between points and projected in the plan. The equation Eq. (15) generalizes the Pythagorean theorem to find the projection of all points on the hyperplane

$$(D'(O'_i, O'_j))^2 = (D(O_i, O_j))^2 - (x_i, x_j)^2, \forall i, j = 1, .., n. \qquad (15)$$

When all the distances are projected on the hyperplane, we can return to the first procedure, finding the projection of the points over the line pivot, so finding a new vector of coordinates. The recursion of the previous procedures is followed until achieve a desired number of dimensions.

4. Materials and Methods

We consider in this paper a comparative analysis of four algorithms MDS described above. Our main objectives were: i) compare the ability of the algorithms in accommodate the similarity information into 2 and 3 dimensional spaces, and ii) verify the preservation of existing patterns in such spaces. For this purpose, we used three comparison criteria:

(1) Stress index: The stress index is used to measure the divergence between the original distance matrix and the Euclidian distance computed from the resulting data points. We used two stress indices given by the Eq. (7) and Eq. (8).

(2) Computational time index: Refers to the computational time (in seconds) used by MDS algorithms measures from the invoked main program moment until the return results moment.

(3) Clustering purity index: This is an index used to assess the proximity of groups found by some method of clustering with respect to the actual classes. The value of purity indicates the percentage of elements that have the majority class in each group. The objective is identify if can be possible found in the mapping data the same separation between different clusters after be used a MDS algorithms. For a dataset expressed in tuples (x_i, cl_i), where x_i indicates the attributes of the data i, and cl_i indicates the belonging class of the data $cl_i \in cl_1, .., cl_m$, the purity of grouping C with clusters $c_1, c_2, ..., c_g$ obtained by some method of grouping is defined

as:

$$C = \frac{\sum_{c_i \in C} max\left(N_{c_i}(cl_1), .., N_{c_i}(cl_m)\right)}{n} \tag{16}$$

where $N_{c_i}(cl_j)$ denotes the number of elements with class cl_j within the group c_i and n denotes the size of data. The index of clustering indicates the percentage of elements that have the majority class in all the groups.

The procedure performed to calculate the described indixes consisted in execute the MDS algorithms for each datasets several times, each time increasing in one the dimensionality of the MDS space. The clustering algorithm chosen in this work was the *K-Means*[6], which requires as input the cluster number. The value used for this parameter is the known number of classes of each dataset. The CMDS algorithm, unlike the other algorithms, is executed only once for each dataset. This is because CMDS always returns the maximum number of possible dimensions (corresponding to positive eigenvalues). Therefore, the calculation of the computational time is performed only once. The outputs of each method and experiments in the dataset correspond to one time vector, two vectors of stress (Stress-1 and Stress-2), and one vector containing the clustering purity index of groups induced. Each element of these vectors corresponds to an index in a given dimensionality.

The performance index was tested in three known datasets: Iris plant, Wisconsin breast cancer and Image segmentation[1]. Table 1 presents a summary of the datasets including the number of data (total numbers x attributes) and the cluster number for each datasets.

Table 1. Summary of attributes for the three datasets used in this work.

	Size initial	Number of clusters
Iris	150x4	3
Cancer	683x9	2
Images	2100x19	7

To show the performance of the presented comparison criteria, we developed a tool in Matlab called **MDSExplorer**[7,8]. In this tool were implemented the experiments, running in a PC AMD Athlon 64 operating at a speed of 3 GHz with 2 GBytes of memory and operating system MS Windows XP.

5. Results

Figure 2 shows the measured values of the criterion of stress **S1** in the three datasets. The results for the stress index **S2** are not showed, since in all the cases this index presented the same behavior than the stress index **S1**. Each figure shows four curves corresponding to the stress of the four studied algorithms as a function of the dimensionality. As expected, it can be noted that in all algorithms MDS the stress values are close to 0 when the dataset is close to its true dimensionality. The algorithm SMACOF presented the lower levels of stress in all the analyzed dimensions and the FASTMAP algorithm presented the highest values of stress. These results suggest that the optimization performed in SMACOF is effective in accommodate more similarity information into few dimensions than the other algorithms.

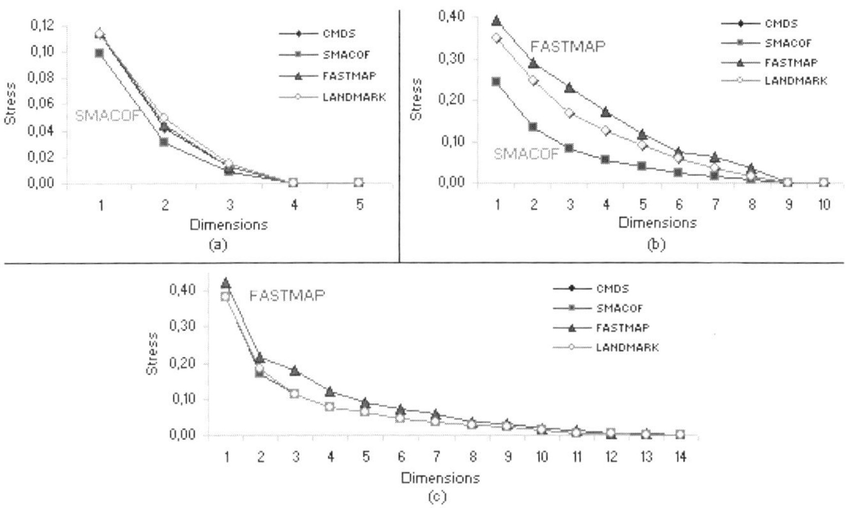

Figure 2. Stress index S1 for the dataset: (a)Iris, (b)Cancer and (c)Images. The x-axis is the dimensionality and the y-axis is the measure of the stress by each dimension.

Figure 3 shows the curves of the computational time for all analyzed algorithms as a function of the number of dimensions. As can be seen, the SMACOF curve, presents computational time significantly larger than the other algorithms, this is due to the iterative nature of SMACOF. Note also that in this algorithm the computational time drops significantly when the dimensionality is close to the true dimensionality of the dataset. This behavior can be due to the mapping optimized for SMACOF is very close to optimal and therefore the convergence is achieved quickly. Moreover, CMDS has the second highest curve of computational time, it is noted that this curve is constant because the CMDS is performed only once and the performance does not depend on the number of dimensions. The computational time of the CMDS was replicated in all dimensions only for comparison purposes. The curves of the algorithms FASTMAP and LANDMARK have the least computational time, however, the FASTMAP presents a computational time increasing with the number of dimensions. LANDMARK curves are smaller in all dimensions, being almost insensitive to the dimensionality.

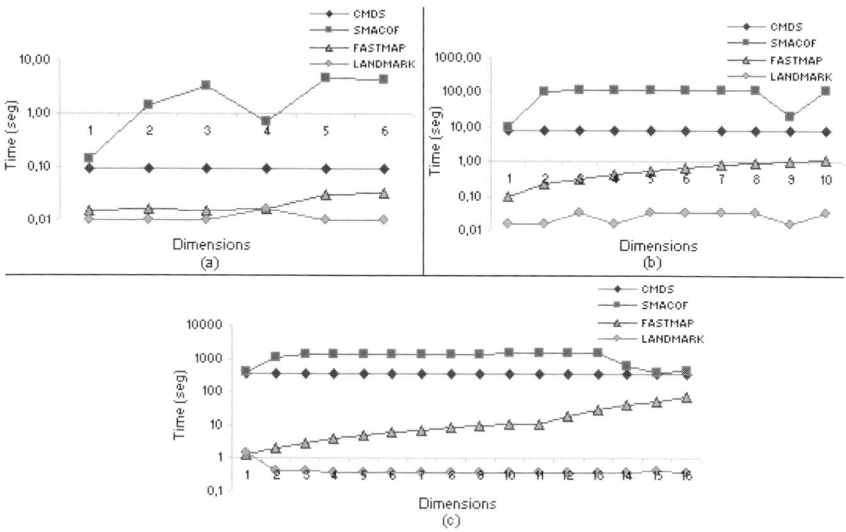

Figure 3. Computational time index for the dataset: (a)Iris, (b)Cancer and (c)Images. The x-axis is the dimensionality and the y-axis is the measure of time in seconds. Note that the curves are shown in logarithmic scale due to large differences among the algorithms in this index

Figure 4 shows the clustering purity index for the analyzed MDS algorithms as a function of the dimensionality of the mapped datasets. It can be noted that CMDS, SMACOF and LANDMARK have very similar values for this index and similar to the clustering purity index computed in the original dataset (data not shown). Only the FASTMAP algorithm presented lower values for this index when few dimensions are taken. This means that the data representation outputted by FASTMAP induce clusters of lower quality (further from reality) when compared to the other

MDS algorithms with the same dimensionality. This results is also consistent with the stress index, where FASTMAP was the least effective in accommodate more similarity information into few dimensions.

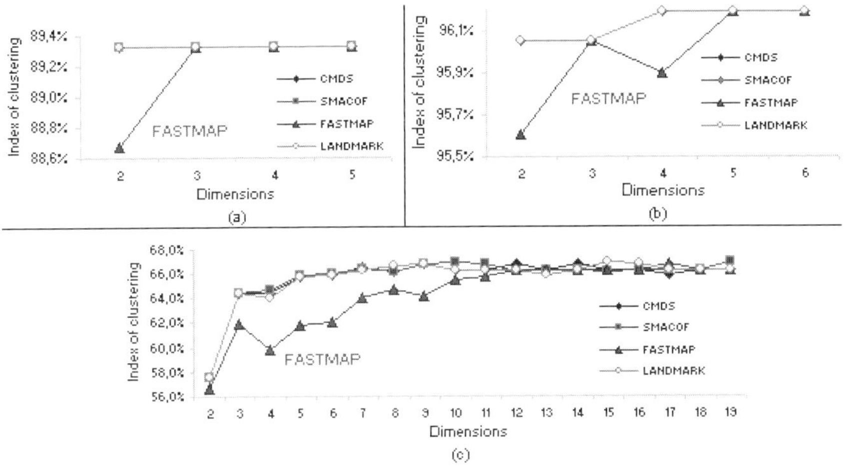

Figure 4. Clustering purity index for the dataset: (a)Iris, (b)Cancer and (c)Images. The x-axis is the dimensionality and the y-axis is the measure in percentage of the clustering purity index.

Summarizing, the results obtained on the studied datasets showed that the stress index vary with respect to the analyzed MDS algorithms. SMACOF was the MDS algorithm that best fits the similarity information in low dimension spaces. Meantime, FASTMAP was the least capable to accomplish such task. With respect to the clustering purity index, CMDS, LANDMARK and SMACOF presented indistinguishable results and similar to the clustering purity index obtained in the original datasets. This result suggest that in fact such MDS algorithms conserve the most important information in the mapped representations, showing the same group structures that in the the original datasets. By other hand, FASTMAP lost some information in the mapped representations that caused the identification of imprecise patterns. With respect to the computational time index, SMACOF presented the largest processing time and LANDMARK the shortest. FASTMAP showed a time index that is notoriously sensitive to the dimensionality of the target space.

Based on the above results and with the intention to get accurate mappings, we select the SMACOF algorithm to map a genotype data that was previously represented with dendrograms and LANDMARK[8]. The results of such mapping are shown to follow.

5.1. *Mapping of genotype data*

We present here a case of study showing the mapping of a genotype dataset. This dataset[14] is formed by RFLP-PCR images corresponding to a Brazilian collection of N_2-fixing bacterial strains belonging to the genus Bradyrhizobium. This symbiotic bacteria is important in agriculture by its capacity to transform the nitrogen of the atmosphere (N_2) into plant usable forms. The dataset is formed by 119 bacterial strains identified by a sequential number. Each strain is described by three lanes, which correspond to the RFLP-PCR analysis of three molecular markers in the ribosomal region 16S using the restriction enzymes: *Cfo I*, *Dde I* and *Msp I*[14].

The Bradyrhizobium dataset was pre-processed according to the procedure described in the works of Ref. 8, 7. Three distance matrices were obtained with such procedure ($D1$, $D2$ and $D3$), which correspond to the RFLP-PCR images in the ribossomal region 16S and restriction enzymes Cfo, Dde and Msp.

Figure 5 shows a 2D visualization of the three mappings performed with the SMACOF algorithm. The quantity of information exhibited in these representations corresponds to the fraction of the 2 higher eigenvalues over the total sum of eigenvalues of the covariance matrix, calculated for each mapping. This values are respectively 66%, 78% and 68% for $D1$, $D2$ and $D3$ mappings. After a visual exploration, we identified 4 groups for $D1$ mapping and 3 groups for $D2$ and $D3$ mapping. The K-Means algorithm was then executed with these numbers of groups. To differentiate the results in the Figure 5, we enclosed the resulting clusters with ovals. The centers of the clusters are indicated with asterisks. The found clusters in the $D1$ and $D2$ mapping were similar to those found in the LANDMARK mapping in Ref. 8. The $D3$ mapping showed some differences. In this mapping we identified 3 clusters, while in the corresponding mapping of Ref. 8 were identified 4 clusters. After a visual analysis we noted that the clustering performed in the SMACOF mapping seems to be more natural than the clustering performed in the LANDMARK mapping (the major cluster (blue) in the SMACOF mapping was divided into two proximal clusters in the LANDMARK mapping). We verify that the quantity of information captured by LANDMARK with two dimensions is 61%, 70%

and 66% for the *D*1, *D*2 and *D*3 mappings respectively. That is to say, in two dimensions the SMACOF mapping represent more information that the LANDMARK mapping.

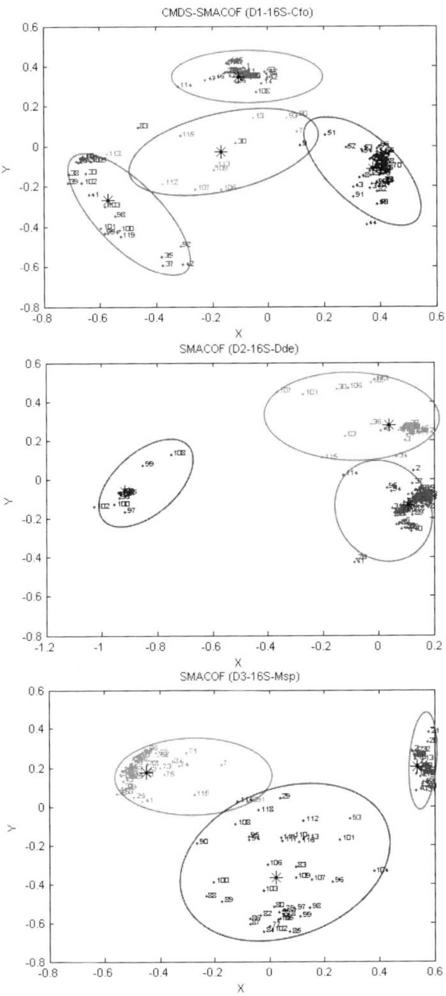

Figure 5. Clustering performed on the SMACOF mappings with K-Means algorithm. These representations contain respectively 66%, 78% and 68% of total information for the D1, D2 and D3 Bradyrhizobium mappings. The ovals indicate the groups.

6. Conclusions

The question of which MDS algorithm is more suitable to map genotype data was addressed in this paper. Four popular MDS algorithms were analyzed in terms of three criteria: their ability to fit the similarity information contained in the data into low dimensional spaces, their ability to preserve true group structures in the mapped space, and their computational costs. The results obtained from three well know datasets showed that SMACOF is the MDS algorithm that best accommodate the similarity information into few dimensions, but at a high computational cost. The preservation of true group structures was performed with the same performance by CMDS, LANDMARK and SMACOF in the three datasets. FASTMAP was the algorithm that presented the worst performance, both in the fitting of the similarity information in low dimensions as in the preservation of true group structures. SMACOF showed to improve the identification of clusters in a real genotype dataset. In situations where precision is not an important issue, the LANDMARK algorithm would be more suitable to map genotype data, since it presented the lowest computational time and an acceptable precision in the mapping.

References

1. A. Asuncion and D. Newman. UCI machine learning repository, 2007.
2. I. Borg and P. Groenen. *Modern Multidimensional Scaling: Theory and Applications*. Springer Press, second edition, 2005.
3. T. Cox and M. Cox. *Multidimensional Scaling*. Probability and Mathematical Statistics, second edition, 1995.
4. J. De Leeuw. Convergence of the majorization method for multidimensional scaling. *Journal of Classification*, 5(1):163–180, 1988.
5. V. de Silva and J. Tenenbaum. Global versus local methods in nonlinear dimensionality reduction. In *In S.Becker, S. Thrun, and K. Obermayer, editors, Proc. NIPS*, volume 15, pages 721–728, 2003.
6. R. O. Duda, P. E. Hart, and D. G. Stork. *Pattern Classification*. A Wiley-Interscience, second edition, 2001.
7. S. Espezua-Llerena. Mapeamento de dados genômicos usando escalonamento multidimensional. Master's thesis, Engenharia Elétrica de São Carlos da Universidade de São Paulo, 2008.
8. S. Espezua-Llerena and C. D. Maciel. Exploratory visualization of rflp-pcr genomic data using multidimensional scaling. In *SIBGRAPI 2008: XXI Brazilian Symposium on Computer Graphics and Image Processing*, pages 05–312. Brazilian Comput Soc, 2008.
9. K. Faloutsos, C.; Lin. Fastmap: A fast algorithm for indexing, data-mining and visualization of traditional and multimedia datasets. In *ACM SIGMOD*,

volume 24, page 163–174, 1995.
10. P. Germano, M. G.and Menna, F. L. Mostasso, and M. Hungria. Rflp analysis of the rna operon of a brazilian collection of bradyrhizobial strains from 33 legume species. *International Journal of Systematic and Evolutionary Microbiology*, 56(1):217–229, 2006.
11. A. Kelemen, A. T. V. Vasilakos, and Y. Liang. Computational Intelligence in Bioinformatics: SNP/Haplotype Data in Genetic Association Study for Common Diseases. *IEEE Transactions on information Technology in Biomedicine*, 13(5):841–847, SEP 2009.
12. S. Kim, L. Shen, and e. a. Saykin, AJ. Data synthesis and tool development for exploring imaging genomic patterns. In *Conference Information: IEEE Symposium on Computational Intelligence in Bioinformatics and Computational Biology*, pages 298–305, 2009.
13. O. Maimom and L. Rokach. *The Data Mining and Knowledge Dicovery Handbook*. Springer, second edition, 2005.
14. S. Milagre, C. Maciel, A. Shinoda, M. Hungria, and J. Almeida. Multidimensional cluster stability analysis from a brazilian bradyrhizobium sp. rflp/pcr data set. *Journal of Computational and Applied Mathematics*, 2009.
15. M. Schroeder, D. Gilbert, J. Van Helden, and P. Noy. Approaches to visualization in bioinformatics: from dendrograms to space explorer. *Information Sciences*, 139(1-2):19–57, 2001.
16. D. W. Scott. *Multivariate Density Estimation: Theory, Practice, and Visualization*. Wiley Series in Probability and Mathematical Statistics, second edition, 1992.
17. P. Sebastiani, N. Timofeev, D. A. Dworkis, T. T. Perls, and M. H. Steinberg. Genome-wide association studies and the genetic dissection of complex traits. *American Journal of Hematology*, 84(8):504–515, AUG 2009.
18. M. Su and C. Chou. A k-means algorithm with a novel non-metric distance. In *5th Joint Conference on Information Sciences (JCIS 2000)*, volume 1-2, pages 417–420, 2000.
19. M. Vlachos, B. Taneri, E. Keogh, and P. S. Yu. Visual exploration of genomic data. In Kok, JN and Koronacki, J and DeMantaras, RL and Matwin, S and Mladenic, D and Skowron, A, editor, *Knowledge Discovery in Databases: PKDD 2007, Proceedings*, volume 4702 of *Lecture notes in artificial intelligence*, pages 613–620. Springer-Verlag Berlin, 2007.

UNIVERSAL FEATURES FOR EXON PREDICTION

DIEGO FRIAS
Department of Natural and Applied Sciences, Bahia State University
Salvador, BA, Brazil

NICOLAS CARELS
Oswaldo Cruz Foundation, Oswaldo Cruz Institute, Laboratory for Functional Genomics and Bioinformatics,
Rio de Janeiro, RJ, Brazil

Nucleotide correlations in coding sequences result from the functional constraints on physico-chemical properties of proteins. These constraints are imprinted in the coding DNA in the form of a purine bias with a purine preference in first position of codons. The resulting codon pattern is RNY (or Rrr) and has been called "ancestral codon pattern". Here, we describe a method that we called UFM (for Universal Feature Measure) for the CDS/intron classification based on the statistics of purine bias and stop codons. The proposed method is species-independent, GC-content independent, does not need prior training nor parameter adjustment and performs well with small DNA fragments >300bp. The results obtained with six model organisms (*A. thaliana, D. melanogaster, P. falciparum, O. sativa, C. reinhardtii and Homo sapiens*) show that for sequences of size >600bp the new classifier achieves a sensitivity > 97% and a specificity > 94% in all species.

1. Introduction

Nucleotide sequencing of genomes (DNA) and transcriptomes (RNA) of plants, animals, and microbes has revolutionized biology and medicine. While much genome projects address full chromosome DNA sequencing, transcriptome projects has generally consisted in the partial sequencing of the RNA pool of targeted cells or tissues resulting in the inventory of RNA fragments as a collection of expressed sequence tags (ESTs). According with the Genome OnLine Database (GOLD) accessed on May 21 of 2010 there are actually 7141 (6566 genome + 343 transcriptome + 231 metagenome + 1 metatranscriptome) sequencing projects including complete, incomplete and starting projects[1]. Inventories of genomic DNA sequences from environmental samples containing thousands of microbial species are referred to as Metagenomes or as Metatranscriptomes if RNA sequences are concerned. At the moment only 19% of the genomic, 35% of the metagenomic and 6.2% of the transcriptomic projects have being completed which means that a huge amount of DNA and RNA sequences shall be treated with bioinformatic platforms in the next years.

Bioinformatic pipelines involve two basic sequential steps: (1) Sequence Processing and (2) Data Analysis. Sequence Processing namely comprises three steps: (1) Read Processing, (2) Contig Assembly, and (3) Sequence Finishing. Data Analysis occurs in two steps: (1) Gene Prediction and (2) Gene Annotation which consists in assigning a function and a cell location (nuclear, cytoplasm, membrane) to the predicted protein.

Gene prediction in eukaryotes is carried out in two steps: (1) Exon finding and (2) Gene assembly. Exon finding is a complex coding/noncoding classification problem that can be addressed looking for variable-length nucleotide/aminoacid patterns (*ab initio* methods) or seeking for sequence similarity in protein libraries[2]. The basic hypothesis behind sequence similarity (SS) methods is that if two (large enough) sequences show significant similarity, they probably shared same ancestry and, therefore, have similar functions. The basic algorithm consists of scoring local or global alignments of target and library sequences (typically stored in relational databases). A number of bioinformatic tools for similarity searches are available[3–5]. SS methods have high specificity or low rate of false genes, except when the reference library is contaminated with spurious genes. In a recent revision of the annotation of the yeast genome, probably the most studied model specie, approximately 8.3% of the 6062 open reading frames (ORFs) which potentially encode proteins of at least 100 amino acids were found to be spurious predictions[6]. Furthermore, SS methods lack the ability to discover new genes having and unpredictable rate of false negatives. Moreover, while the size of sequence databases is growing faster than the speed of their processing, the rate of SS-based annotation of new function is growing only slowly[7].

The basic premise behind *ab initio* exon finding methods is that the nucleotide ordering derived from the biological message carried by coding sequences (CDS) is responsible for conserved patterns that can be described by a set of proper features. According to the features used, these methods can be grouped into three families:

(1) **Positional nucleotide correlation:** The methods in this family explore the fact that the arrangement of nucleotides in codons in-

troduces a period-3 and other short-range correlations that can be measured with the amplitude and/or phase by Fourier Transform[8, 10, 21, 23] or/and Average Mutual Information which is able to sense a decrease in the entropy of the coding sequences[9, 23, 24]. A great advantage of these methods is their independence from the biological species considered which allows the skipping of training or learning steps. However, they have some basic drawbacks that limit their application: (a) the coding potential is the same in the six reading frames which complicates the coding frame detection and (b) the sensitivity decreases strongly when the size of the sequence decreases below 400 bp.

(2) **Codon and nucleotide statistics:** The methods in this family use a quite large number of features which are mostly based on trinucleotide (4^3) and hexanucleotide (4^6) frequencies. Domain specific (exon, intron, intergenic) models (maximum likelihood matrixes[16−22] or neural networks[14,15]) are "learned" using training data sets. After (cross-)validation, the models are simultaneously applied to each input sequence (ORF or DNA fragment captured by an sliding window) and the higher scoring model is selected for final gene prediction. Models for number of specific species or group of genetically similar organisms are already available. The most accurate gene predictors are the generalized hidden Markov models (GHMMs). A number of tools have been developed for gene prediction (e.g. GENSCAN[16]; EasyGene[17]; GeneMark.hmm[18]18; TWINSCAN[19]; GLIMMER[20]; and AUGUSTUS[13]). However, bias in the models due to error propagation from training sets may produce spurious genes[2]. Thc rate of specificity of these methods is improved when they are combined with methods of cross-species comparison at genomic level such as in TWINSCAN and AUGUSTUS. There is another method called Z-curve[25−28] that belongs to this family. It uses as primary features codon-positional frequencies of nucleotides (P_{nj} - frequency of nucleotide n=A,G,C,T at codon position *j=1,2,3*) and dinucleotides (P_{n1m2}, and P_{n2m3} - frequency of nucleotides n and m at first and second and at second and third codon positions, respectively). These 44 primary features are transformed into a vector of 33 secondary features. Using training data sets model-vectors are calculated for exons and introns and then applied for classification of ORFs. The classifier is the Euclidean

distance between the model and the input sequence vector.

(3) **Codon compositional pattern:** The methods in this family explore the existence of compositional bias in the codons of most genes. Particularly, a preference for purine (R=A,G) in the first position of codons and predominance of pyrimidine (Y=T,C) in the third codon position that is called RNY pattern. Here N indicates any bases from ACGT pool. The prevalence of RNY in CDS is assumed to be an evidence of an ancestral genetic code[11]. Shepherd in 1981 used this regularity for exon prediction[11]. More recently, Nikolaou and Almirantis (2004) proposed the Codon Structure Factor (CSF)[12] that is based on codon-position nucleotide frequencies and on the frequencies of trinucleotides based on the RNY pattern. For each RNY trinucleotide the ratio $P_{\mathrm{R1N2Y3}}/(P_{\mathrm{Y1}}P_{N2}P_{R3})$ is calculated and added to CSF. Given a certain threshold for CSF, lets denote it as τ_{CSF}, the input sequence is classified as coding if CSF $> \tau_{\mathrm{CSF}}$. Like the first family of *ab initio* methods, these methods do not require training step and are, at some extent, species-independent.

Here, we review on CDS prediction based on compositional pattern and propose a new feature set and two classification procedures. The results are compared with those of [12] for sequence size varying from 50 bp to 600 bp.

2. Materials and Methods

We only considered CDSs experimentally validated through peer review publications in order to avoid the possible contribution of systematic annotation errors. CDSs were from six model species covering the complete range of GC levels in third positions of codons (GC3) and sequence complexity in eukaryotes. These species were: Plasmodium *falciparum* (CDS=197, GC3=0-30%), Chlamydomonas *reinhardtii* (CDS=102, GC3=60-100%), Arabidopsis *thaliana* (CDS=1,206, GC3=25-65%), Oryza *sativa* (CDS=401, GC3=25-100%), Drosophila *melanogaster* (CDS=1,262, GC3=40-85%) and Homo *sapiens* (CDS=1,199, GC3=30-90%).

We built datasets of CDS fragments (sequences in coding frame +1) of the 6 model species with fixed sizes ranging from 50 to 600 bp, increasing in 50 bp each time. The fixed-size CDSs were extracted both from the beginning (5' side) and from the end (3' side) of the genes, in order to take into account that codon usage varies along the genes.

Aiming to test the ability of the method in detecting the coding strand, we built data sets with CDSs in frame -1 with the reverse complement sequences of the coding sequences in the reference data sets.

To test the heuristic features for exon/intron classification a dataset of introns of A. *thaliana* (n=5301), D. *melanogaster* (n=18749) and H. *sapiens* (n=2030) retrieved from http://hsc.utoledo.edu/bioinfo/eid/index.html was built. Again fixed-sized datasets with sequence length varying between 50 and 600 bp were built by cutting pieces of specific lengths from the 5' and 3' sides.

Two different approaches to the problem were reported earlier. The first approach, published in [29], used 9 basic features. Such features were combined to form 5 derived features, which later were combined to form 2 linear classifiers, one used for the coding frame determination and the other for intron classification. The second approach was published in [30] and used only 3 of the previous basic features, combining them into a single measure of the coding potential of the sequence. The second approach naturally derived from the first.

In this work a basic feature F assigns a positive real value $f \in \Re^+$ to any input DNA fragment $S = n_1 n_2 n_3, ..., n_{L-2} n_{L-1} n_L$ comprised of L nucleotides $n_i \in \{A, G, C, T\}$, $i = 1, 2, \ldots, L$, that is, $F : S \rightarrow f$. Basic features are used to build linear discriminators φ called derived features. Here, we assume that the fragment S does not contain more than one complete ORF in any frame and that coding ORFs in different frames do not overlap each other. The latter holds in higher eukaryotes, but not in species with compact genomes, like some bacteria and viruses. The derived features φ are designed to take their maximum values at the coding ORF in the DNA fragment S.

2.1. *Basic features*

Our method investigate the coding potential in the six frames of S. Let S_k denote the sequence S given in the reading frame *k= +1, +2, +3, -1, -2, -3.* The score of the *j-th* feature F_j for the input sequence S_k is denoted by $F_j^k = F_j(S_k)$. We introduced nine basic features listed in Table 1 where P_{XYZ} is the frequency of the nucleotide triplet XYZ and P_{Xi} is the frequency of nucleotide X in codon position *i=1,2,3*.

An extensive statistical study with the six model species showed that generally in CDS (coding frame k =*+1|-1)*:

(1) F1>F2 and F1>F3 as seen in Figure 1 (LEFT),

Table 1. Basic features based on nucleotide and stop codon frequencies

	Basic features	
$F_1 = P_{A1}P_{G1}$	$F_4 = P_{C1}P_{G1}$	$F_7 = P_{C3}P_{G1}P_{A2}$
$F_2 = P_{A2}P_{G2}$	$F_5 = P_{C2}P_{G2}$	$F_8 = P_{C2}P_{G3}P_{A1}$
$F_3 = P_{A3}P_{G3}$	$F_6 = P_{C1}P_{G2}P_{A3}$	$F_9 = P_{TAA} + P_{TAG} + P_{TGA}$

(2) F6<F7 and F6<F8 as seen in Figure 1 (RIGHT), and
(3) F4>F5 when the GC content exceeds 55% (result not shown).

The ninth basic feature is the overall frequency of stop codons and then F_9=0 by definition in the coding part of the genes. Moreover, such regularities rarely hold in non-coding frames ($k \neq +1|-1$) and in non-coding sequences. Notice that the two first regularities above are related to the existence of a purine gradient in CDS. This set of basic features was used to form linear discriminators maximized by the coding sequences in the coding frame, as described below.

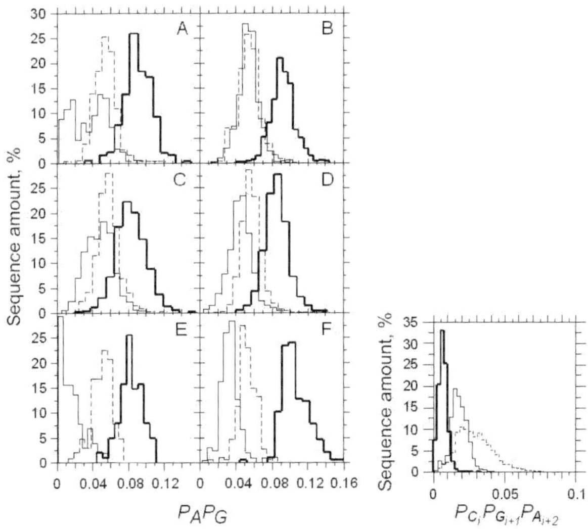

Figure 1. **LEFT:** Distribution of F_1 (bold), F_2 (dashed) and F_3F (thin) in *O. sativa* (A), *A. thaliana* (B), *H. sapiens* (C), *D. melanogaster* (D), *C. reinhardtii* (E) and *P. falciparum* (F). Notice that in most cases $F_1 > F_2 > F_3$. **RIGHT:** Distribution of F_6 (bold), F_7 (dashed) and F_8 (thin) in the six model species grouped together. Notice that in most cases $F_6 < F_8 < F_7$.

2.2. *Derived features*

Any derived feature φ_j must satisfy three conditions:

A. If is coding in frame $c \Rightarrow \varphi_j^c > \tau_j$, where τ_j is a threshold for feature φ_j to be estimated. Failing to fulfill this condition causes a CDS to be classified as non-coding which counts as a case of false negative (FN). When this condition holds (and also the condition C below), it is counted as a true positive (TP).

B. In the case that S is non-coding (intron), it is expected that $max_k({\varphi^k}_j) \le \tau_j$, for all reading frame k. Failing to fulfill this condition causes an intron to be classified as coding which counts as a false positive (FP). When this condition holds, it is counted as a true negative (TN).

C. If S is coding in frame $c \Rightarrow \varphi_j^c > \varphi_j^k, \quad \forall k \neq c$. Failing in this condition causes an error in the coding frame determination and is counted as false negative (FN) independently of the result of condition A, as indicated above.

The performance of the exon predictor as usual is expressed in terms of the sensitivity $S_n = TP/(TP + FN)$ and specificity $S_p = TN/(TN + FP)$. We introduced the harmonic mean of the specificity and sensitivity $F - score = 2S_nS_p/(S_n + S_p)$[45], which is used for estimating the optimal threshold τ_j for the classifier φ_j.

2.2.1. *First approach*

In the first approach the nine basic features in Table 1 were combined to form five derived features as summarized in Table 2 below.

Table 2. Derived features in the first approach.

Features		
$\varphi_1 = 1 - F_9$	$\varphi_3 = F_7 + F_8 - 2F_6$	$\varphi_5^* = F_4 - F_5$
$\varphi_2 = 1 - F_6$	$\varphi_4 = 2F_1 - F_2 - F_3$	$(*)\,\text{ifGC} > 55\%$

The two first derived features were used to determine the frame with highest coding potential. We defined the frame-dependent measure ${m^k}_f = {\varphi^k}_1 + {\varphi^k}_2$ which was calculated for the six frames of the input sequence.

According with the condition A above, the putative coding frame c is assumed as the frame that maximizes m_f, that is, $m^c{}_f = \max_k\left(m_f^k\right)$. In the experiments the coding frame of the input sequences was +1 or -1 to simplify the analysis. Thus, whenever $c \neq \pm 1$, we counted a false negative result.

Once the putative coding frame is determined, the next task is to evaluate the coding potential of the sequence S_c, that is to apply the condition C. The coding potential is calculated as $m_{CDS}^c = \phi_1^c + \phi_3^c + \phi_4^c$, except in the case where the GC content of the input sequence > 55%, in which ϕ_5^c is added to m_{CDS}^c. Given a threshold τ_{CDS}, the input sequence is considered coding if $m_{CDS}^c \geq \tau_{CDS}$. Otherwise it is considered non-coding.

In the experiment, the prediction was compared with the nature of the input sequence, counting a TP or an FN result when the input sequence was a CDS and a TN or an FP result when the input sequence was an intron.

2.2.2. *Second approach*

The second approach, called Universal Feature Method (UFM), differs from the previous one in three aspects: (1) Only three basic features, F_1, F_6 and F_9 are used, (2) A single coding potential measure $m_{UFM}^k = F_1^k/(F_6^k + F_9^k + \omega)$ is calculated for the six reading frames, where $\omega = 0.01$ is a constant included to avoid division by zero, and (3) The classification condition takes into account the change of m_{UFM}^k among reading frames instead of its value itself. The classification condition is now written in the form: if $\max_k\left(m_{UFM}^k\right) - \min_k\left(m_{UFM}^k\right) \geq \tau_{UFM}$ then the input sequence is coding and the coding frame is given by the frame that maximizes m_{UFM}^k, that is, $m_{UFM}^{\aleph} = \max_k\left(m_{UFM}^k\right)$. Otherwise the input sequence is classified as non-coding.

2.3. *Reference methods*

We used two methods for success rate comparison. The success rate of the first UFM approach was compared to that of the ORFFinder method[31]. ORFFinder simply select the largest ORF found in the six reading frames of the given sequence, as being the coding one, rejecting overlapping ORFs and ORFs with size smaller than a prefixed minimum ORF size. The results of this method without ORF size limitation can be simulated by maximizing the derived feature φ_1^k(Table 2). The second UFM approach was compared with a method proposed in [20] based on the Codon Structure Factor (CSF). CSF also explores the conserved preference for codons containing purine in the first codon position and pyrimidine in the last codon position (RNY). For comparison we implemented the CSF method and estimated the average optimal threshold using the data sets of the six model species addressed here.

3. Results and Discussion

According to [11], we found that there is a gradient in the probability of purine towards the first codon position in all 6 species, as shown in Figure 1 (LEFT). Therefore, we denoted this purine bias by Rrr. The purine probability $P_{Aj}P_{Gj}$ was, on average, $P_{A1}P_{G1} = 9.08E-2$, $P_{A2}P_{G2} = 5.47E-2$ and $P_{A3}P_{G3} = 3.90E-2$. The average error in the estimates was ~7.0E-4 (Table 3). Both values are remarkably conserved among distant species whatever their average GC level.

Table 3. Purine porbability ($P_A P_G$) at the 3 codon positions.

Species	samples	1st	2nd	3th
A. thaliana	1206	0.0930±0.0004	0.0550±0.0004	0.0550±0.0003
O. sativa	401	0.0910±0.0008	0.0540±0.0006	0.0360±0.0010
H. sapiens	1199	0.0840±0.0005	0.0580±0.0004	0.0480±0.0004
D. melanogaster	1262	0.0860±0.0004	0.0580±0.0003	0.0450±0.0004
C. reinhardtii	102	0.0840±0.0013	0.0510±0.0012	0.0170±0.0013
P. falciparum	197	0.1070±0.0012	0.0520±0.0007	0.0330±0.0007
Total/Averages	4367	0.0908±0.0008	0.0547±0.0006	0.0390±0.0007

Despite its extreme AT rich composition, P. *falciparum* also shows the Rrr bias (Fig. 1A - F). It is interesting to note that in contrast to Adenine, Guanine does not show correlation between codon positions 1 and 2 in any of the 6 species (result not shown).

Other interesting regularity shown in Figure 1 (RIGHT) is that the product of the frequencies of nucleotides C, G and A, (taken in that order) is significantly lower than the product of the frequencies of the same nucleotides when their position within the codon is changed in a circular way, that is, G,A,C and A,C,G. The overlap between $P_{G1}P_{A2}P_{C3}$ and

$P_{A1}P_{C2}P_{G3}$ is only 7% of the CDS samples of the 6 species considered together. Therefore, the feature φ_3 (Table 2) is maximized in the coding frame (+1) in 93% of the CDS. Notice that the minimum triple-product of nucleotide frequencies $F_6 = P_{C1}P_{G2}P_{A3}$, is obtained for a succession of nucleotides opposite to the purine gradient, that is YRR. However, other circular triple-products of nucleotide frequencies opposite to the purine gradient did not show such a significant bias.

3.1. *First approach*

We calculated the success rate of coding frame detection using only stop codons frequency ($\varphi_1 = 1 - F_9$) as classifier, that is using the same principle than in ORFFinder[31]. We tested with CDSs in coding frame +1 and -1 with sizes varying from 50 bp to 600 bp. The success rate in this case was defined as the percentage of cases when the smallest number of stop codons was found in the coding frame of the input sequence.

Figure 2 shows that coding frame detection is easier in AT-rich than in GC-rich sequences (O. *sativa* and C. *reinhardtii*), but also that the depletion in stop codon frequency introduced by the coding frame is not enough for an accurate diagnosis of coding ORFs. In Figure 3 we show the results of the addition of one more feature ($\varphi_2 = 1 - F_6$) for detection of the coding frame, under the same conditions of Figure 3. A significant improvement in accuracy and in independence of the GC content can be observed. For CDS with size $>$ 200bp the success rate is $>$ 90% in all species considering coding regions in both positive and negative strands.

We then tested the ability of the coding potential m_{CDS} to reject introns. Firstly, we plotted the distributions of m_{CDS} for CDSs in frame +1 and for introns of A. *thaliana*, D. *melanogaster* and H. *sapiens* in the six frames (Figure 4). For the purpose of clarity, we grouped the CDSs of the 6 species all together. Secondly, we noticed that the better discrimination between CDSs and introns is achieved with a threshold $\tau_{CDS} = 1.05$ which is constant in the size interval of 200bp to 600 bp, as indicated with a vertical line in Figure 5. The overlapping area in Figure 5 concerns the sequences for which the intron/exon classification cannot be trusted. Thirdly, we calculated the rates of wrong classifications with the chosen threshold. The false negative rate (CDS classified as intron) varied from 10% at 200bp to 7% at 600 bp, while the false positive rate (intron classified as CDS) varied with the specie. For introns of A. *thaliana* the false positive rate varied between 8% at 200bp and 0% at 600bp. For introns of H. sapiens and D.

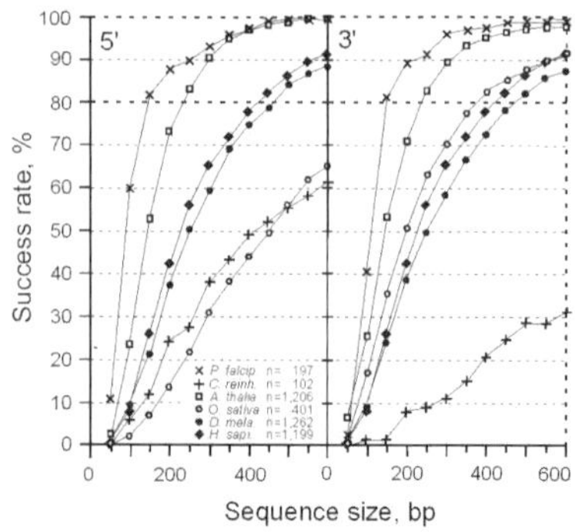

Figure 2. Classification of the coding frame with stop codon frequency (φ_1) for input CDSs in frame +1 (left) and -1 (right). The CDS size veries between 50 and 600 bp. Species: *P. falciparum* (X), *C. reinhardtii* (+), *A. thaliana* (square), *O. sativa* (circle), *D. melanogaster* (filled circle) and *H. sapiens* (filled diamond).

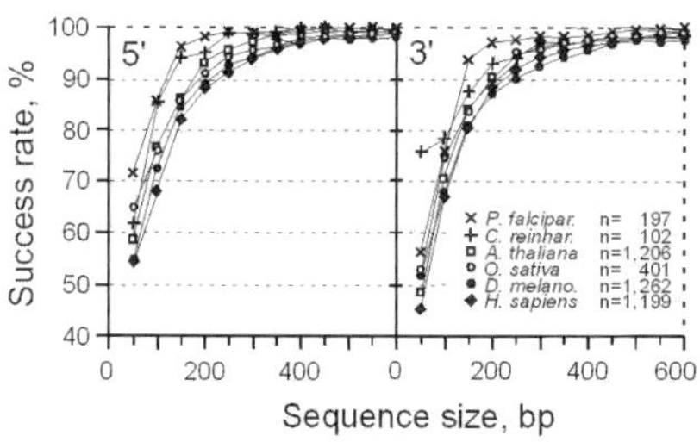

Figure 3. Classification of the coding frame maximizing the measure ($m_f = \varphi_1 + \varphi_2$) for input CDSs in frame +1 (left) and -1 (right). The CDS size veries between 50 and 600 bp. Species: *P. falciparum* (X), *C. reinhardtii* (+), *A. thaliana* (square), *O. sativa* (circle), *D. melanogaster* (filled circle) and *H. sapiens* (filled diamond).

melanogaster the false positive rate varied between 15% at 200bp and 3% at 600bp. Notice that the false positive rate decreases more rapidly than the false negative rate when the size of the input sequence increases.

3.2. *Second approach*

We built data sets of 500 sequences of equal sizes for CDS and introns of *H. sapiens*, *D. melanogaster* and *A. thaliana*, in order to normalize the experiment.

These data sets were then used for the calculation of the distribution of the Codon Structure Factor (CSF) and of the coding potential m_{UFM} as a function of the sequence length, in order to determine an optimal

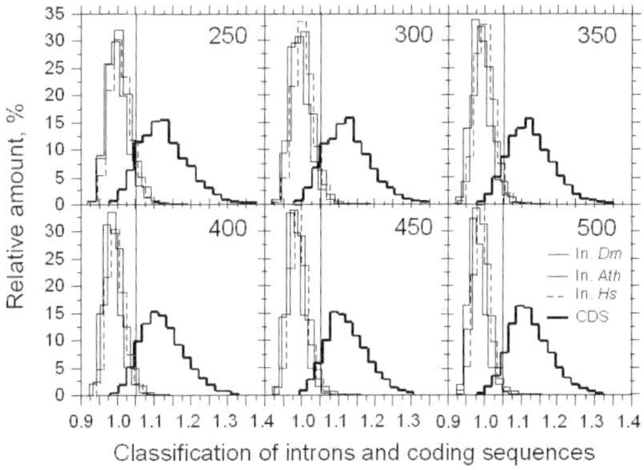

Figure 4. Distribution of the coding potential m_{CDS} in 4.367 CDS of six model species (bold) and in 26.080 introns (In) of three species: *A. thaliana* (Ath, n=5.301, plain line), *D. melanogaster* (Dm, n=18.749, thin line) and *H. sapiens* (*Hs*, n=2.030, dashed line). The coding potential in introns was calculated among the 6 frames. The sequence size varied between 250bp and 500bp. The intron distributions are centered at $m_{CDS} = 0.95$ while the CDS distribution at 1.10. The plain line (vertical) at 1.05 is the chosen threshold for coding/noncoding classification.

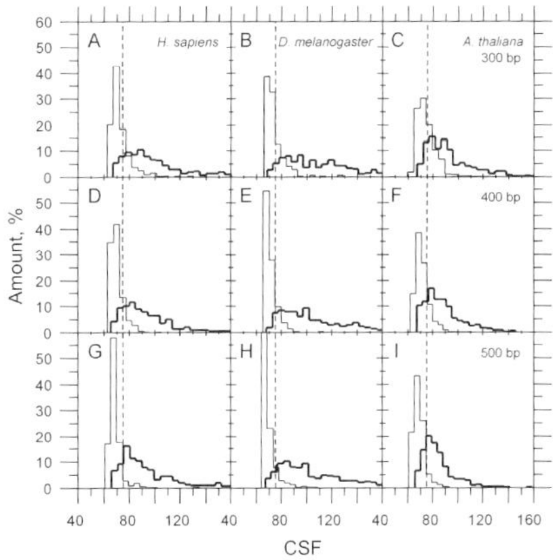

Figure 5. CSF for coding sequences (bold) and introns (thin) of *Homo sapiens* (1st*col.*), *Drosophila melanogaster* (2nd*col.*) and *Arabidopsis thaliana* (3rd*col.*) at 300 (1st*row*), 400 (2nd*row*) and 500 bp (3rd*row*) by CSF. The vertical dashed line separates coding (right) from noncoding (left).

classification threshold in both cases. Doing this, we estimated $\tau_{CSF} = 75.0$ and $\tau_{UFM} = 1.0$ as the thresholds that maximizes the average F-score in the sequence length interval considered (see Figures 5 and 6).

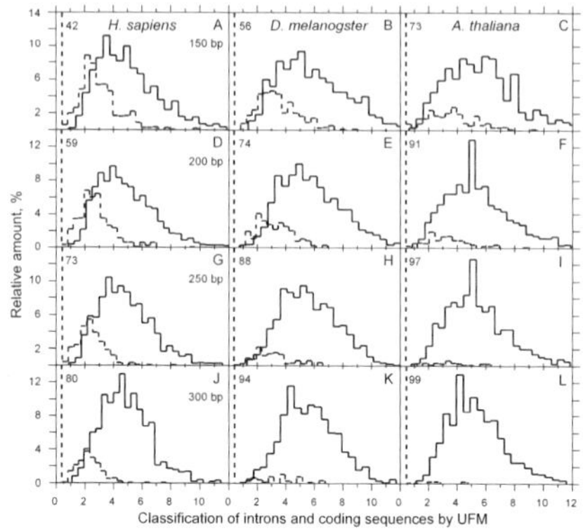

Figure 6. Classification of small exons (solid line) and introns (dashed line) of *H. sapiens* ($1^{st}col.$), *Drosophila melanogaster* ($2^{nd}col.$) and *A. thaliana* ($3^{rd}col.$) at 150 ($1^{st}row$), 200 (2nd row), 250 ($3^{rd}row$) and 300 bp ($4^{th}row$)by UFM. Specificity (%) for τ_{UFM} = 1.0 indicated at top-left corner.

Table 4. Comparative analysis of CDS/Intron classification by CSF ($\tau_{CSF} = 75$) and $UFM(\tau_{UFM} = 1)$.

		SCF			UFM		
Species	Size, bp	Sn^1	Sp^2	$F - score^3$	Sn^4	Sp^5	F-score
H. sapiens	300	788.8	74.6	81.1	100.0	76.0	86.4
	400	86.6	87.6	87.1	100.0	88.0	93.6
	500	85.2	93.6	89.2	100.0	93.0	96.4
	600	84.2	93.4	88.6	100.0	97.4	98.7
D. melanogaster	300	95.2	71.4	81.6	99.8	97.4	98.6
	400	95.8	82.8	88.8	99.8	97.8	98.8
	500	93.4	89.6	91.5	100.0	98.6	99.3
	600	94.8	92.6	93.7	100.0	98.6	99.3
A. thaliana	300	90.8	59.6	72.0	100.0	100.0	100.0
	400	82.2	78.8	80.5	100.0	100.0	100.0
	500	78.6	90.8	84.3	100.0	100.0	100.0
	600	78.8	93.0	85.3	100.0	100.0	100.0

Note: (^{1}) Sensitivity (%) of CSF, (^{2}) Specificity (%) of CSF, (3) F-score (%) = $2SnSp/(Sn + Sp)$. (^{4}) Sensitivity (%) of UFM, (^{5}) Specificity (%) of UFM for.

Finally we calculated the sensitivity, specificity and F-score of UFM and CSF with the chosen thresholds. The results are shown in Table 4 and Figure 7.

The F-scores of UFM were higher than those of CSF in all three species with differences of 8%, 11% and 24%, on average, in *H. sapiens*, *D. melanogaster* and *A. thaliana*, respectively (Table 4, Figure 7). The perfor-

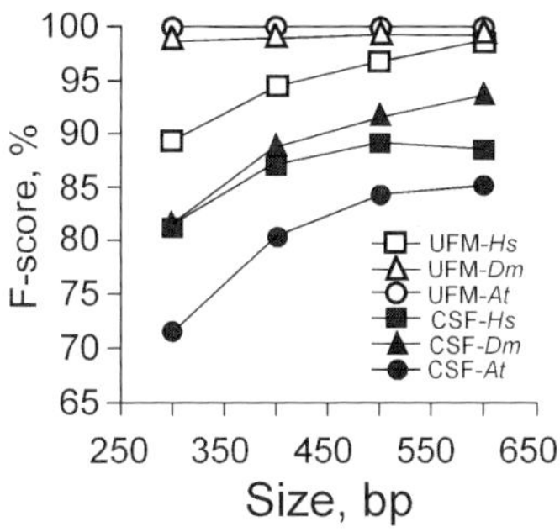

Figure 7.

mance of UFM was found to be higher in *A. thaliana* and *D. melanogaster* than in *H. sapiens* (Figure 7) suggesting fundamental differences in the intron composition of *H. sapiens* compared to the other two species. However, convergence between CDS/intron classification among the three species was reached at sequence size > 600 bp with a classification rate > 97%. By contrast, CDS/intron classification with CSF was higher for *D. melanogaster* and *H. sapiens* than for *A. thaliana* and was still < 95% at 600 bp without significant convergence trend (Figure 7).

Notice that both Sn and Sp of UFM increase with sequence size for all species, which is a strong evidence of the independence of its threshold of both sequence size and species and suggests that it is a robust classifier.

4. Conclusions

The features analyzed in this study allow the improvement of sensitivity and specificity of CDS vs intron classification at small ORF sizes with respect to the CSF method. The optimal thresholds for all the linear discriminators studied were the same among the six model species considered. Thus, UFM is species-independent (does not require species-related parameter setting) which makes it appropriated for anonymous genomes annotation. The different success rates of CDS/intron classification between *A. thaliana*, on one hand, and *H. sapiens*, *D. melanogaster*, on the other hand, are apparently due to intrinsic difference of base composition. The difference of GC level between introns and CDS was found to be higher, on average, in *A. thaliana*[32] (5% to 15-30%), than in *H. sapiens*, *D. melanogaster*[33] (5%). In addition, the vast majority of plant introns are GC-poor[34], which is not the case in *H. sapiens* and *D. melanogaster*. The results show that UFM is an accurate and species-independent coding ORF predictor for sequences > 300bp.

Acknowledgments

This research was supported by the Brazilian FIOCRUZ/CAPES (CDTS) Program providing researcher fellowships to N. Carels.

References

1. N. Kyrpides, *Bioinformatics*, **15**:773 (1999).
2. J. Wang et al., *Eur. J. Biochem.*, **268**:4261 (2004).
3. S.F. Altschul et al., *Nucleic Acids Res.*, **25**(17):3389 (1997).
4. J.H. Badger and G.J Olsen, *Mol. Biol. Evol.*, **16**:512 (1999).
5. D. Frishman, A. Mironov, *Nucleic Acids Res.*, **26**:2941 (1998).
6. M. Kellis et al., *Nature*, **423**:241 (2003).
7. C. Camacho et al., *BMC Bioinformatics*, **10**:421 (2009).
8. E.N. Trifonov and J.L. Sussman, *Proc. Natl. Acad. Sci. USA*, **77**:3816 (1980).
9. I. Grosse, H. Herzel et al., *Phys. Rev. E*, **61**:5624 (1999).
10. D. Anastasiou, *Bioinformatics*, **16**:1073 (2000).
11. J.C.W. Shepherd, *Proc. Natl. Acad. Sci.*, **78**:1596 (1981).
12. C. Nikolaou and Y. Almirantis, *J. Mol. Evol.*, **59**:309 (2004).
13. M. Stanke and S. Waack, *Bioinformatics*, **19**(Suppl 2):215 (2003).
14. E.C. Uberbacher and R.J. Mural, *Proc. Natl. Acad. Sci. USA*, **88**:11261 (1991).
15. K.J. Hoff et al., *BMC Bioinformatics*, **9**:217 (2008).
16. C. Burge and S. Karlin, *J. Mol. Biol.*, **268**:78 (1997).
17. T.S. Larsen and A. Krogh, *BMC Bioinformatics*, **4**:21 (2003).
18. J. Besemer and M. Borodovsky, *Nucleic Acids Res.*, **27**:3911 (1999).
19. I. Korf, P. Flicek et al., *Bioinformatics*, **15**, S140 (2001).
20. A.L. Delcher et al., *Bioinformatics*, **23**:673 (2001).
21. S. Tiwary et al., *CABIOS*, **13**:263 (1997).
22. D. Kotlar and Y. Lavner, *Genome Res.*, **13**:1930 (2003).
23. R.B. Farber et al., *J. Mol. Biol.*, **226**:471 (1992).
24. I. Grosse et al., *Pacific Symposium on Biocomputing*, **5**:611 (2000).
25. J. Besemer et al., *Nucleic Acids Res.*, **29**:2607 (2001).
26. R. Zhang and C.T. Zhang, *J. Biomol. Struct. Dyn.*, **11**:767 (1994).
27. M. Yan, M., Z.S. Lin and C.T. Zhang,*Bioinformatics*, **14**:685 (1998).
28. F. Guo et al., *Nucleic Acids Res.*, **31**:1780 (2003).
29. N. Carels, R. Vidal and D. Frías, *Bioinform Biol Insights*, **3**:37 (2009).
30. N. Carels and D. Frías, *Bioinform Biol Insights*, **3**:141 (2009).
31. D.L. Wheeler et al., *Nucleic Acids Res.*, **28**(1):10 (2000).
32. N. Carels, P. Hatey et al., *J Mol Evol*, **46**:45 (1998).
33. O. Clay, S. Caccio et al., *Mol Phylogenet Evol*, **5**:2 (1996).
34. N. Carels and G. Bernardi, *Genetics*, **154**:1819 (2000).

MODELING AND ANALYSIS OF BIOFILMS

MARGARITA M. GONZÁLEZ-BRAMBILA, HÉCTOR PUEBLA
Departamento de Energía, Universidad Autónoma Metropolitana, Azcapotzalco.
Av. San Pablo No. 180, Reynosa Tamaulipas,
D. F., 02200, México.
E-mail: mmgb@correo.azc.uam.mx

FELIPE LÓPEZ-ISUNZA
Departamento de IPH, Universidad Autónoma Metropolitana, Iztapalapa,
Av. San Rafael Atlixco No. 186, Iztapalapa,
D.F., 09340, México. E-mail: felipe@xanum.uam.mx

In this work we are addressing the mathematical modeling of microbial biofilms and studying the effect of the biofilm structure on the concentration profiles inside the biofilm thickness. Two dynamic mathematical models were developed: (i) the first model considers the biofilm as a homogeneous solid, (ii) the second model considers the biofilm like a heterogeneous material composed by two phases, one solid phase enclosed cells and extracellular polymers, and a liquid phase containing the liquid outside the biofilm flowing inside channels and porous, which interchanges mass with the solid phase. Numerical simulations show the prediction capabilities and comparison of the proposed models.

1. Introduction

In nature, most of microorganisms live forming biofilms (Costerton, 1999). Biofilms constituent cells remain together by an extracellular matrix of polysaccharides (EPS), proteins and some nucleic acids, produced by the cells (Branda, et al., 2005; Lasa, 2006). In biofilms microorganisms live either in a symbiotic relationship an in competition for substrates and space, however biofilms provide protection from a wide array of environmental dangerous like antibiotics, predators, shear stress and even human immune system (Singh, et al., 2000; Leid, et al., 2005).

Biofilms can be formed by mixed or pure cultures; they exist in a great variety of environments, from water pipes, to indwelling devices in hospital patients. These facts have been motivated research on microbial biofilms to understand the mechanisms leading to physical persistence of microbes on surfaces and their resistance to antibiotics. The persistence of microorganisms in biofilms provides a reservoir for these microbes that can be advantageous in biofilms reactors; however, biofilms are disadvantageous in clogging of catheters and pipes, creating drag in ships, etc.

In most environmentally biofilms grow at the interface of an aqueous phase and a solid surface used as support by the cells. In all biofilms substrates need to be available to the cells. These compounds can be present in aqueous phase, such as in drinking water pipes or medical catheters, or be present over the solid surface like in minerals or insoluble organic matter (cellulose, protein) (Romeo, et al., 2008).

Biofilms are characterized physically by three parameters: medium composition, thickness biofilm and diffusional mass transfer of substrates. These characteristics determine the species in the biofilms, their spatial arrangement inside, and the dynamic balance between these organisms (Costerton, et al., 1995).

The extracellular polymers are responsible for the biofilms integrity and form a water-logged porous structure like a sponge, where the soluble substrates can move freely, but in which the solid substrates and the cells are trapped (Widerer and Characklis, 1989). The substrates arrive at the biofilm-liquid interface by convective flow and they are transported inside biofilm by diffusion and convection. The biofilm behaves as a viscoelastic material recovering its shape after being deformed by the shear stresses due to external flow, although it also promotes the convective transport inside the biofilm (Christensen and Characklis, 1990; Klapper et al., 2002).

To study the effects of the biofilms structure over mass transport and kinetic reaction two dynamic mathematical models were developed. The first model describes a homogeneous biofilm in a membrane biofilm reactor, composed by a solid phase, the biofilm, (mass transfer by diffusion); in contact with an external liquid phase, the liquid in the bioreactor (mass transfer by convection). The second model considers the biofilm as a heterogeneous material, like an sponge, with porous and channels filled with liquid.

Both models were compared with experimental results obtained in an experimental device shown in Figure 1. A single tube membrane biofilm reactor was used with a biofilm attached to a silicon rubber membrane tube held in the glass reactor. The bioreactor was connected to a continuous stirred tank reservoir. The silicon membrane is impermeable to the dissolved substrate. The system operates in a continuously recycled way between the membrane module and the stirred tank. During the recycling process a flow of oxygen is fed through the membrane tube. The oxygen diffuses through both the membrane wall and the biofilm. Air was bubbled into the reservoir tank.

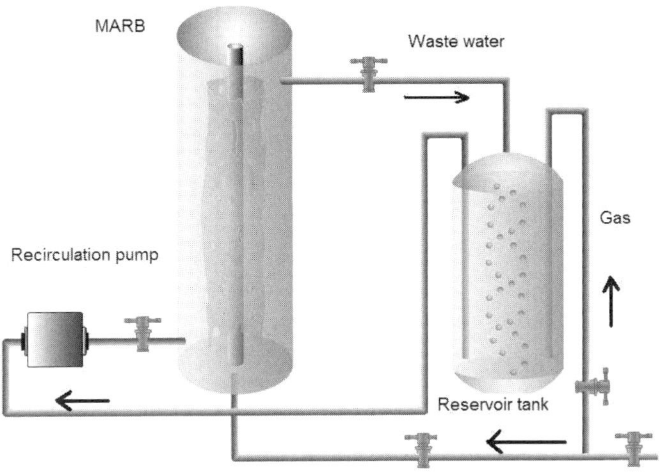

Figure 1. Schematic of the membrane-attached biofilm reactor (MABR) system.

2. Homogeneous mathematical model

The homogeneous model of the membrane bioreactor assumes that the biofilm is a homogeneous material formed by cells and EPS. Dissolved oxygen diffuses from the wall membrane, arrive to the membrane-biofilm interface, and diffuse toward the biofilm. The substrate dissolved in the liquid in the external part of the biofilm, arrive by convection from the liquid to the liquid-biofilm interface, then it diffuses inside biofilm. In biofilm the cells consume the oxygen and the substrate to growth, to maintain alive and to produce EPS. Figure 2 shows a schematic of the homogeneous biofilm in the bioreactor. The reaction system is modeled as two interconnected ideally stirred tanks: the reservoir tank containing the carbon source (S), oxygen is continuously sparged in the reservoir tank; the second tank is the bioreactor with the membrane tube with the biofilm attached, the reaction takes place only inside the biofilm, because all the cell are tramped in the biofilm. The cells consume substrate and oxygen to growth, to maintenance and to produce EPS. The production of biomass and EPS cause the biofilm grow giving rise to a moving boundary at l_b.

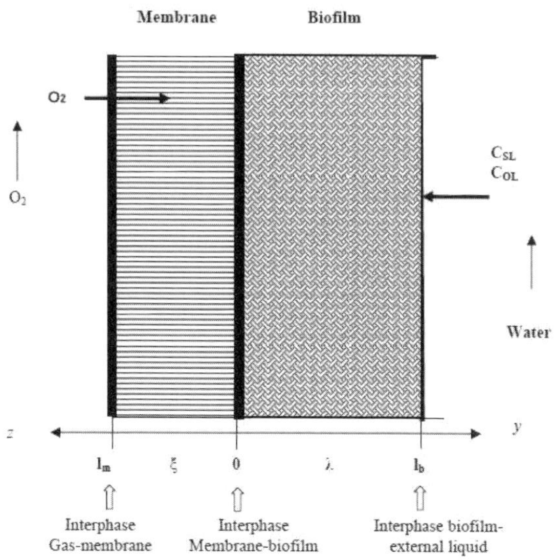

Figure 2. Schematic of the membrane-attached biofilm reactor showing: the membrane with thickness l_m through which oxygen permeates; the homogeneous biofilm with thickness l_b, and the external liquid phase.

The model assumes that there is only one carbon source, only one single species forming the biofilm, cell activity inside biofilm is the same at all times; outside the biofilm the cell activity is negligible. Mass transfer inside biofilm takes place by diffusion according to Fick Law. The external liquid phase interchanges mass by convention with the biofilm, and mass transfer coefficients remain constant during the operation of the bioreactor. Cell death and biomass detachment are neglected.

2.1. *Mass balances in biofilm*

The mass balances in biofilm for substrate (S), oxygen (O) and biomass (X) following the above assumptions are given by equations 1, 2 and 3. They consider the diffusion (D_{efS}, D_{efO}) from the interface into the biofilm, and consume of substrate and oxygen to grow (R_X), to maintenance (km/s) and to EPS production (R_{PEC}). The substrate and oxygen concentration in the homogeneous biofilm model are function of time and position inside biofilm:

$$\frac{\partial S}{\partial t} = D_{efS}\frac{\partial^2 S}{\partial y^2} - \frac{1}{Y_{x/S}}R_x - k_{m/S}x - Y_{PEC/S}R_{PEC} \tag{1}$$

$$\frac{\partial O}{\partial t} = D_{efO}\frac{\partial^2 O}{\partial y^2} - \frac{1}{Y_{x/O}}R_x - k_{m/O}x - Y_{PEC/O}R_{PEC} \tag{2}$$

where D_{efS} is the effective diffusion coefficient of substrate in biofilm, D_{efO} is the effective diffusion coefficient of oxygen in biofilm; $Y_{x/s}$ is the yield coefficient of biomass/substrate; $Y_{x/O}$ is the yield coefficient of biomass/oxygen; $k_{m/O}$ is the maintenance coefficient for oxygen, $k_{m/S}$ is the maintenance coefficient for substrate; y is the axial coordinate in biofilm, $Y_{pec/S}$ is the yield coefficient of substrate to EPS; $Y_{pec/O}$ is the yield coefficient of oxygen to EPS, and t is time. The Monod kinetics for biomass growth with a double limiting substrate (carbon source and dissolved oxygen) is:

$$R_x = \mu_{\max} x \left(\frac{S}{K_S + S}\right) \cdot \left(\frac{O}{K_O + O}\right) \tag{3}$$

where μ_{max} is the maximum specific constant growth rate, K_S is the saturation constant in Monod kinetics for substrate, K_O is the saturation constant in Monod kinetics for oxygen and x is the biomass concentration. Biofilm thickness grows with time due to the production of both biomass and EPS, and is described by a Stefan type of moving boundary (Crank, 1984); the effect of the curvature of the tubular membrane is neglected:

$$\rho_{bp}\frac{dl_b}{dt} = \frac{V_b}{A_{ext}}\langle R\rangle_b = l_b\langle R\rangle_b \tag{4}$$

where ρ_{bp} is the biofilm density, V_b is the volume of the biofilm, l_b = biofilm thickness A_{ext} is the external area of the biofilm and R_b is the average rate biofilm growth of the whole biofilm:

$$\langle R\rangle = \frac{1}{l_b}\int_0^{l_b}\left(1 + Y_{PEC/x}\right) R_x dl_b \tag{5}$$

2.2. *Mass balances in the liquid phase in biofilm reactor*

The substrate and oxygen concentrations in the liquid in the bioreactor are given by the input and output flow concentrations, minus the interfacial mass transport in the liquid-biofilm interphase.

$$V_R \frac{dS_L}{dt} = F\left(S_{LT} - S_L\right) - V_R k_l a_v \left(S_L - S\right) \tag{6}$$

$$V_R \frac{dO_L}{dt} = F\left(O_{LT} - O_L\right) - V_R k_l a_v \left(O_L - O\right) \tag{7}$$

where V_R = volume of the bioreactor, S_L = substrate concentration in the liquid in the bioreactor, S_{LT} = substrate concentration in the liquid in the reservoir tank, O_L = oxygen concentration in the liquid in the bioreactor, O_{LT} = oxygen concentration in the liquid in the reservoir tank, and F is the flow.

2.3. *Mass balances in the liquid tank*

The mass balance for substrate in the liquid in the reservoir tank is:

$$V_R \frac{dS_{LT}}{dt} = F\left(S_L - S_{LT}\right) \tag{8}$$

where V_T = Volume of the reservoir tank.

The mass balance for oxygen includes the contributions to mass transfer from the sparging air into the reservoir stirred tank plus the interchange between this tank and the bioreactor is:

$$V_R \frac{dO_{LT}}{dt} = F\left(O_L - O_{LT}\right) + V_T k_{lbub} a_{vbur} \left(O_{eq}^* - O_{LT}\right) \tag{9}$$

where k_{lbub} = interfacial mass transfer coefficient liquid-bubble, a_{vbub} = volumetric area of bubble, O_{eq}^* = equilibrium oxygen concentration.

2.4. *Mass balance in membrane*

For the permeable membrane the mass balance for dissolved oxygen is:

$$\frac{\partial O_m}{\partial t} = D_{efO_m} \frac{\partial^2 O_m}{\partial z^2} \tag{10}$$

where O_m = oxygen concentration in the membrane, D_{efOm} = effective diffusion coefficient of oxygen in membrane, z = axial coordinate in membrane.

2.5. *Initial and boundary conditions*

The initial conditions (t = 0) are:

(1) Dissolved oxygen concentration at gas membrane interface is equal to oxygen concentration in the gas phase.
(2) Dissolved oxygen and substrate concentration inside biofilm are cero ($O_0 = S_0 = 0$).
(3) Dissolved oxygen concentration at liquid in the reservoir tank is the equilibrium concentration O^*_{eq}.
(4) Dissolved oxygen in the liquid in bioreactor is $0.5 * O^*_{eq}$
(5) Substrate concentration in the liquid in reservoir tank and in the liquid in bioreactor are initial substrate concentration (S_0)

The boundary conditions are:

(1) At the membrane-biofilm interface: z = y = 0, continuity of flux and concentration of oxygen is required:

$$\frac{\partial O}{\partial y} = H_b \frac{\partial O_m}{\partial z} \tag{11}$$

(2) The membrane is impermeable to substrate S, then:

$$\frac{\partial S}{\partial z} = 0 \tag{12}$$

(3) At the liquid-biofilm interface (y = l_{bf}), continuity of fluxes for substrate and dissolved oxygen are:

$$-D_{efS} \frac{\partial S}{\partial y} = k_l \left(S_L - S\right) \tag{13}$$

$$-D_{efO} \frac{\partial O}{\partial y} = k_l \left(O_L - O\right) \tag{14}$$

3. Heterogeneous mathematical model

The heterogeneous model considers that the biofilm is formed by two phases: one is the solid phase containing clusters made by EPS housing the cells, which are interconnected by pores and channels filled with liquid, which form the second phase. Both phases interact with the external liquid phase. Experimental observations (de Beer et al., 1994, Lewandowski, 1998) have shown two different dissolved oxygen concentration profiles measured in pore and cells cluster in a biofilm. These pores and channels exchange mass with the clusters, as wells as with the external liquid phase. Figure 3 shows the schematic representation of the heterogeneous biofilm in the same membrane bioreactor considerer above.

3.1. *Mass balances in biofilm*

The mass balances for dissolved oxygen and substrate in the EPS-cluster phase in biofilm are respectively:

$$\in_p \frac{\partial O}{\partial t} = D_{efO} \frac{\partial^2 O}{\partial y^2} + \in_p k_{l\,\mathrm{int}} a_{v\,\mathrm{int}} (O_{li} - O) - \frac{1}{Y_{x/O}} R_x - m_{O/X} x - Y_{PEC/O} R_{PEC} \tag{15}$$

$$\in_p \frac{\partial S}{\partial t} = D_{efS} \frac{\partial^2 S}{\partial y^2} + \in_p k_{l\,\mathrm{int}} a_{v\,\mathrm{int}} (S_{li} - S) - \frac{1}{Y_{x/S}} R_x - m_{S/X} x - Y_{PEC/S} R_{PEC} \tag{16}$$

where $\in_p$= solid fraction of biofilm, $k_{l\,\mathrm{int}}$= internal mass transfer coefficient in liquid-solid biofilm interface, $a_{v\,\mathrm{int}}$= specific area per unit volume of liquid phase of biofilm, O_{li}= dissolved oxygen concentration in the liquid contained in porous and channels. The mass balances for dissolved oxygen and substrate in porous and channels in the biofilm are:

$$(1 - \in_p) \frac{\partial O_{li}}{\partial t} = D_{efO} \frac{\partial^2 O_{li}}{\partial y^2} - (1 - \in_p) k_{l\,\mathrm{int}} a_{v\,\mathrm{int}} (O_{li} - O) \tag{17}$$

$$(1 - \in_p) \frac{\partial S_{li}}{\partial t} = D_{efS} \frac{\partial^2 S_{li}}{\partial y^2} - (1 - \in_p) k_{l\,\mathrm{int}} a_{v\,\mathrm{int}} (S_{li} - S) \tag{18}$$

The biomass growth kinetics and the moving boundary are the same used in the homogenous biofilm model.

3.2. *Mass balances in the liquid phase in biofilm reactor*

The mass balances in the liquid phase in biofilm reactor consider the mass transport toward both phases of the biofilm, the solid and the liquid phase:

$$V_R \frac{dO_L}{dt} = F (O_{LT} - O_L) - V_R k_{l\ ext} a_{v\ ext} [\in_p (O_L - O) + (1 - \in_p)(O_L - O_{li})] \tag{19}$$

$$V_R \frac{dS_L}{dt} = F\left(S_{LT} - S_L\right) - V_R k_{l\ ext} a_{v\ ext}\left[\in_p \left(S_L - S\right) + \left(1 - \in_p\right)\left(S_L - S_{li}\right)\right] \quad (20)$$

where $k_{l\ ext}$= external mass transfer coefficient in biofilm-liquid interface in bioreactor, $a_{v\ ext}$= specific area per unit volume of biofilm.

3.3. *Mass balances in the liquid tank and in membrane*

The mass balances in the liquid in the reservoir tank and in the membrane wall are the same used in the homogeneous model.

3.4. *Initial and boundary conditions*

Initial conditions are equals than in homogeneous model. The boundary conditions are:

(1) At the membrane-biofilm interface: z = y = 0, oxygen diffuses toward the two phases of the biofilm:

$$-D_{ef\ O_m} \frac{\partial O_m}{\partial z} = \in_p D_{ef\ O} \frac{\partial O}{\partial y} + \left(1 - \in_p\right) D_{ef\ O_{li}} \frac{\partial O_{li}}{\partial y} \quad (21)$$

where $D_{ef\ O_m}$= effective diffusion coefficient of oxygen in membrane.

(2) At the membrane-biofilm interface: z = y = 0, hc substrate is impermeable to the membrane:

$$\frac{\partial S}{\partial y} = 0 \quad (22)$$

$$\frac{\partial S_{li}}{\partial y} = 0 \quad (23)$$

(3) At the liquid-biofilm interface (y = l_b), for the solid phase of the biofilm:

$$-\in_p D_{ef\ O} \frac{\partial O}{\partial y} = k_{l\ ext}\left(O_L - O\right) \quad (24)$$

$$-\in_p D_{ef\ S} \frac{\partial S}{\partial y} = k_{l\ ext}\left(S_L - S\right) \quad (25)$$

(4) At the liquid-biofilm interface ($y = l_{bf}$), for the liquid phase of the biofilm:

$$-(1-\in_p)D_{ef\ O_l}\frac{\partial O_{li}}{\partial y} = k_{l\ ext}\left(O_L - O_{li}\right) \tag{26}$$

$$-(1-\in_p)D_{ef\ S_l}\frac{\partial S_{li}}{\partial y} = k_{l\ ext}\left(S_L - S_{li}\right) \tag{27}$$

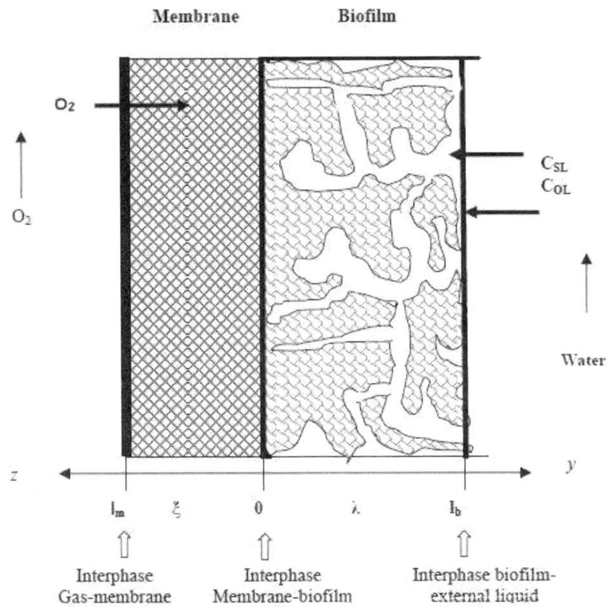

Figure 3. Schematic of the membrane-attached biofilm reactor showing: the membrane with thickness l_m through which oxygen permeates; the heterogeneous biofilm containing EPS-clusters surrounded by pores and channels, with thickness l_b, and the external liquid phase.

4. Numerical solution

For the numerical solution the method of orthogonal collocation was used (Villadsen and Michelsen, 1978), including the calculation of the moving boundary due to the growth of the biofilm thickness (Crank, 1984). This was done by making the dimensionless biofilm thickness equal to unity at all times; this front fixing method introduces an extra-convective term in the mass balances, which is given by a concentration gradient term multiplied by the dimensionless rate of biofilm growth (Lpez-Isunza et al., 1997). The modified set of mass balances were solved using 14 interior collocation points for the biofilm and seven interior collocation points for the membrane. The resulting set of non-linear ODE was solved with a fourth-order Runge-Kuta method.

Using substrate and dissolved oxygen measurements from batch experiments, transport and kinetic parameter in the model were estimated using the Marquardt non-linear estimation method (Meeter, 1965; Draper and Smith, 1996). These parameter are the effective diffusion coefficients, the interfacial mass transfer coefficients, the maximum growth rate, the Monod's saturation constants and the initial biomass concentration in the biofilm.

5. Results and Discussion

In previous works (González-Brambila, *et al.*, 2006; González-Brambila and López-Isunza, 2008), the model and experimental results were compared, and successful predictions of the observed concentrations measurements were obtained using the two models developed. In this work both model results are presented. These results consider that oxygen is flowing inside the membrane, and that there is not air bubbling in the reservoir tank.

In Figure 4 the homogenous model predictions for oxygen concentration profiles in the biofilm thickness are presented. At the beginning oxygen crosses over the membrane wall and arrive to the membrane-biofilm interface. After 0.5 hours oxygen begin to diffuse through the biofilm, then the oxygen concentration in biofilm increases with time. At the same time oxygen is been consumption by the cells. According with the homogenous model the concentration inside biofilm increases continuously and at 10 hours the oxygen from the membrane arrives to the biofilm-liquid interface.

Figure 5 illustrates the homogeneous model predictions for the substrate profile concentration. At the beginning the substrate diffuses form the liquid - biofilm interface inside the biofilm, and the concentration inside biofilm increases, however at the same time the substrate concentration in the liquid is decreasing and the concentration inside the biofilm decreases too.

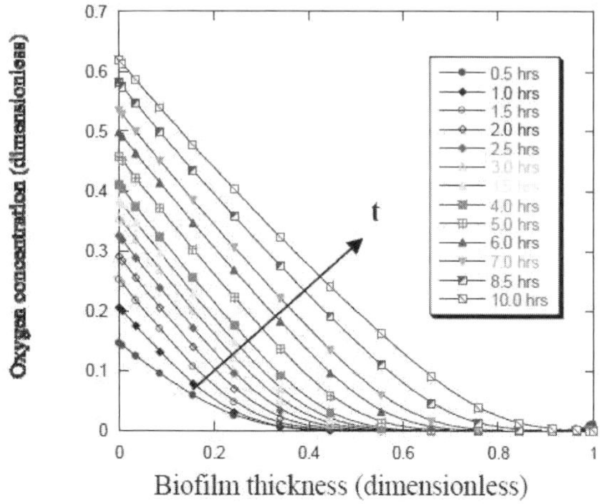

Figure 4. Homogeneous model: predicted oxygen concentration profiles within the biofilm.

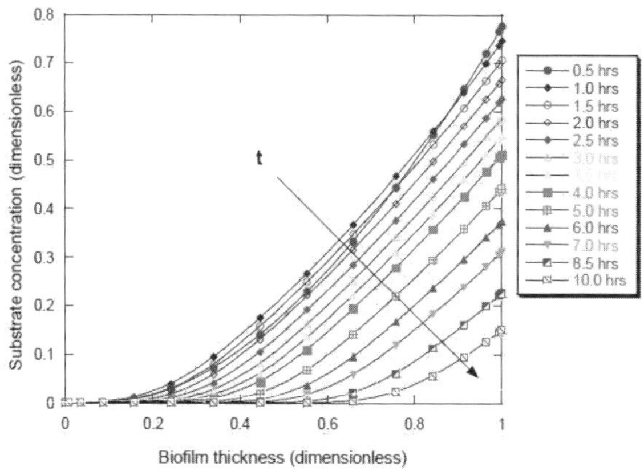

Figure 5. Homogeneous model: predicted substrate concentration profiles within the biofilm.

According with Monod kinetics with double limiting substrate, the reaction rate is a function of the substrate and oxygen concentration inside the biofilm. These concentrations are function of the position in the biofilm thickness, and then there is a different reaction rate in each position in the biofilm thickness. Figure 6 shows these different reaction rates at each position in biofilm thickness. The reaction rates are greater near the liquid

- biofilm interface because the substrate concentration is higher. In agreement the substrate concentration is decreasing near the membrane biofilm interface, the reaction rate near this interface is decreasing too.

Figures 7 to 11 present the heterogeneous model prediction in the biofilm thickness. Figures 7 shows the oxygen profile concentrations in the solid phase of the biofilm where it is consumed by the cells, and Figure 8 illustrates the oxygen profile concentration in porous and channels in the biofilm. The shapes of the profiles are very different, the oxygen concentration is higher in porous and channels because there is not cells in porous and then there is not consumption. However in both figures the oxygen concentration profiles increases with time and reaches the liquid biofilm interface.

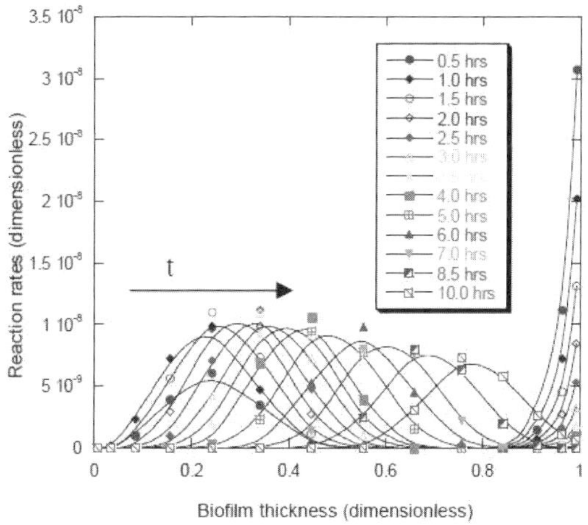

Figure 6. Homogeneous model: predicted reaction rate profiles within the biofilm

Figures 9 and 10 illustrate the heterogeneous model substrate concentration profiles throw time, in the solid phase and in porous and channels, in the biofilm thickness respectively. In both figures the concentration profiles decreases on time, due the substrate concentration in liquid decreases with time by the cells consumption. This is different compared with oxygen profiles because the oxygen is supply continuously and substrate is fed intermittently. In this case, oxygen profiles concentrations are higher in porous and channels than inside the cells clusters.

Finally Figure 11 shows the heterogeneous model reaction rates profiles throw time. At the beginning there are two maximum reaction rates, one of them is near the membrane-biofilm interface due the higher oxygen concentration, and the other one is near the biofilm liquid interface due the higher substrate concentration. As time increases the reaction rates decreases due the liquid substrate consumption.

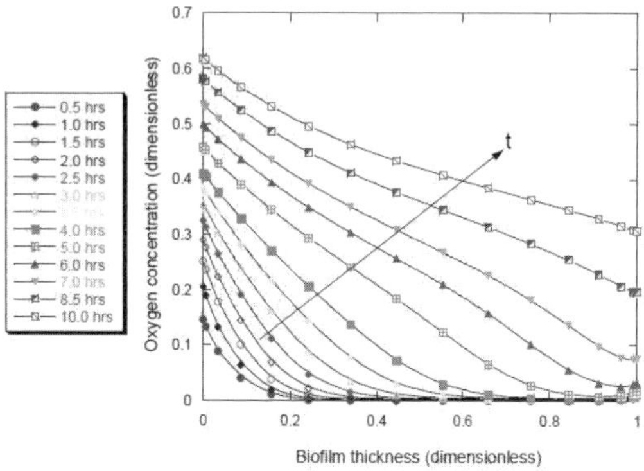

Figure 7. Heterogeneous model: predicted oxygen concentration profiles within the solid phase of the biofilm.

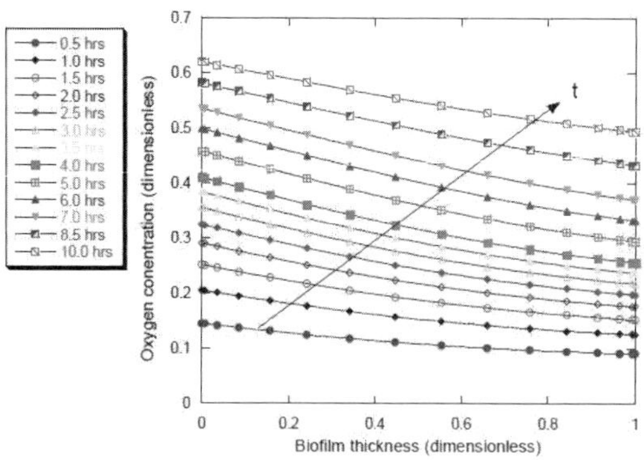

Figure 8. Heterogeneous model predicted oxygen concentration profiles through the time within porous and channels in the biofilm.

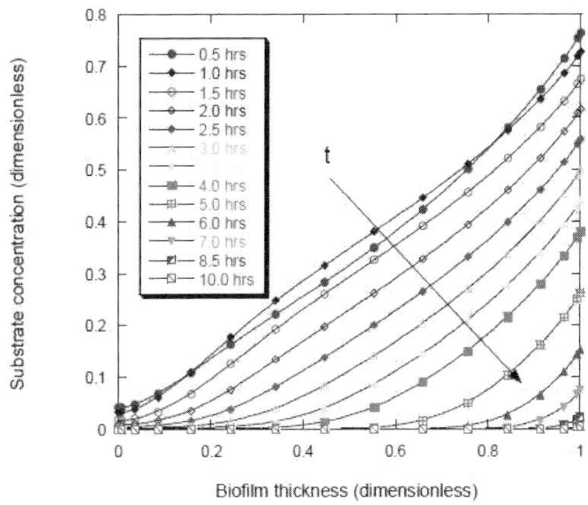

Figure 9. Heterogeneous model: predicted substrate concentration profiles within the solid phase in the biofilm.

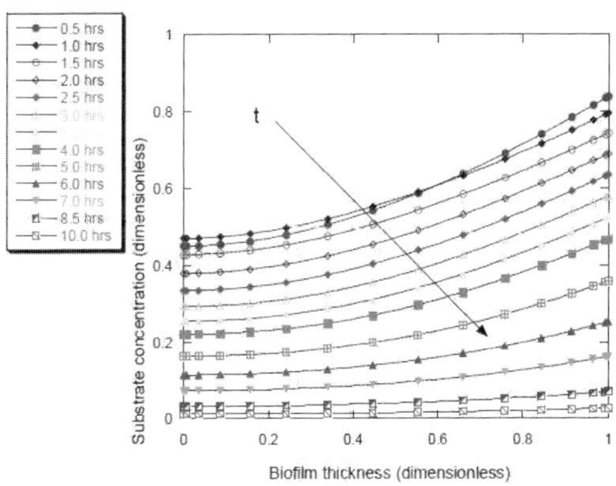

Figure 10. Heterogeneous model: predicted oxygen concentration profiles within porous and channels in the biofilm.

6. Conclusions

The modeling of microbial biofilms is important to understand the mechanisms leading to physical persistence of microbes on surfaces and their resistance to antibiotics. In this work we have proposed homogeneous and heterogeneous mathematical models of microbial biofilms in a membrane-attached biofilm reactor. Although two models predict the experimental results obtained in the liquid phase outside the biofilm, successfully, the pre-

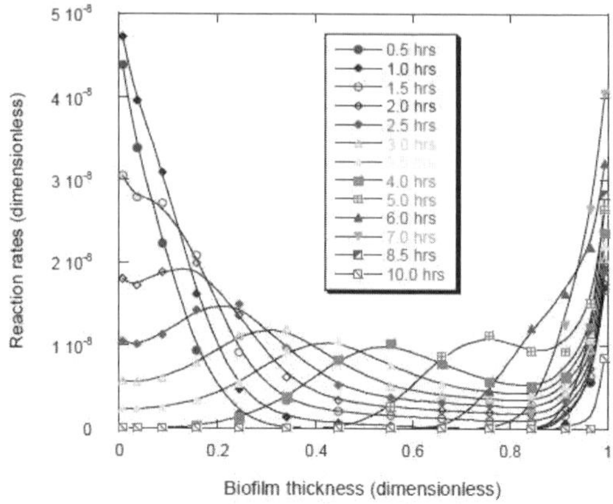

Figure 11. Heterogeneous model: predicted reaction rate profiles within the solid phase of the biofilm.

dictions inside de biofilm thickness are very different. Homogeneous model predicts lower profiles concentration due oxygen and substrate come into the biofilm only by diffusion in a solid biofilm. On the other hand, heterogeneous model predict higher reaction rates along all the biofilm thickness due oxygen and substrate come into the biofilm by diffusion in the solid phase and by convection in porous and channels, then the concentration gradient between clusters and channels promote the mass transfer from channels toward clusters. Homogeneous model predicts that the cells near the membrane - biofilm interface do not receive substrate almost time, leading to microorganism dead and then detached from the membrane, which in our previous experiments not occur. The heterogeneous model predictions are according with experimental results reported in the literature obtained in porous and cluster in the biofilm thickness.

Acknowledgments

This work was partially supported by Programa de Mejoramiento de Profesores, PROMEP, Secretara de Educación Pública, México.

References

1. Branda S.S., Vik S., Friedman L., Kolter R., *Trends Microbiol* **13**, 20 (2005).
2. Christensen B.E., Characklis W.G. Biofilms, *Willey* (1990).

3. Costerton J.W., Lewandowski Z., Caldwell D.E., Korber D.R., Lappin- Scott H.M., *Ann Review Microbiology*, **49**, 711 (1995).
4. Costerton J.W., Stewarte P.S., Greenberg E.P., *Science*, **284**, 1318 (1999).
5. Crank, J., *Free and moving boundary problems.* **Oxford**, (1984).
6. Draper N.R., Smith H., *Applied regression analysis.* Willey (1966).
7. González-Brambila M., López-Isunza F., *Int J of Chem Reactor Eng*, **A40**, (2008).
8. González-Brambila M., Monroy O., López-Isunza F., *Chem Eng Sci*, **61**, 5268, (2006).
9. Klapper I., Rapp C.J., Cargo R., Purvedorj B., Stoodley P., *Biotech and Bioengineering*, **80**, 289 (2002).
10. Lasa, I., *Int. Microbiol.*, **9**, 21(2006).
11. Leid J.G., Willson C.J., Shirtliff M.E., Hassett D.J., Parsek M.R., Jeffers A.K., *J Immunol*, **175**, 7512 (2005).
12. López-Isunza F. Larralde Corona, C.P., Viniegra González G., *Chem Eng Sci*, **38**, 877 (1997).
13. Meeter D.A., *Computer code from University of Wisconsin*, (1965).
14. Romeo, T., *Bacterial Biofilms.*, **Springer** (2008).
15. Singh P.K., Schaefer A.L., Parsek M.R., Moninger T.O., Welsh M.J., Greenber E.P., *Nature*, **407**, 762 (2000).
16. Villadsen J., Michelsen M.L., *Solution of differential equation models by polynomial approximation*, Prentice Hall. (1978).
17. Widerer P.A., Characklis W. F., *Structure and function of biofilms*, Willey (1989).

QUALITATIVE ANALYSIS OF CHEMICAL BIOREACTORS BEHAVIOR*

AMERICA MORALES DIAZ

Robótica y Manufactura Avanzada, Cinvestav Saltillo,
km 13 Carr. Saltillo-Monterrey Ramos Arizpe, Coahuila 25900, Mxico

LOURDES DIAZ JIMENEZ, SALVADOR CARLOS HERNANDEZ

Recursos Naturales y Energticos, Cinvestav Saltillo,
km 13 Carr. Saltillo-Monterrey Ramos Arizpe, Coahuila 25900, Mxico

Chemical bioreactors are commonly used in pharmaceutical production, wastewater treatment and odor compounds production for cosmetics and food industry, among other processes. These systems present nonlinear behavior due to the complex reactions taking place in the process of transforming raw materials to specific products. For that reasons, finding an analytical solution and doing a global analysis to study the properties of that kind of processes is not an easy task. Instead, a qualitative analysis by graphic methods is often considered in order to provide insights of the processes; in example, phase portrait and bifurcation analysis. In this work, a phase portrait analysis is performed for the case of chemical bioreactors for organic wastes transformation. The influence of process variables such as substrate input, dilution rate and death rate are analyzed. This study provides information about the operation points that allows the process to obtain high efficiency, the operating conditions which lead the process to instability, and the physical parameter limits. The analysis is done trough numerical simulations in specialized software.

1. Introduction

1.1. *Chemical bioreactors*

A chemical reactor is a recipient where chemical reactions take place under controlled conditions; they are designed in order to optimize the reactions. Two elements are clearly identified inside the reactor: reactants (material to be transformed) and products. Besides, a catalyst (material used to modify the velocity of the reactions) can be present. When a microorgan-

*This work is supported by grants conacyt 0125851, p51739-r, 105844.

ism is involved in the transformation process, the recipient is known as bioreactor or biochemical reactor [1-3]. Several applications of bioreactors have been identified in different technological and scientific domains such as pharmaceutical production, odor compounds production for cosmetics and food industry and wastewater treatment [4].

In this paper, an anaerobic wastewater treatment process is studied since the production of effluents is a recurrent problem in several cities. Anaerobic processes are efficient to transform high organic loads in a biogas which can be used as an alternative energy source since it is composed of methane and carbon dioxide. Organic complex molecules are progressively degraded by different anaerobic bacteria population in four successive stages; a particular task is performed in each stage and then, each one has particular dynamics. Some operating conditions, such as large variations on pH, temperature and organic load can inhibit or stop the anaerobic digestion [5-7].

In the anaerobic wastewater treatment systems, several bacteria species are involved; however, in order to simplify the mathematical representation and the dynamic analysis of the processes, two compendiums of bacteria are considered in this paper: the acidogenic and the methanogenic. Depending on the organic components of the effluents to be treated, three operational regimes are taking into consideration: batch, continuous and semi-continuous (fedbacth). In Figure 1 a scheme of these operation modes is presented.

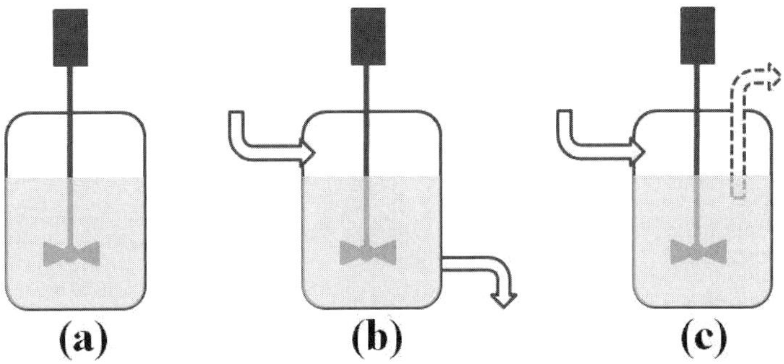

Figure 1. Different operating bioreactor regimes: (a) batch, (b) continuous and (c) semi continuous or fed-batch.

In the batch operation mode, an initial organic load is set inside the bioreactor and after some time, the process is stopped, the load is totally removed and a new experiment can be started. The removal rate is time varying and produces a high consumption of the substrates. This implies the bioreactor is volume dependent. In the continuous process, a continuous flow of the substrate is fed; also an output flow is keeping during all the process. Commonly, the input and output flows are equal and constant. The fedbatch is similar to the batch reactor; the difference is that, in the fedbatch, an addition or removal of substrate is done at certain interval time depending on the process application.

1.2. *Fedbatch bioreactors*

Nowadays, it has been an increase in the use of the fedbatch processes because this kind of reactors allows higher organic matter to be removed in comparison with batch and even with a continuous process [8]. Moreover, fedbatch regime allows the quality of products to be improved. In addition to better yields and selectivity, gradual addition or removal assists in controlling temperature particularly when the net reaction is highly exothermic. Thus, the use of a fedbatch reactor intrinsically permits more stable and safer operation than in a batch regime. In industrial applications, fedbatch processes are used in order to avoid bacteria inhibition caused by an excessive production of specific components. The advantage in comparison with continuous reactors is that fedbatch processes are easier to manipulate and to control. Since the operation mode implies a continuous change on the substrate, the systems is not able to reach a complete steady state. Then, the study of optimization and control strategies of fedbatch reactors is an active research subject in this domain [9-12].

2. Modeling batch and fedbatch reactors

Mathematical models are an interesting tool in order to analyze the dynamical behavior of systems, to predict the process evolution by means of numerical simulations and to develop control strategies which allow processes to be enhanced. In the case of bioreactors, the model should consider the essential phenomena (biological, hydrodynamic, physicochemical) and the main variables (biomasses, substrates) in order to guarantee a reliable representation of the process. Then, mass balance and transportation are usually considered [13-15].

As said before, in this paper two bacteria populations are taking into account for modeling a bioreactor operating in batch and fedbatch mode as follows:

$$\begin{aligned}
\frac{dS_1}{dt} &= -k_3\mu_1 x_1 + uS_{1_0} \\
\frac{dx_1}{dt} &= \mu_1 x_1 - \mu_{r_1} x_1(1-u) \\
\frac{dS_2}{dt} &= -k_1\mu_2 x_2 + k_4\mu_1 x_1 + uS_{2_0} \\
\frac{dx_2}{dt} &= \mu_2 x_2 - \mu_{r_2} x_2(1-u)
\end{aligned} \tag{1}$$

$$\begin{aligned}
u &= 0 \quad batch \\
u &= \vec{u} \quad fedbatch
\end{aligned} \tag{2}$$

where S1 and S2, corresponds to the organic matter concentration (g/L) and volatile fatty acid concentration (mmol/L), respectively; x_1 and x_2 are the acidogenic and the methanogenic bacteria concentration, both in (g/L); $\overline{u} = 0$ indicates the batch operation of the bioreactor; $\overline{u} = D$ is the dilution rate (h^{-1}) for the semi continuous operation; the constants k_1 and k_4 represent the production yield of bacteria and μ_{r_1} ,μ_{r_2}, represent a death rate. S_{1_0} (g/L) and S_{2_0} (mmol/L) are the fed substrate concentration in the inlet for the semi continuous operation; μ_1 and μ_1 are the bacterial growth rate and corresponds, respectively, to the Monod and Haldane expressions as follows:

$$\begin{aligned}
\mu_1 &= \frac{\mu_{m_1} S_1}{K_{S_1} + S_1} \\
\mu_2 &= \frac{\mu_{m_2} S_2}{K_{S_2} + S_2 + K_I S_2^2}
\end{aligned} \tag{3}$$

where μ_{m_1}, μ_{m_2} are the maximal growth rates for x_1 and x_2, respectively; K_{S_1}, K_{S_2} are the saturation coefficients for x_1 and x_2, respectively; and K_I is the inhibition coefficient for x_2.

In equation (1), the natural behavior of the death rate represents the inactive bacteria; this inactivity can be caused by death or extraction from the bioreactor when fresh load is introduced in the semi continuous operation. The term $(1 - \mu)$ represents a positive effect induced by the fresh load for the semi continuous operation. That means, the flow injected by the dilution rate μ promotes the bacterial growth instead of the inactive bacteria. However, for practical effects either growth or death rates can be considered depending on the operational range in the dilution rate.

The numerical parameters of the system (1) are presented in Table 1. These values correspond to a bioreactor used for wastewater treatment and they are taken from results reported in the literature [16].

Table 1. Bioreactor parameters.

Parameter	Value	Units
k_1	268	mmol g^{-1}
k_2	42.14	-
k_4	116.5	mmol g^{-1}
K_{S_1}	8.85	gL^{-1}
K_{S_2}	23.2	mmol
K_I	0.0039	$mmol^{-1}$
μ_{m_1}	1.2	d^{-1}
μ_{m_2}	0.74	d^{-1}
μ_{r_1}	0.1	d^{-1}

3. Model analysis

This section is devoted to analyze the properties of the mathematical model presented in section 2. For this purpose, graphical methods are considered: phase portrait and bifurcation analysis.

3.1. *Phase portrait analysis*

This method is useful to elucidate dynamic properties of a nonlinear system without linearization. In fact, a global analysis can be done considering a large operation interval. Phase portrait analysis consists in calculate and draw (in the so called phase plane) the trajectories of a system considering different initial conditions. From these trajectories a qualitative analysis can be done in order to obtain information concerning the process behavior and to predict future trajectories [17, 18].

Concerning the bioreactor (1), a first series of simulations is performed to draw the phase portraits of each state variable. By considering a nominal dilution rate of $\overline{u} = 2.6h^{-1}$, the time evolution for the substrates and biomasses involved in equation (1) is presented through Figures 2 - 5. During the first 50 h the process operates in batch mode; the initial conditions for the substrate and bacteria concentration were modified randomly, remaining positive.

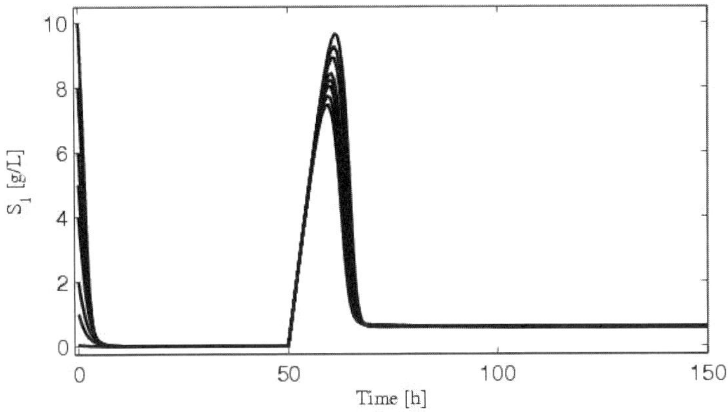

Figure 2. Phase portrait for substrate degradation in a fedbatch bioreactor: acidogenesis stage.

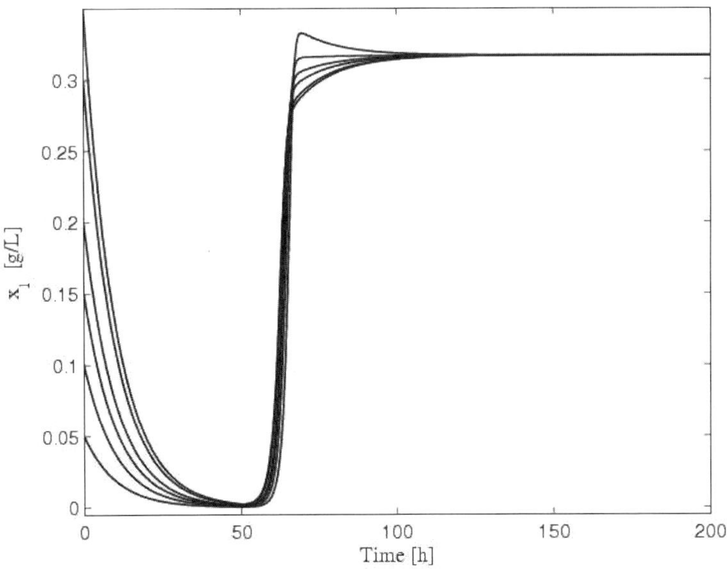

Figure 3. Phase portrait for bacteria evolution in a fedbatch bioreactor: acidogenesis stage.

As can be seen, all variables (substrates and biomasses) decrease leading the system to an equilibrium point; substrates are degraded and then the respective graphics tends to zero. On the other side, since there is not input substrate to the reactor, bacteria do not have enough substrate to grow and then they tend to an equilibrium point close to zero. After this period, a certain quantity of substrate is taken from the reactor and new organic matter is reloaded, therefore, substrate concentration increase as shown in Figures 2 and 4 around t = 50 h. At the same time, bacteria grow since new substrate is available to be degraded. Bacteria keep growing until to reach a maximal value which is sufficient to treat the additional substrate; this implies the bacteria keep constant and substrate decrease. Around t = 150 h, the system arrives to a steady state.

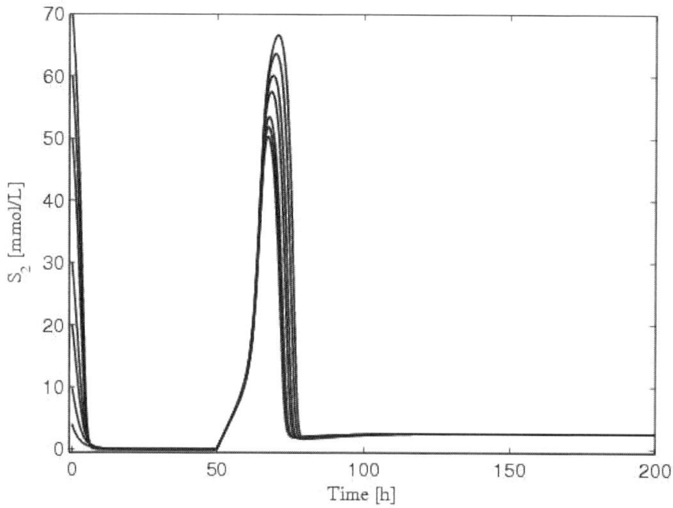

Figure 4. Phase portrait for substrate degradation in a fedbatch bioreactor: methanogenesis stage.

Although several initial conditions were simulated, the same equilibrium point was achieved in both operating conditions (batch and fedbatch). From these simulations, it can be concluded that the fedbatch process allows more organic matter to be removed. Also, for this specific case, the batch operation mode should be less than 50 h; though organic matter is removed the concentration of bacteria is reduced, which could represent a risk since this situation could lead the system to washout (absence of active bacteria inside the reactor). In order to study the effect of the dilution rate variations and of the death rate, the equilibrium point of the system (1), considering

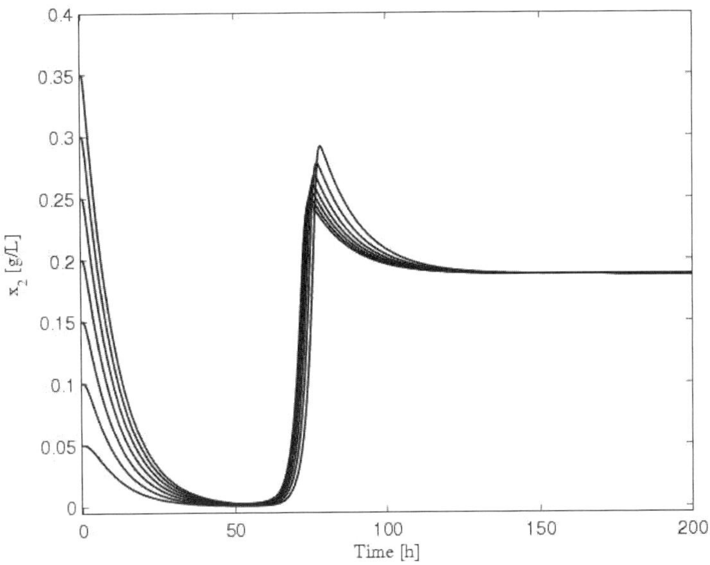

Figure 5. Phase portrait for bacteria evolution in a fedbatch bioreactor: methanogenesis stage.

the kinetic expressions (3) a bifurcation analysis is performed and presented in next section.

4. Bifurcation analysis

Bifurcation theory studies the behaviour of a system when one parameter changes. A continuous nonlinear process has fixed points or solutions for differential equations that describe the system. Commonly, these solutions are parameter dependents. These fixed points or solutions determine some properties about stability. Hence, when changes on the parameters occur the fixed points can change the stability properties. These qualitative changes in the dynamics are called bifurcations, and the parameters values at which they occur are called bifurcation points. Bifurcations are important scientifically, because they provide information about transitions and instabilities as some parameters varies [19, 20]. From this information, control strategies can be developed. In this paper, the objective of bifurcation analysis is to determine the effect of different parameters on the considered fedbatch bioreactor.

First, the equilibrium point of the system (1) is computed as follows:

$$
\begin{aligned}
0 &= -k_3\mu_1 x_1 + uS_{1_0} \\
0 &= \mu_1 x_1 - \mu_{r_2} x_1(1-u) \\
0 &= -k_1\mu_2 x_2 + k_4\mu_1 x_1 + uS_{2_0} \\
0 &= \mu_2 x_2 - \mu_{r_2} x_2(1-u)
\end{aligned}
\tag{4}
$$

Replacing the expression of biomass growth rate μ_1 and μ_2 (described by eq. 3) in (4), it is possible to compute the value of each state at the equilibrium $(S_1^*, S_2^*, x_1^*, x_2^*)$. Then, the next expressions are obtained:

$$S_1^* = \frac{k_{S_1}\mu_{r_1}(1-u)}{\mu_{m_1} - \mu_{r_1}(1-u)} \tag{5}$$

$$x_1^* = \frac{uS_{1_0}}{k_3\mu_1} \tag{6}$$

$$S_2^* = \frac{-b \pm \sqrt{b^2 - 4ac}}{2a} \tag{7}$$

$$x_2^* = \frac{k_4\mu_1 x_1^* + uS_{2_0}}{k_1\mu_2} \tag{8}$$

where:

$$a = 1; \quad b = \frac{\mu_{r_2}(1-u) - \mu_{m_2}}{k_I\mu_{r_2}(1-u)}; \qquad c = \frac{k_{s_2}}{k_I}$$

Since the Haldane kinetic (μ_2) considers the saturation and inhibition of bacteria population degrading volatile fatty acid, a quadratic equation is deduced in steady state. It implies there are two solutions for S_2 and consequently two for methanogenic bacteria (X_2).

It can be seen that equations (5) to (8) are functions of several parameters; two of these parameters are identified to be variable: the dilution rate (D) and the death rate (μ_{r_1} ,μ_{r_2}), both of them have influence on the solutions of the system equations; therefore, they are considered as bifurcation parameters. For this reason, the variation of each of them in the equilibrium equations (4)-(7) are studied via simulations in order to show that influence. Since only the fedbatch operation involves the effect of the dilution rate in all the states, these solutions are considered. When D is changing, the death rate is considered constant and equal for acidogenesis and methanogenesis process, i.e. $\mu_{r_1} = \mu_{r_2} = 0.10$.

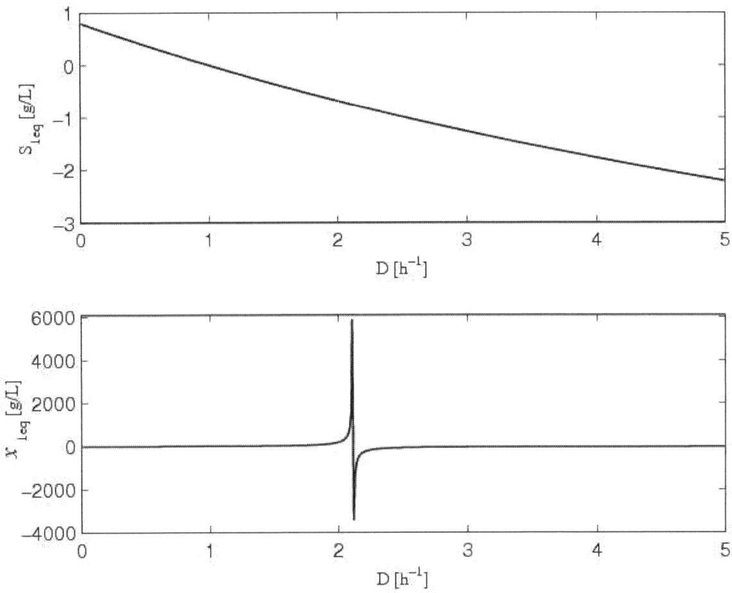

Figure 6. Bifurcation diagram for S_1^*, x_1^* with dilution rate $\bar{\mu} = D$.

In Figure 6, for $D = 2h^{-1}$ a change in the process behavior is noted; in the case of the concentration of biomass x_1 a jump between the growth and decreased rate is remarked. Before this limit, the substrate concentration experiences a fast decreased and even reach negative values; this does not have a physical sense. Therefore, it is recommendable to operate the bioreactor with $D = 1h^{-1}$.

On the other side, volatile fatty acid and the methanogenic bacteria concentration have two solutions, these are presented on Figure 7. This figure shows that between $0 < D \leq 1h^{-1}$ a bifurcation is presented for both biomass and substrate. Therefore, this imposes that the dilution rate should be $D \leq 1h^{-1}$ for an adequate operation of the process in fedbatch mode.

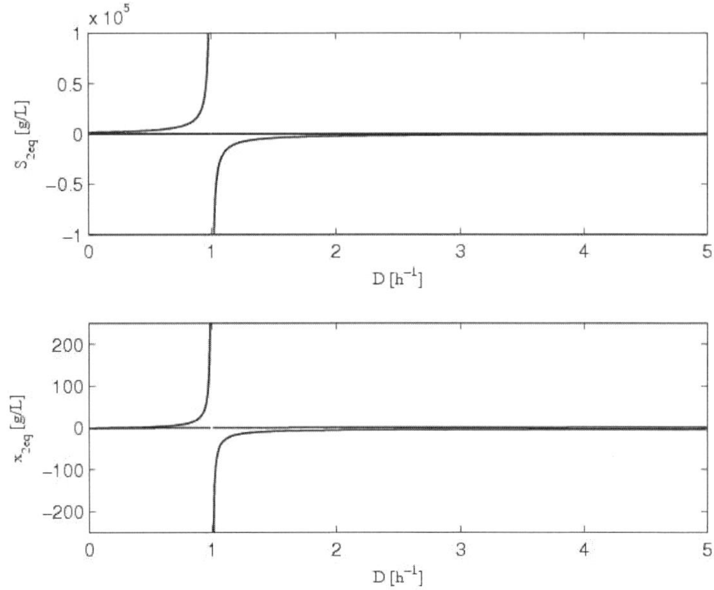

Figure 7. Bifurcation diagram for S_2^*, x_2^* with dilution rate $\bar{\mu} = D$.

In order to determine the effect of death rate, other series of simulations is done considering death rate variations meanwhile the dilution rate remains constant ($D = 0.26h^{-1}$). The obtained results are presented in Figures 8-9.

As expected, an increase on the death rate causes an increase of the organic matter due to a decrease of acidogenic bacteria, as shown in Figure 8. This phenomenon is known as washout and this is not a desirable situation in real processes, since there are not active microorganisms inside the bioreactor and the substrate cannot be transformed.

For the range between $0 < \mu_r < 0.65$ two solutions are clearly identified in the methanogenesis stage: firs one for high growth of bacteria and production of substrate, second one for lower growth and production (Figure 9). The lower production corresponds to a physical feasible situation; the higher solution is related only to a numerical scenario.

Remark 1: the behavior for the acidogenesis is expected to be monotonically due to the equilibrium equations. The term (1-μ) induces a competition mechanism between the growth and death rates.

Remark 2: the methanogenesis stage implies a saddle node bifurcation

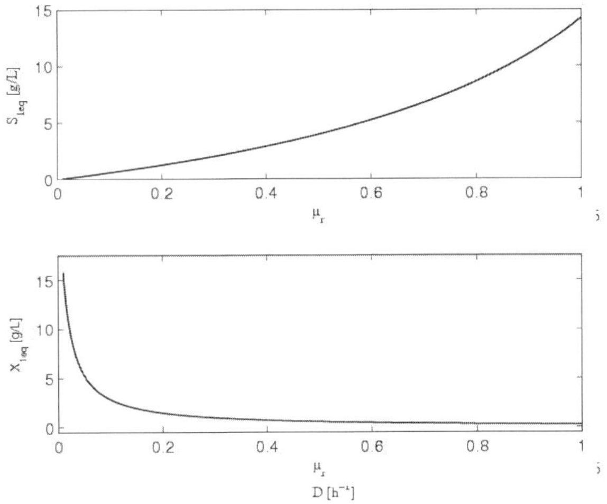

Figure 8. Bifurcation diagram for S_1^*, x_1^* for several values of $\mu_{r_1} = \mu_{r_2} = \mu_r$.

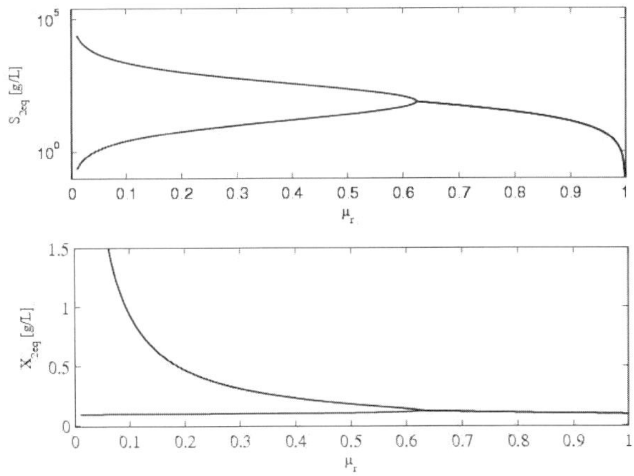

Figure 9. Bifurcation diagram for S_2^*, x_2^* for several values of $\mu_{r_1} = \mu_{r_2} = \mu_r$.

derived from the Haldane kinetics.

5. Conclusion

Two graphic methods were employed in order to analyze the dynamical behavior of an anaerobic process which is operated in batch and fed batch regime. Phase portraits are used to determine the effect of initial operating conditions of the process, mainly in the transition from batch to fedbatch; meanwhile, bifurcation analysis is developed to study the effect of two parameters on the process performances: dilution rate and death rate. From the phase portraits analysis, it is possible to conclude that the process reach a same equilibrium point independently of the initial conditions. Also, the fedbatch process allows more organic matter to be removed. Also, with the phase portrait analysis is possible to determine the time interval which is adequate to operate the process in batch mode before to switch the operation to fedbatch mode. On the other side, bifurcation analysis allows the user to identify the operating limits for the bifurcations parameter where the fedbatch process works adequately.

References

1. H.S. Fogler, *Elements of Chemical Reaction Engineering* **Prentice Hall**, U.S.A. (2005).
2. K. Riet, J. Tramper, *Basic biorector design,* **Marcel Dekker Inc.**, U.S.A. (1991).
3. G. Liden, *Bioprocess. Biosyst. Eng.*, **24**, 273 (1998).
4. R. K. Bajpai, R. H. Luecke, *In Chemical Engineering Problems in Biotechnology. M. L. Shuler, Eds American Institute of Chemical Eng.*, **1**, U.S.A., 301 (2001).
5. P.L. McCarty, *Public works*, **9(10)**, 107 (1964).
6. J. Mata-Alvarez, S. Mace, P. Llabres, *Bioresour. Technol.*, **74**, 3 (2000).
7. P.F. Pind, I. Angelidaki, B.K. Ahring, K. Stamatelatou, G. Lyberatos, *Adv. Biochem. Eng./Biotechnol.*, **82**, 135 (2003).
8. G. P. Longobardi, *Bioprocess. Eng.*, **10**, 185 (1994).
9. P. K. Shukla, S. Pushpavanam, *Chem. Eng. Sci.*, **53(2)**, 341 (1998).
10. A. S. Soni, R. S. Parker, *Ind. Eng. Chem. Res.*, **43**, 3381 (2004).
11. A. Tholudur, W. F. Ramirez, *Biotechnol. Progr.*, **12(3)**, 302 (2008).
12. J.R. Banga, E. Balsa-Canto, C.G. Moles, A.A. Alonso, *Proc. Ind. Acad. Sci.*, **1**, (2002).
13. J. F. Andrews, *Water Res.*, **8**, 261 (1974).
14. S. J. Parulekar, H. C. Lim, *Adv. Biochem. Eng./Biotechnol.*, **32**, 207 (1985).
15. G. Wang, E. Feng, Z. Xiu, *J. Process Control*, **18**, 458 (2008).
16. N.I. Marcos, M. Guay, D. Dochain, T. Zhang, *J. Process Control*, **14**, 317 (2004).
17. J-J. E. Slotine, W. Li, *Applied Nonlinear Control.*, **Prentice-Hall Inc**, U.S.A. (1991).
18. I. Karafyllis, G. Savvoglidis, L. Syrou, K. Stamatelatou, C. Kravaris, G. Lyberatos, *Global Stabilization of Continuous Bioreactors, American Institute of Chemical Engineers - Annual Meeting, Sn. Francisco*, **U.S.A.** (2006).

19. S.H. Strogatz, *Nonlinear dynamics and Chaos, Westview Press*, **U.S.A.** (1994).
20. S. Shen, G. C. Premier, A. Guwy, R. Dinsdale, *Nonlinear Dyn.*, **48**, 391 (2007).

A PEDIGREE ANALYSIS INCLUDING PERSONS WITH SEVERAL DEGREES OF SEPARATION AND QUALITATIVE DATA

CHARLES E. M. PEARCE, MACIEJ HENNEBERG

School of Mathematical Sciences, School of Medical Sciences,
The University of Adelaide,
Adelaide, SA 5005, Australia
E-mail: charles.pearce, maciej.henneberg@adelaide.edu.au

One application of genetic analysis that has received considerable attention involves the probable descent of present–day individuals from Thomas Jefferson by his slave, Sally Hemings. This analysis was made possible through the use of a rare Y chromosome. In this chapter we address the possible descent of present–day individuals from a prominent Australian statesman, Charles Cameron Kingston (1850–1908), by two partners, G. McCreanor and M.P. Holt. Despite the obvious parallel, the current problem would appear to be more complex because of the absence of a Y chromosome marker: Kingston and his wife were childless and DNA typing based on an exhumed bone of Kingston failed to produce a Y chromosome result. Further, only four loci of Kingston which gave results were common to the 17 for which typing was done for three putative descendants. The problem is both of some historical interest and an example of the complications when working from a split data set.

1. Introduction

Pedigree analysis is an area of genetics of rapid growth encompassing a considerable variety of problems, such as the genetic determinants of medical family traits, the reconstruction of genealogies from genetic data, the extinction of genes in small populations, and the ancestry of alleles or prediction of their future distribution. Characteristically these problems involve the analysis of data on large and complex genealogies. An excellent account of the basic underlying ideas is given by Elizabeth Thompson[11], who provides as applications genealogies for the Yanomama Indians of Brazil, a Mennonite–Amish genealogy, the ancestral origins of current polymorphic genes in the population of Tristan da Cunha, a dermatoglyphic trait in a Habbanite isolate and lymphoreticular malignancies in western Newfoundland. An interesting general algebraic approach has been presented by Hilden[8]. For a recent treatment of theoretical foundations of pedigree analysis, see Ginsberg *et al.*[7]

By contrast, use of X and Y chromosomes often enables the resolution of questions involving determination of paternity and maternity, possibly over lineages of appreciable length. Application to historical and prehistoric questions has popularized these ideas. We note in particular the question of lines of descent from Thomas Jefferson by his slave Sally Hemings[1,2,3,5,6] (which is discussed in a substantial literature, genetic and historical), and the long sweep of human history covered in Oppenheimer[9] and Sykes[10].

This chapter concerns a parallel question, the possible descent of a number of present–day individuals from Charles Cameron Kingston (1850–1908). CCK was a prominent Australian statesman known particularly for his role in drafting the Australian Constitution and for his period as Premier of South Australia, which included his bringing the vote to women in the state of South Australia. While he and his wife were childless, it has been suggested that he fathered children by more than two other partners.

There are strong traditions of descent from CCK in the families of a putative daughter, the second G. McCreanor, by one partner (the first G. McCreanor), and of a putative son, W.M. Kaiser, by another partner, M.P. Holt. CCK's public life was extensively documented and his movements and whereabouts are not inconsistent with the family traditions. There is also an inheritable morphology of a rare ear shape in these families that was possessed by CCK. This gives strong support to the existence of a family link.

The Kingston Research Group, which includes a number of putative descendants, has sought to investigate the matter further through genetic evidence. DNA from CCK was obtained by disinterrment and extraction of a bone sample and involved typing of alleles at several segregated autosomal loci. See Table 1. DNA was also suppled by B, a son of W.M. Kaiser, and by a brother and sister, C and D, grandchildren of the second G. McCreanor. DNA typing was obtained for B, C, D at 17 autosomal loci, including four of those at which typing was successful for CCK. See Table 2.

The content of Table 1 was derived by the Victorian Institute of Forensic Medicine, Monash University. NR indicates no result obtained. That of Table 2 was produced by the Institute of Medical and Veterinary Science, Adelaide. Table 3 records population frequencies for the alleles that feature in Tables 1 and 2. The entries in Table 3 were supplied by The Adelaide Institute of Medical and Veterinary Science. When available, values derived

Table 1. CCK: DNA typing

locus	allele 1	allele 2
CSF1PO	10	12
THO1	7	8
F13A01	3.2	5
FES	11	12
D8S639	NR	NR
FABP	10	11
D7S460	NR	NR
vWA	NR	NR
Penta E	NR	NR
Penta D	9	11

Table 2. DNA typing for C, D and B

locus	C		D		B	
	1	2	1	2	1	2
FES	10	12	10	12	10	13
F13A01	6	7	6	7	3.2	4
D8S1179	13	16	13	16	13	13
D21S11	30	31.2	30	31.2	28	30
D7S820	10	12	10	12	9	10
CSF1PO	11	12	11	12	13	14
D3S1358	16	17	15	17	17	17
TH01	7	9.3	7	9.3	7	9.3
D13S317	8	12	12	12	11	12
D16S539	12	13	12	12	12	12
D2S1338	23	25	16	22	20	24
D19S433	13	14	13	13	14	16
vWA	17	19	17	17	17	18
TPOX	9	11	8	9	8	9
D18S51	14	18	14	18	10	15
D5S818	11	11	11	11	9	11
FGA	19	25	20	24	21	23

Table 3. Australian allele frequencies

FES	10	12	13			
	0.269	0.251	0.048			
F13A01	3.2	4	5	6	7	
	0.071	0.021	0.250	0.323	0.286	
D8S1179	13	16				
	0.3156	0.0303				
D21S11	28	30	31.2			
	0.1504	0.2472	0.1017			
D7S820	9	10	12			
	0.1543	0.2631	0.1543			
CSF1PO	10	11	12	13	14	
	0.293	0.278	0.343	0.045	0.015	
D3S1358	15	16	17			
	0.2691	0.2561	0.1940			
TH01	7	8	9.3			
	0.183	0.141	0.306			
D13S317	8	11	12			
	0.1259	0.3026	0.2811			
D16S539	12	13				
	0.333	0.140				
D2S1338	16	20	22	23	24	25
	0.055	0.135	0.023	0.120	0.108	0.095
D19S433	13	14	16			
	0.233	0.353	0.053			
vWA	17	18	19			
	0.2719	0.2032	0.0868			
TPOX	8	9	11			
	0.533	0.120	0.253			
D18S51	10	14	15	18		
	0.0089	0.1582	0.1495	0.0776		
D5S818	9	11				
	0.0364	0.3739				
FGA	19	20	21	23	24	25
	0.0605	0.1394	0.1787	0.1423	0.1352	0.0804

Figure 1. Photograph of C.C. Kingston showing his ear.

by Weir *et al.*[12] were used. The remainder derived from samples sent to Molecular Pathology for routine pathology testing.

As we see in the following section, the dispersal of the DNA data amongst B, C, D and CCK makes the standard likelihood ratio method (see Fang and Hu[4]) not ideally suited to analysis of our problem. The presence of qualitative information is also a complication. Our approach uses a robust statistical hypothesis testing technique.

In Section 3 we propose our procedure to overcome the difficulties of the likelihood ratio method. Detail of the technique appears in Section 4. In Section 5 we present the results of its implementation. We conclude in Section 6 with a discussion.

A summary of the results of this study was presented publicly by the second–named author at Old Parliament House, Adelaide, on May 11, 2010.

The meeting was organized by The Kingston Research Group and attended by the media.

2. Likelihood Ratio Comparisons

We note that the DNA typing for CCK, which was over a century old, was done by a different laboratory from that for B, C and D, resulting in only four of the loci being typed for both the old and current DNA. We may readily verify that as a consequence strong likelihood ratio comparisons are not available. For example, consider let us compare B with CCK.

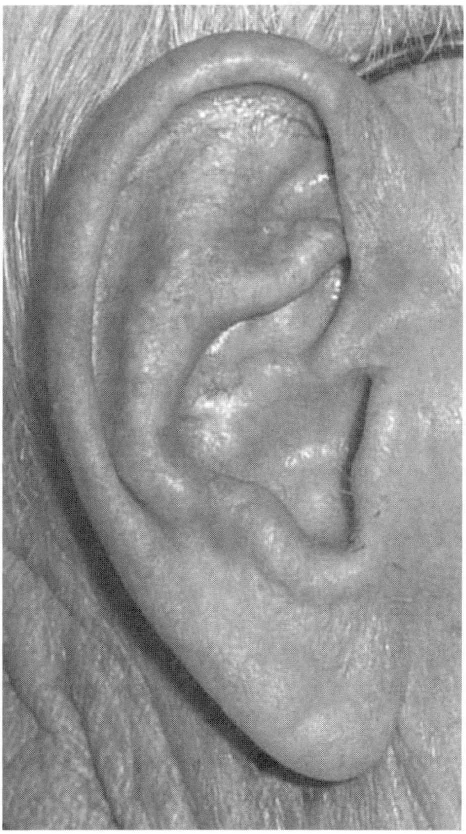

Figure 2. Photograph of B showing his ea.

Suppose first that, in accordance with tradition, CCK is grandfather of

B. Then B has probability 1/4 of inheriting any particular allele from CCK. From the tables, we have that the probability of B having the relatively rare 3.2 allele of F13A01 (in homozygous or heterozygous form) is

$$\frac{1}{4} + \frac{3}{4}\left[1 - (1 - 0.071)^2\right] = 0.3527.$$

Table 4. $x(n)$, $x_4(n)$ & $v(n)$ values (reference: B)

locus	allele 1	allele 2	$x(n)$	$x_4(n)$	$v(n)$
FES	10	13	0.317	0.4024	2
F13A01	3.2	4	0.092	0.2055	0
D8S1179	13	13	0.3156	0.4012	2
D21S11	28	30	0.3976	0.4729	2
D7S820	9	10	0.4174	0.4902	2
CSF1PO	13	14	0.060	0.1775	0
D3S1358	17	17	0.1940	0.2948	2
TH01	7	9.3	0.489	0.5529	2
D13S317	11	12	0.5837	0.6357	2
D16S539	12	12	0.333	0.4164	2
D2S1338	20	24	0.243	0.3376	0
D19S433	14	16	0.406	0.4803	1
vWA	17	18	0.4751	0.5407	2
TPOX	8	9	0.653	0.6964	2
D18S51	10	15	0.1484	0.2549	0
D5S818	9	11	0.4103	0.4840	2
FGA	21	23	0.3210	0.4059	0

Under the supposition that B and CCK are unrelated, the corresponding probability is 0.1370. For the 7 allele of TH01, the corresponding probabilities are 0.4994 and 0.3325, respectively.

On the other hand, again on the assumption that CCK is grandfather to B, the probability that B has neither of the alleles 10 and 12 for CSF1PO held by CCK is

$$\frac{1}{2}\left[1 - (1 - 0.293 - 0.343)^2\right] = 0.4338.$$

The corresponding probability under the assumption that CCK and B are unrelated is

$$1 - (1 - 0.293 - 0.343)^2 = 0.8675.$$

Similarly for FES we derive 0.4439 and 0.8878, respectively.

Thus if CCK is grandfather to B, the probability that B shares the given F13A01 and TH01 alleles of CCK, but neither of the latter's CSF1PO or FES alleles, is

$$0.3527 \times 0.4994 \times 0.4338 \times 0.4439 = 0.0339.$$

The corresponding probability when B and CCK are unrelated is

$$0.1370 \times 0.3325 \times 0.8675 \times 0.8878 = 0.0351.$$

So despite the presence of the strongly suggestive 3.2 and 7, we are unable to advance a strong probability likelihood ratio argument for relatedness based on the four common loci.

3. Analysis Strategy

Most of the available DNA information relates to B, C and D, for whom the joint DNA typing is suggestive. We shall proceed via a comparison of B, C and D. There are three natural hypotheses for relationships with CCK.

- C and D are descendants of CCK, but B is not.
- B is a descendant of CCK, but C and D are not.
- B, C and D are all descended from CCK.

We may in fact treat these together. We shall adopt the null hypothesis H_0 that B is unrelated to the sib pair C, D. It will turn out that the evidence against H_0 is very significant. Indeed, the genetic similarity between B on one hand and C and D on the other is remarkably high even under the assumption that C and D are five degrees removed from B. Here we adopt the standard usage that parent and child are one degree removed, grandparent and child two degrees, and so on. We note that two children with only one parent in common are also two degrees removed from each other. It will therefore be inappropriate to seek a more remote connection than through CCK. We are thus able to combine the more complete typing of the recent DNA with the appreciable non-DNA evidence to obtain a strong argument for common descent through CCK.

Our analysis is somewhat complicated by the connection between C and D. We treat this issue in the following section.

4. Degrees of Separation & Sib Probabilities

We treat B as a reference person. To test statistically whether another person is related to B, we examine the extent of similarity of their DNA typing.

Suppose B has alleles a_1, a_2 (possibly $a_1 = a_2$) at a specified autosomal locus, say locus n. We denote by p_1, p_2 respectively the probability that a randomly selected allele in the locus n population is a_1, a_2. If a_1, a_2 are distinct, we specify the reference set for locus n as $E = \{a_1, a_2\}$. If $a_1 = a_2$, we take $E = \{a_1\}$. We define $x(n) = P(E)$, so $x(n) = p_1 + p_2$ when $a_1 \neq a_2$ and $x(n) = p_1$ when $a_1 = a_2$.

We refer to a person as belonging to class 0, 1 or 2 for reference B and locus n according as
(0) Neither of their alleles at the locus belongs to E. A randomly chosen person will belong to this class with probability $[1 - x(n)]^2$.
(1) Precisely one allele belongs to E. For a randomly chosen person, this occurs with probability $2x(n)[1 - x(n)]$.
(2) Both alleles belong to E. For a randomly chosen person, this occurs with probability $[x(n)]^2$.

If a randomly selected person is replaced by one at m degrees of removal from B, the value $x(n)$ in (0)–(3) is replaced by

$$x_m(n) = \frac{1}{2^{m-1}} + \left[1 - \frac{1}{2^{m-1}}\right] x(n),$$

which is greater than $x(n)$. In the limit $m \to \infty$ we have $x_m(n) \to x(n)$, giving the results of the preceding paragraph.

We now address the situation of a pair of sibs, with one parent chosen at random from the population and the other m degrees of separation from B and separating the children from B, so that the children are $m + 1$ degrees separated from B. Denote by $V_m(n)$ the number (0, 1 or 2) of sibs having one or more elements of E at locus n.

For $V_m(n) = 0$, we have either

- two class 0 parents, in which case $V_m(n) = 0$ is automatic;
- one class 0 and one class 1 parent, when $V_m(n) = 0$ occurs with probability 1/2; or
- two class 1 parents, when $V_m(n) = 0$ occurs with probability 1/4.

Adding the component probabilities for these three cases provides

$$
\begin{aligned}
P(V_m(n)=0) &= (1-x_m(n))^2(1-x(n))^2 + [2x_m(n)(1-x_m(n)) \\
&\quad \times(1-x(n))^2 + (1-x_m(n))^2\cdot 2x(n)(1-x(n)]\cdot\frac{1}{2} \\
&\quad +2x_m(n)(1-x_m(n))\cdot 2x(n)(1-x(n))\cdot\frac{1}{4} \\
&= [1-x_m(n)][1-x(n)]. \qquad (1)
\end{aligned}
$$

For $V_m(n)=1$, we have either

- one class 0 and one class 1 parent, in which case $V_m(n)=1$ occurs with probability 1/2;
- two class 1 parents, in which case $V_m(n)=1$ occurs with probability 1/2;
- one class 0 and one class 2 parent, in which case $V_m(n)=1$ is automatic;

or

- one class 1 and one class 2 parent, in which case $V_m(n)=1$ occurs with probability 1/2.

We derive

$$
\begin{aligned}
P(V_m(n)=1) &= \big[(1-x_m(n))^2\cdot 2x(n)(1-x(n)) \\
&\quad +(1-x(n))^2\cdot 2x_m(n)(1-x_m(n))\big]\,\frac{1}{2} \\
&\quad +2x_m(n)(1-x_m(n))\cdot 2x(n)(1-x(n))\cdot\frac{1}{2} \\
&\quad +\big[(1-x_m(n))^2(x(n))^2 + (1-x(n))^2 x_m(n)\big] \\
&\quad +\big[2x_m(n)(1-x_m(n))\cdot x(n)^2 \\
&\qquad +2x(n)(1-x(n))x_m(n)^2\big]\,\frac{1}{2} \\
&= x(n)+x_m(n)-2x(n)x_m(n). \qquad (2)
\end{aligned}
$$

For $V_m(n)=2$, we have either

- two class 2 parents, in which case $V_m(n)=2$ occurs with probability 1/4;
- one class 1 and one class 2 parent, in which case $V_m(n)=2$ occurs with probability 1/2; or
- two class 2 parents, in which case $V_m(n)=2$ is automatic.

Hence

$$\begin{aligned} P(V_m(n) = 2) &= 2x_(n)(1 - x(n)) \cdot 2x_m(n)(1 - x_m(n)) \cdot \frac{1}{4} \\ &\quad + \left[2x(n)(1 - x(n)) \cdot x_m(n)^2 \right. \\ &\quad \left. + 2x_m(n)(1 - x_m(n))x(n)^2\right] \frac{1}{2} \\ &\quad + x_m(n)^2 x(n)^2 \\ &= x(n)x_m(n). \end{aligned} \tag{3}$$

From (1)–(3) we deduce that

$$E(V_m(n)) = P(V_m(n) = 1) + 2P(V_m(n) = 2) = x(n) + x_m(n) \tag{4}$$

and

$$\begin{aligned} E(V_m(n)^2) &= P(V_m(n) = 1) + 4P(V_m(n) = 2) \\ &= x(n) + x_m(n) + 2x(n)x_m(n), \end{aligned}$$

so

$$\begin{aligned} \mathrm{var}(V_m(n)) &= E(V_m(n)^2) - \left[E(V_m(n))^2\right] \\ &= x(n) - x(n)^2 + x_m(n) - x_m(n)^2. \end{aligned} \tag{5}$$

For the situation with both parents chosen randomly from the population, the number $V(n)$ of sibs having one or more elements of E at locus n is obtained by letting $m \to \infty$ in (4) and (5). We derive

$$E(V(n)) = 2x(n), \quad \mathrm{var}(V(n)) = 2x(n)[1 - x(n)]. \tag{6}$$

5. Analysis & Results

We adopt the hypothesis H_0 that C and D are unrelated to B. We have the similarity measure

$$V = \sum_n V(n)$$

for the genetic similarity of the pair C, D to B at the loci typed. We denote by v, $v(n)$, respectively, the actual values of V, $V(n)$ for the pair C, D. Table 4 gives $v = 23$ coincidence sites for E. Since the loci are segregated, the values $V(n)$ are independent and we have from (6) that

$$E(V) = 2\sum_n x(n), \quad \mathrm{var}(V) = 2\sum_n x(n)[1 - x(n)].$$

Under H_0, we have from (6) that $E(V) = 11.71$ and $\text{var}(V) = 7.08$, so the distribution of V has standard deviation 2.66.

For testing H_0, we shall need to determine the probability that $V \geq v$. The distribution of $V(n)$ has a probability generating function

$$F_n(z) = \sum_{i=0}^{2} P(V(n) = i)z^i,$$

a quadratic polynomial in z. Since the loci are segregated, the $V(n)$ are independent and so the distribution of $\sum_n V(n)$ has probability generating function

$$F(z) := \prod_n F_n(z),$$

which for 17 loci will be a polynomial

$$F(z) = \sum_{j=0}^{34} c_j z^j$$

in z of degree 34. We have

$$P(V \geq v) = \sum_{i=v}^{34} c_j,$$

which may be evaluated through use of a short computing programme.

However, a simple approximation is available. By independence of the $V(n)$, the central limit theorem gives that V may be approximated by a normal distribution with mean 11.71 and standard deviation 2.66. We remark that although the distribution of each individual $V(n)$ is in general skewed, the normal approximation is not. Also, the standard errors in the determined values of allele frequencies are such that it is not physically meaningful to distinguish the normal approximation from the theoretical distribution.

With integer correction, the value $v = 23$ is located

$$\frac{22.5 - 11.71}{2.66} = 4.05$$

standard deviations above the mean. The probability of getting as high a value as this by chance is approximately 0.00003. This represents a deviation from what is expected under H_0 that is highly significant. The genetic similarity between the pair C, D and B is thus too great to be reasonably attributable to chance.

How close a relationship is indicated between B and C, D? Suppose we adopt a null hypothesis that one parent of C, D is four degrees removed from B and that the other is unrelated to B. We may carry out a calculation similar to the previous one using (4), (5) in place of (6). We derive that

$$V_4 := \sum_n V_4(n)$$

may be taken as following approximately a normal distribution with mean 13.11 and standard variation 2.72. The integer correction gives that having 23 coincidence sites corresponds to a location 3.46 standard deviations above the mean. This level or more occurs by chance with probability approximately 0.0003. Accordingly we should expect the parent of C, D related to B to be at most four degrees removed from B. Given the evidence of oral tradition, what is known of the family trees, and the morphology of the ear, this gives strong support for CCK as a common ancestor to B, C, D.

6. Discussion

In the previous section we found strong support for descent of B, C and D from CCK. The argument is robust and makes full use both of non-DNA information and of the DNA information we have for B, C and D. It is not restricted by the fact that the DNA information we have about CCK is very limited. Its main deficiency is that it does not use the DNA information we do have about CCK.

By contrast the likelihood ratio method utilizes the DNA information about CCK we have at the four crucial loci CSF1PO, TH01, F13A01 and FES, but neglects most of the DNA information we have concerning B, C and D. One could modify our portmanteau assumption technique to incorporate the typing of the DNA of CCk, though such an approach would appear not to be entirely satisfactory or natural. What is needed is more DNA typing.

The Kingston Research Group believes that there may be further descendants of CCK by other partners, which may in itself initiate further work. The Group also hopes that more DNA evidence may become available, though it recognizes that further DNA may be neither readily available nor inexpensive. However it seems premature to believe that the present study will prove to be the last word on the subject. Preliminary information involving one further partner has come to hand during the last stages of the writing of this chapter.

Acknowledgements

The authors wish to acknowledge Dadna Hartman of the Victorian Institute of Forensic Medicine for work done on the CCK DNA analysis, and Zbigniew Rudzki, Scott Hamish and Karen Ambler of the IMVS, Adelaide, for the analyses of DNA from B, C and D and for providing background allele frequency data. We also thank Malcolm Simpson, who funded the entire project, and John Bannon, who led The Kingston Research Group and organized and chaired the public meeting at Old Parliament House.

References

1. E. Check, Jefferson's descendants continue to deny slave link, *Nature* **417**, 16 May 2002, 213–213.
2. S. T. Corneliussen, Jury out on Jefferson's alleged descendants, *Nature* **418**, 11 July 2002, 124–125.
3. G. Davis, The Thomas Jefferson paternity case, *Nature* **397**, 7 January 1999, 32–32.
4. W. K. Fang and Y. Q. Hu, *Statistical DNA forensics: theory, methods and computation*, John Wiley & Sons, Chichester (2008).
5. E. A. Foster, M. A. Jobling, P. G. Taylor, P. Donnelly, P. de Knijff, R. Mieremet, T. Zerjal and C. Tyler–Smith, Jefferson fathered slave's last child, *Nature* **396**, 5 November 1998, 27–28.
6. E. A. Foster, M. A. Jobling, P. G. Taylor, P. Donnelly, P. de Knijff, R. Mieremet, T. Zerjal and C. Tyler–Smith, Reply: The Thomas Jefferson paternity case, *Nature* **397**, 7 January 1999, 32–32.
7. E. Ginsberg, I. Malkin and R. C. Elston, *Theoretical aspects of pedigree analysis*, Ramot Publ., Tel-Aviv Univ. (2006).
8. J. Hilden, GENEX – An algebraic approach to pedigree probability calculus, *Clinical Genetics* **1**, 319–348 (1970).
9. S. Oppenheimer, *Out of Eden*, Constable, London (2003). K. J. Navara, Humans at tropical lattitudes produce more females, *Biol. Lett.* **5** (2009), 524–527.
10. B. Sykes, *The Seven Daughters of Eve*, Bantham, London (2001).
11. E. A. Thompson, *Pedigree Analysis in Human Genetics*, The Johns Hopkins University Press, London and Baltimore (1986).
12. B. S. Weir, A. Bagdonavicius, B. Blair, C. Eckhoff, C. Pearman, P. Stringer, J. Sutton, J. West, L. Wynen, Allele frequency data for Profiler Plus loci in Australia, *J. Forensic Sci.* **49** (5) 1121–1123 (2004).

EFFECTS OF MOTILITY AND CONTACT INHIBITION ON TUMOR VIABILITY: A DISCRETE SIMULATION USING THE CELLULAR POTTS MODEL

JONATHAN LI

St. Margaret's Episcopal School
San Juan Capistrano, CA 92697, USA

JOHN LOWENGRUB

Department of Mathematics, University of California at Irvine
Irvine, CA 92697, USA

We analyze the effects of cell migration and contact inhibition on tumor growth using the Cellular Potts Model (CPM). Motility, cell-to-cell adhesion, contact inhibition, and cell compressibility factors are incorporated into the model. We find that increased motility has a direct effect on the growth rate of a tumor. Cell lines with greater motility overcome the attractive forces of cell-to-cell adhesion and have more space to proliferate. In addition, contact inhibition amplifies the effect of motility. Strict contact inhibition penalizes clumped cells by halting their growth, giving motile cells a greater advantage. The model also shows that cells with less response to contact inhibition are more invasive. This raises questions on the effectiveness of some chemotherapy treatments, which may actually select for these more invasive cells. We also explore inherent problems in the CPM and compensate for them in the model.

1. Introduction

Tumor growth begins with the genetic mutations in oncogenes or tumor suppressor genes. Most oncogenes are the result of a mutation or overexpression of proto-oncogenes which regulate cell growth and proliferation. Oncogenes cause normal cells to bypass apoptosis and proliferate instead. Uncontrolled growth follows, leading to the formation of a tumor (Todd and Wong, 1999; Yokota, 2000; Corce, 2008). Tumor suppressor genes also regulate the growth of cells and repair mistakes in DNA copying. If necessary, these genes can initiate apoptosis. Tumor suppressor genes are very successful in preventing tumor development and thus pose a risk when they are mutated (Sherr, 2004). Tumor development is a complex process which depends on a variety of factors. Our model focuses on the two basic properties, cell migration and contact inhibition, although others are incorporated into the simulation. Cell migration is described by three processes: the extension of a side of a cell, the contraction of the cell body, and the detachment of the trailing side. Motility can be influenced by cellular interactions, chemical signaling, and tissue function (Friedl and Brcker, 2000). Contact inhibition is the restriction on growth or motility when a cell is in contact with other cells. This can result from signaling from cell-to-cell junctions such as the cell adhesion molecule NECL-5 (Takai *et al.*, 2008).

Both *in vitro* and *in vivo* testing have their advantages and disadvantages when studying tumor growth. *In vivo* experiments are more realistic but it is difficult to pinpoint all factors contributing to an observed behavior. On the other hand, *in vitro* experimentation allows for a more controlled environment but is obviously less realistic than performing *in vivo* experiments. Another method used to observe tumor development is mathematical modeling. Modeling allows the user to have a completely controlled environment with set parameters. Factors are varied one at a time, making changes to tumor growth easier to observe. Modeling also allows for growth to be measured more quantitatively. Previous studies have considered the effect of motility and proliferation on tumor development (Fidler, 1989) including the use of mathematical modeling (Thalhauser *et al.*, 2010; Phan, 2010).

2. The Model

We model tumor cells using the Cellular Potts Model (CPM), which is a lattice-based computational modeling method to simulate the behavior of cells (Izaguirre, *et al*, 2004). The simulation is implemented in COMPUCELL3D (CC3D), a C++ modeling environment that provides the foundation for CPM simulations (Cickovski, *et al*, 2007). Parameters for the model are shown in Table 1.

In CC3D, the simulation starts with a rectangular lattice grid with each lattice point assigned an index, $\sigma(\vec{i})$. A cell is composed of lattice points of the same cell type, $\tau(\sigma(\vec{i}))$. All cells are assigned effective energies which control cell behavior and interactions.

CC3D, which uses a modified Metropolis algorithm, is ruled by many index-copy attempts. In each attempt, a lattice point, $\vec{i}$, is compared with a neighbor, $\vec{J}$. If the two pixels lie in the same cell, $\sigma(\vec{i}) = \sigma(\vec{j})$, no indexes are copied. If they are from different cells, $\sigma(\vec{i}) \neq \sigma(\vec{j})$, there is a

Table 1. List of Model Parameters

$\sigma(\vec{i})$	Index of pixel at $\vec{i}$
$\tau(\sigma(\vec{i}))$	Cell type at pixel $\vec{i}$
J	Adhesion energy
λ_v	Volume constraint constant
v_σ	Volume of cell
V_T	Target volume
λ_S	Surface area constraint constant
S_σ	Surface area of cell
S_T	Target surface area
db	Doubling volume
ts	Time step for motility
$\vec{\lambda}_{mot}$	Vector describing motility
r_c	Measures amount of free surface area
CI	Contact Inhibition Constant

probability that the index $\sigma(\vec{i})$ will be copied to $\sigma(\vec{j})$. A successful copy is called an index-copy. One Monte Carlo Step (MCS) consists of one index-copy attempt for each pixel. The probability of an index-copy is given by the Boltzmann acceptance function:

$$p(copy) = \begin{cases} 1, & \Delta H < 0 \\ e^{-\frac{\Delta H}{T_m}}, & 0 < \Delta H \end{cases}$$

The parameter T_m directly affects the chance that an index-copy attempt will be successful. A high T_m will increase the number of accepted attempts and essentially increase cell motility. Thus, T_m is often related to the temperature of the system in analogy with physical systems. ΔH is the change in effective energy, which is a combination of the Hamiltonians (energies) for cell-to-cell adhesion, surface and volume constraints, and motility:

$$\Delta H = \Delta H_{adhesion} + \Delta H_{surface} + \Delta H_{volume} + \Delta H_{motility}$$

CC3D incorporates adhesion energies between cells, which has the form:

$$H_{adhesion} = \sum_{\substack{i,j \\ neighbors}} J\left(\tau\left(\sigma\left(\vec{i}\right)\right), \tau\left(\sigma\left(\vec{j}\right)\right)\right)\left(1 - \delta\left(\sigma\left(\vec{i}\right), \sigma\left(\vec{j}\right)\right)\right)$$

where J is the boundary energy. Though not a main focus of this paper, the adhesion energies also play an important role in contact inhibition and motility. The Hamiltonian for the volume constraint is:

$$H_{volume} = \sum_{\sigma} \lambda_v (v_\sigma - V_T)^2$$

where λ_v is the inverse compressibility of the cell, v_σ is the volume, and V_T is the target volume. Thus, when the volume is much larger or much smaller

$$H_{surface} = \sum_{\sigma} \lambda_s (s_\sigma - S_T)^2$$

Cell migration is modeled using the ExternalPotential plugin in CC3D designed to achieve directional movement. Each cell is assigned a vector, $\vec{\lambda}_{mot}$. The Hamiltonian is then:

$$\Delta H = \vec{\lambda}_{mot} \cdot \vec{s}$$

where $\overrightarrow{S}$ is the spin flip direction. In the simulation, migration is varied using the time step, that is, the number of MCS between migration attempts, and $\overrightarrow{\lambda_{mot}}$. Every time step, $\overrightarrow{\lambda_{mot}}$ is set to $< m, 0 >$, $< -m, 0 >$, $< 0, m >$,$< 0, -m >$ or $< 0, 0 >$, where m is the magnitude, $|\overrightarrow{\lambda_{mot}}|$. Each vector has equal probability of being chosen and promotes movement left, right, up, down, or not at all. Simulations show that altering the time step had a much smaller effect than changing the magnitude so most of the models are run at a constant time step. This model for motility is a variation of random walks.

Note that this plugin does not directly have complete control over the movement of cells. It only contributes to the total energy. If $|\overrightarrow{\lambda_{mot}}|$ is large enough, then the change in energy will be large and the cell will have a higher probability of moving in the $\overrightarrow{\lambda_{mot}}$ direction. Hence, the plugin has no bearing on the actual velocity of the cells. Instead, by altering $|\overrightarrow{\lambda_{mot}}|$, the probability of accepting index-copy events is changed, which in turn affects the average speed of cells.

All cells in the simulation are modeled to undergo mitosis once the cell volume, v_σ reaches the doubling volume, db. Mitosis simply divides the parent cell into two cells with equal volumes. Cells reach doubling volume by increasing their target volumes. In this model, the target volumes are increased by 1 every 10 MCS although this is changed by factoring contact inhibition into the simulation. Contact inhibition is modeled by determining the ratio:

$$r_c = \frac{SA\ in\ contact\ with\ other\ cells}{total\ SA\ of\ cell}$$

If r_c is below a certain value, CI, the target volume is increased by 1. However, if r_c is above that value, the cell is in contact with other cells and will not grow. Note that contact inhibition is essentially turned off when $CI = 100\%$ since the ratio can never be greater than 1. The high ratio implies that the cell is surrounded by other cells and has little room to proliferate. In this model, contact inhibition only affects proliferation, not migration.

3. Results

3.1. *The Simulation*

The simulation is conducted using CC3D in a *100* x *100* lattice grid. A frame of width 1 from the grid boundary (the yellow boundary in Figure 2.a) is placed on the perimeter of the grid so the actual size is *98* x *98*. The adhesion energy between the tumor cells and the frame is the same as the adhesion energy between tumor cells to prevent them from adhering to the edges of the grid. The initial cell has volume 18 and starts in the center of the grid. Parameters including contact inhibition (CI), compressibility of cells ($\lambda_1 \upsilon$), migration magnitude $(\overrightarrow{\lambda_{mot}})$, and cell-to-cell adhesion (J) are also incorporated into the model. Tumor growth is measured by recording the sum of the volumes of all the cells in Microsoft Excel and MATLAB is used to fit growth curves to the data. The data from CC3D is fitted to a generalized logistic curve:

$$X^{'}(t) = \alpha X(1 - \left(\frac{X}{K}\right)^{p})$$

which has the solution:

$$X(t) = K\left(1 - e^{-\alpha tp} \cdot \frac{x_0^p - K^p}{x_0^p}\right)^{-\frac{1}{p}}$$

where $K = 98^2 = 9604$ is the carrying capacity, α describes the initial growth rate, p describes the intermediate growth rate, and $r_0 = 18$ is the initial cell volume. The growth rate is denoted as (α, p).

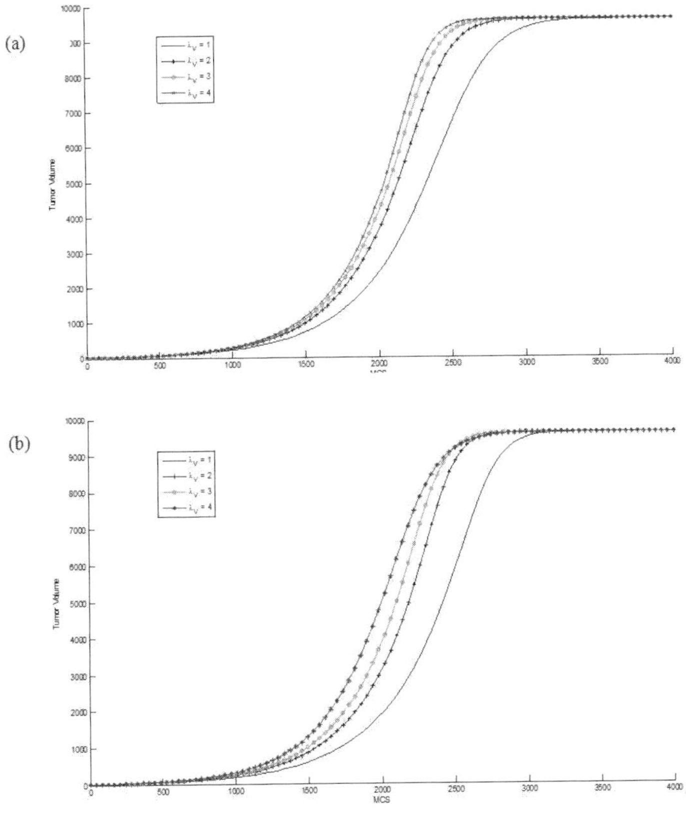

Figure 1. Best fit curves of tumor volume as a function of time for $\lambda_v = 1, 2, 3, 4$. (a) time step (ts)= 50 and $|\vec{\lambda}_{mot}|$= 10. Calculated growth rates: (0.00245, 2.470), (0.00266, 3.169), (0.00272, 3.604), and (0.00276, 4.025) for $\lambda_v = 1, 2, 3$, and 4 respectively. (b) ts = 50 and $|\vec{\lambda}_{mot}| = 50$. Calculated growth rates: (0.00233, 4.000), (0.00258, 4.456), (0.00268, 3.768), and (0.00285, 2.454) for $\lambda_v = 1, 2, 3$, and 4 respectively.

3.2. *Compressibility of Cells*

The parameter λ_v is used to alter the compressibility of the cells. Cells with a high λ_v have a large Hamiltonian and will have less deviation from the target volume. In Figure 1, motility is held constant by keeping the time step and magnitude constant. Also, contact inhibition was not incorporated into the simulation, allowing cells to undergo a constant rate of growth; each cell's target volume increased by 1 every 10 MCS. Figure 1 shows a correlation between λ_v and proliferation with a high λ_v promoting a faster growth rate. The initial growth rate, α , increases only slightly from 0.00245 to 0.00266 to 0.00272 to 0.00276, which implies that λ_v has only a small effect in the early proliferation stages of the tumor growth. This is expected since early in growth, cells have room to grow and are not affected by compression very much. The intermediate growth constant shows a steeper increase from 2.470 to 3.169 to 3.604 to 4.025. This correlation can be explained by considering the cells in the center of the tumor. Because all cells are constantly growing, cells in the center experience pressure from other cells. As center cells get compressed, this creates a deviation between their actual volumes and their target volumes. Thus, cells with larger λ_v are more resistant to compression and be more likely to keep up with their growing target volumes.

Figure 2 illustrates the compression patterns in tumor growth. Initial parameters are set at λ_v= 2 and $|\vec{\lambda}_{mot}| = 0$ (no motility). Cells are color-coded based on compression which is measured by the ratio of the actual cell volume and the target volume. Hence, cells with a low ratio are compressed. In Figure 2, the cells are assigned colors in the following manner:

$$color = \begin{cases} dark\ \ gray, & 0.8 < ratio \leq 1 \\ light\ \ gray, & 0.6 < ratio \leq 0.8 \\ white, & 0.4 < ratio \leq 0.6 \end{cases}$$

Note that at $MCS = 2000$ (Figure 2.a), light gray cells predominate in the center while dark gray cells lie mostly on the perimeter of the tumor. This shows that cells in the center get compressed more than cells on the boundary. Furthermore, at $MCS = 2500$ (Figure 2.b), cells become even more compressed and white, highly compressed cells appear in the center, while dark gray cells still exist where there is room to grow.

Figure 2.c has the same conditions as the previous setup except λ_v is changed from 2 to 4. At $MCS = 2000$, there is noticeably less compression, evident by only a few light gray cells. Figure 2.a shows much more compression at the same time frame. Since λ_v is higher in Figure 2.c, cells maintain volumes closer to the target volume.

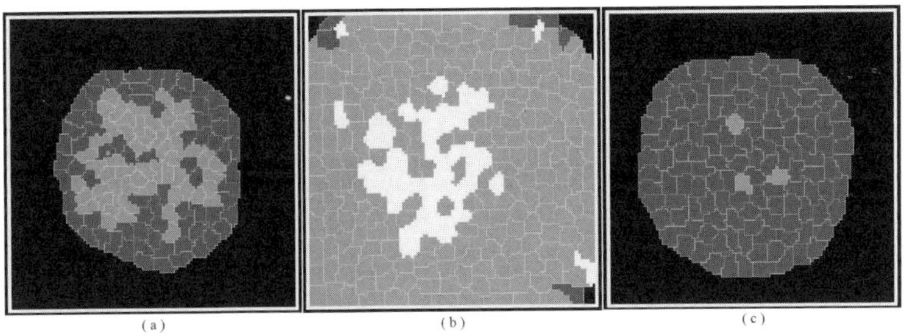

Figure 2. Snapshots of the tumor with $|\vec{\lambda}_{mot}| = 0$, ts = 50, CI = 0. (a) MCS = 2000 and λ_v= 2. (b) MCS = 2500 and λ_v= 2. (c) MCS = 2000 and λ_v= 4.

3.3. *Contact Inhibition*

In Figure 3, motilities (50,10), (50,50), (100,10), (100,50) were tested for $\lambda_v = 1, 2, 3, 4$ ($\lambda_v = 1$ and *2* shown in Figures 3.a and 3.b respectively). Also, $CI = 100\%$ so cells did not experience contact inhibition. For each λ_v, the black curve grew faster than the $\oplus$-curve and the $\circ$-curve grew faster than the $\otimes$-curve. Thus, when there is no contact inhibition, cells with smaller magnitudes grow faster than those with larger ones. This is counter-intuitive because magnitude correlates with cell movement such that a larger magnitude corresponds to more motile cells, and one would expect more migration to lead to a faster growth rate. This can be explained by the inherent design of CC3D, which uses Hamiltonians to dictate index-copies. When the magnitude increases, the Hamiltonian for migration also increases. As migration plays a bigger role, the surface and volume constraints are loosened because migration compensates for the energy. Hence, cell volumes can deviate more from the target volume, a behavior analogous to lowering λ_v. As previously shown, lower λ_v values yield slower tumor growth.

Next, we consider the effect of contact inhibition on cell proliferation. Figure 4 shows results for $\lambda_v = 2$ when motility is (50,10),(50,50),(50,100) and (a)$CI = 25$, (b)$CI = 50$, and (c)$CI = 75$. In all three cases, growth is faster for larger migration magnitudes, in contrast to the results from Figure 2. Thus, when tumor cells do have contact inhibition, the migration amplitude plays a greater role in tumor growth. These results can be attributed to the two factors that can inhibit growth: i) cell compressibility and ii) contact inhibition. Note that both can be avoided if tumor cells are well-separated and have space to grow. Large migration magnitudes help overcome cell-to-cell adhesion and cause cells to separate. This effect

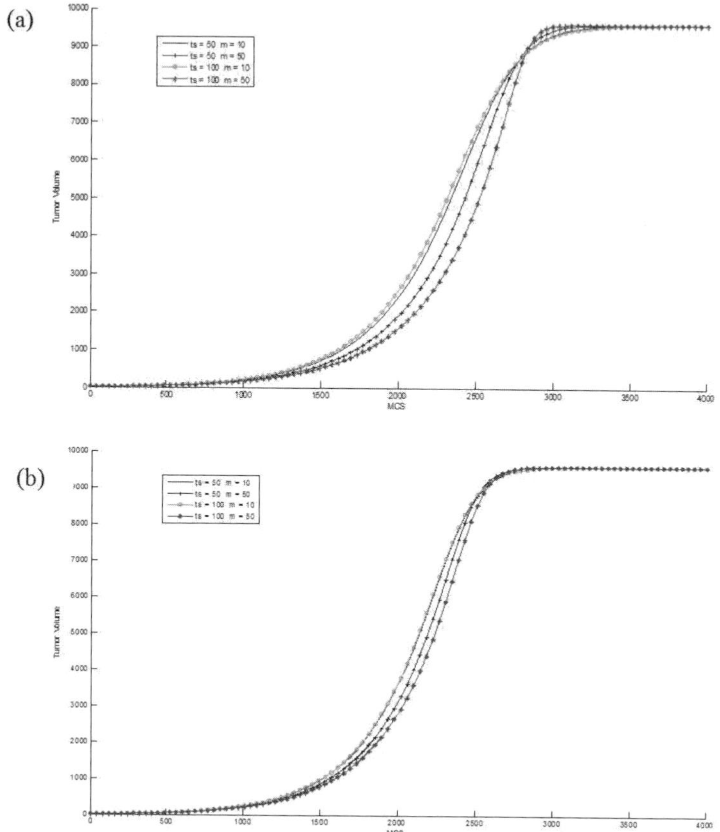

Figure 3. Tumor volume as a function of time for (ts, m) = (50, 10), (50, 50), (100, 10), and (100, 50). (a) $\lambda_v = 1$. Calculated growth rates: (0.00245, 2.470), (0.00233, 4.000), (0.00249, 2.238), and (0.00223, 8.272) respectively. (b) $\lambda_v = 2$. Calculated growth rates: (0.00265, 3.169), (0.00258, 4.456), (0.00266, 3.180), and (0.00252, 5.160) respectively.

is most prominent in Figure 4 when there is contact inhibition since the simulation penalizes tumors with low motility by slowing growth. Without contact inhibition (Figure 3), the only factor that can inhibit growth is cell compressibility.

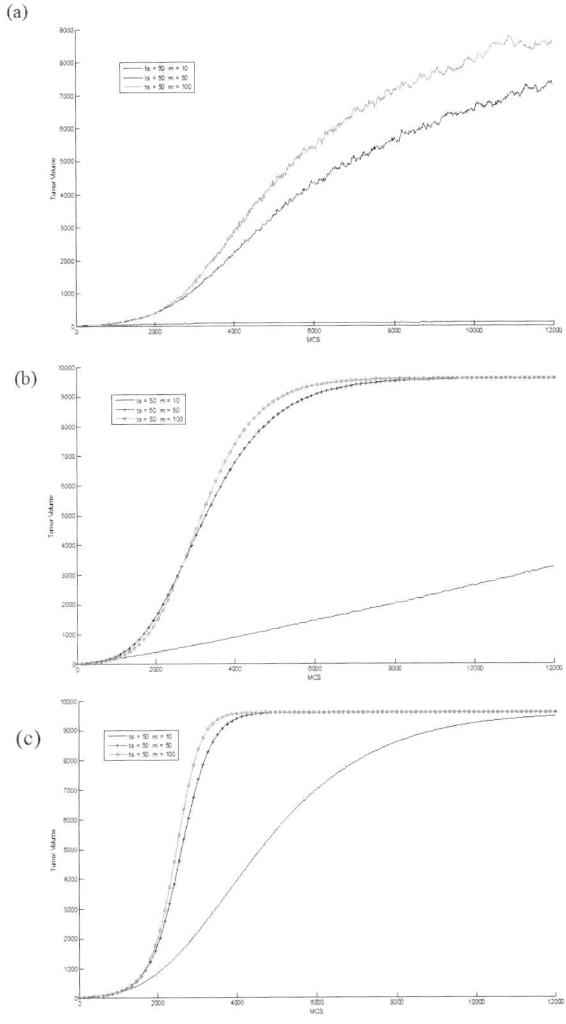

Figure 4.

4. Conclusion

Previous research has shown the effects of proliferation rates and migration on tumor growth. Higher proliferation rates correlate with faster invasion times. Higher migration probabilities create more space for cells in the center of the tumor to grow and proliferate (Fedotov, 2007). However, little research has been done on the effects of contact inhibition, motility, and cell compressibility. Here, a mathematical model was developed to study the effects of contact inhibition on tumor growth. By increasing the contact inhibition restraint, clustered cells were less likely to proliferate. Cells in the center of the tumor were completely surrounded by homotypic cells and thus were not able to grow.

Further, the model suggested that contact inhibition amplifies the effects of cell motility. Contact inhibition added a penalty to compressed cells by halting their growth. Thus, cells in the inside of the tumor were effectively quiescent and did not contribute to the overall growth of the tumor. The growth was also fitted to the generalized logistic curve which described initial growth rate and the intermediate growth rate.

These results also call into question the effectiveness of cancer therapies that involve high cell death rates such as chemotherapy. Previous models have shown that mass die-off places a selective pressure on more motile cells. These cells were able to quickly fill in spaces freed by cell death and proliferate. Each round of die-off increased the proportion of motile cells (Thalhauser *et al.*, 2010). Similarly, high cell death rate therapies could select for cells with a particularly low response to contact inhibition, leading ultimately to more invasive cells. As this would be counterproductive, this could be justification to develop new treatments to specifically target cells with a low response to contact inhibition. As with all mathematical models, this model also has a few limitations. First, motility may have unwanted effects on proliferation. For example, migration is modeled by decreasing the energy in the proposed direction. Unfortunately, this can interfere with growth because cells are less likely to follow their target volumes. Because migration and cell compressibility factor into the same Hamiltonian, migration has an undesired effect on the λ_v of cells. Further research will explore possible ways to counter the effect. Also, the model for contact inhibition is still somewhat crude and, later may be refined to also restrict cell migration and incorporate cell death due to compression.

Acknowledgments

We would like to thank Dr. Maciek Swat of the University of Indiana for providing technical support in the use of the CC3D simulation software.

References

1. Cickovski, T., Ars, K. Alber, M. S., Izaguirre, J. A., Swat M., Glazier, J. A., Merks, R. M., Glimm, T., Hentschel, H. G., and Newman, S. N., *Computing in Science & Engineering*, **July/August**, (2007).

2. Corce, C. M., *N. Engl. J. Med.*, **358 (5)**, 502-11, (2008).
3. Fedotov, S. and Iomin, A., *Physical Review Letters*, **98**, (118101): 1-4, (2007).
4. Fidler, I. J., *Cytometry*, **10**, 673-680, (1989).
5. Friedl, P. and Bröcker, E.B., *Cell. Mol. Life Sci.*, **57**, 41-46, (2000).
6. Izaguirre, J. A., Chaturvedi, R., Huang, C., Cickoviski, T., Coffland, J., Thomas, G., Forgacs, G., Alber, M., Henstschel, G., Newman, S. A., and Glazier, J. A., *Bioinformatics*, **20**, 1129-1137, (2004).
7. Phan, D. D., *A discrete cellular automaton model demonstrates cell motility increases fitness in solid tumors*, **University of California**, Irvine, (2010).
8. Sherr, C. J., *Cell.*, **116**, 235-246, (2004).
9. Takai, Y., Miyoshi J., Ikeda W., and Ogita H., *Nature Reviews: Molecular Cell Biology*, **9**, 603-615, (2008).
10. Thalhauser, C. J., Lowengrub, J. S., Stupack, D., Komarova, N., *Selection in spatial stochastic models of cancer: Migration as a key modulator of fitness*, **Biology Direct**, (2010).
11. Todd, R. and Wong, D.T., *Oncogenes. Anticancer Res.* **19**, (6A): 4729-46, (1999).
12. Yokota, J., *Carcinogenesis*, **21**, (3): 497-503, (2000).

INDEX